Urban Sustainability and Energy Management of Cities for Improved Health and Well-Being

Roberto Alonso González-Lezcano
Universidad CEU San Pablo, Spain

IGI Global
PUBLISHER of TIMELY KNOWLEDGE

A volume in the Advances in Environmental Engineering and Green Technologies (AEEGT) Book Series

Published in the United States of America by
IGI Global
Engineering Science Reference (an imprint of IGI Global)
701 E. Chocolate Avenue
Hershey PA, USA 17033
Tel: 717-533-8845
Fax: 717-533-8661
E-mail: cust@igi-global.com
Web site: http://www.igi-global.com

Library of Congress Cataloging-in-Publication Data

Names: González-Lezcano, Roberto Alonso, 1971- editor.
Title: Urban sustainability and energy management of cities for improved health and well-being / Roberto Alonso Gonzalez Lezcano, editor.
Description: Hershey, PA : Engineering Science Reference, an imprint of IGI Global, [2022] | Includes bibliographical references and index. | Summary: "This book explores opportunities and challenges urban communities face, as they seek to become sustainable systems embedded in their diverse and complex social and environmental contexts"-- Provided by publisher.
Identifiers: LCCN 2022001270 (print) | LCCN 2022001271 (ebook) | ISBN 9781668440308 (h/c) | ISBN 9781668440315 (s/c) | ISBN 9781668440322 (ebook)
Subjects: LCSH: Municipal engineering. | Sustainable urban development. | Cities and towns--Energy conservation. | Communities--Energy conservation.
Classification: LCC TD159 .U734 2022 (print) | LCC TD159 (ebook) | DDC 628--dc23/eng/20220321
LC record available at https://lccn.loc.gov/2022001270
LC ebook record available at https://lccn.loc.gov/2022001271

This book is published in the IGI Global book series Advances in Environmental Engineering and Green Technologies (AEEGT) (ISSN: 2326-9162; eISSN: 2326-9170)

British Cataloguing in Publication Data
A Cataloguing in Publication record for this book is available from the British Library.

Advances in Environmental Engineering and Green Technologies (AEEGT) Book Series

Sang-Bing Tsai
Zhongshan Institute, University of Electronic Science and Technology of China, China & Wuyi University, China
Ming-Lang Tseng
Lunghwa University of Science and Technology, Taiwan
Yuchi Wang
University of Electronic Science and Technology of China Zhongshan Institute, China

ISSN:2326-9162
EISSN:2326-9170

Mission

Growing awareness and an increased focus on environmental issues such as climate change, energy use, and loss of non-renewable resources have brought about a greater need for research that provides potential solutions to these problems. Research in environmental science and engineering continues to play a vital role in uncovering new opportunities for a "green" future.

The **Advances in Environmental Engineering and Green Technologies (AEEGT)** book series is a mouthpiece for research in all aspects of environmental science, earth science, and green initiatives. This series supports the ongoing research in this field through publishing books that discuss topics within environmental engineering or that deal with the interdisciplinary field of green technologies.

Coverage

- Alternative Power Sources
- Policies Involving Green Technologies and Environmental Engineering
- Sustainable Communities
- Radioactive Waste Treatment
- Electric Vehicles
- Contaminated Site Remediation
- Green Transportation
- Renewable Energy
- Industrial Waste Management and Minimization
- Biofilters and Biofiltration

IGI Global is currently accepting manuscripts for publication within this series. To submit a proposal for a volume in this series, please contact our Acquisition Editors at Acquisitions@igi-global.com or visit: http://www.igi-global.com/publish/.

The Advances in Environmental Engineering and Green Technologies (AEEGT) Book Series (ISSN 2326-9162) is published by IGI Global, 701 E. Chocolate Avenue, Hershey, PA 17033-1240, USA, www.igi-global.com. This series is composed of titles available for purchase individually; each title is edited to be contextually exclusive from any other title within the series. For pricing and ordering information please visit http://www.igi-global.com/book-series/advances-environmental-engineering-green-technologies/73679. Postmaster: Send all address changes to above address.

Titles in this Series

For a list of additional titles in this series, please visit: www.igi-global.com/book-series

Handbook of Research on Water Sciences and Society
Ashok Vaseashta (International Clean Water Institute, USA & Transylvania University of Brasov, Romania & Academy of Sciences of Moldova, Moldova) Gheorghe Duca (Institute of Chemistry, Moldova State University, Moldova & Academy of Science of Moldova and Romanian Academy, Romania.) and Sergey Travin (Federal Research Center for Chemical Physics and N. N. Semenov of the Russian Academy of Sciences, Russia)
Engineering Science Reference • © 2022 • 728pp • H/C (ISBN: 9781799873563) • US $465.00

Optimization and Decision-Making in the Renewable Energy Industry
Figen Balo (Fırat University, Turkey) Ayşe Topal (Nigde Omer Halisdemir University, Turkey) Ezgi Demir (Sumer Robotics Engineering and Consultancy Ltd., UK) and Alptekin Ulutaş (Sivas Cumhuriyet University, Turkey)
Engineering Science Reference • © 2022 • 304pp • H/C (ISBN: 9781668424728) • US $245.00

Improving Integrated Pest Management for Crop Protection and Food Security
Muhammad Haseeb (Florida A&M University, USA) Lambert H.B. Kanga (Florida A&M University, USA) and Jawwad A. Qureshi (University of Florida, USA)
Engineering Science Reference • © 2022 • 315pp • H/C (ISBN: 9781799841111) • US $195.00

Prevention and Management of Soil Erosion and Torrential Floods
Slobodan Milutinović (University of Niš, Serbia) and Snežana Živković (Faculty of Occupational Safety, University of Niš, Serbia)
Engineering Science Reference • © 2022 • 335pp • H/C (ISBN: 9781799884590) • US $195.00

Applications of Nature-Inspired Computing in Renewable Energy Systems
Mohamed Arezki Mellal (M'Hamed Bougara University, Algeria)
Engineering Science Reference • © 2022 • 326pp • H/C (ISBN: 9781799885610) • US $225.00

Physiology, Genomics, and Biotechnological Applications of Extremophiles
Aparna B. Gunjal (Dr. D.Y. Patil Arts, Commerce, and Science College, Pune, India) Rebecca Thombre (The University of Kent, UK & Blue Marble Space Institute of Science, USA) and Javid A. Parray (Department of Higher Education, Government Degree College, Srinagar, India)
Medical Information Science Reference • © 2022 • 473pp • H/C (ISBN: 9781799891444) • US $345.00

IGI Global
PUBLISHER of TIMELY KNOWLEDGE

701 East Chocolate Avenue, Hershey, PA 17033, USA
Tel: 717-533-8845 x100 • Fax: 717-533-8661
E-Mail: cust@igi-global.com • www.igi-global.com

Editorial Advisory Board

Table of Contents

Detailed Table of Contents

There is a two-way interaction between the climate and the city. Significant thermal differences may appear between one area and another in big urban areas with diverse urban fabric. This research determines the bioclimatic characterisation of all city neighbourhoods, focusing on the case study of the city of Málaga (Spain) using the definition of urban bioclimatic areas (UBA). The methodology used is based on selecting key factors (geographic and climatic) and in multi-criteria analysis, applied at territorial and urban scale, by geographic information systems (GIS). The results obtained show the microclimatic thermal differences in the city that may help to develop specific urban proposals to achieve the Sustainable Development Goals: SDG.3-Health and well-being, SDG.11-Sustainable cities and communities, and SDG.15-Life of terrestrial ecosystems, within the framework of action proposed by the New Urban Agenda 2030 and heathier neighbourhoods.

Data collected by the United Nations (UN) suggest that infections arise after a close contact with infected people through nasal and oral secretions released when an infected person coughs, sneezes, or talks. In public transportation buses, an important amount of people gather every day for long periods of time, making the air pollution inside these transportation systems a major risk for transmission. Therefore, the objective of this chapter is to know criteria and strategies for a conceptualization of an air-extraction system inside public transportation buses, based on the detailed study applicability of the 1) product-user interaction (technical data, dimensional relationship, and evaluation); 2) creative process, ideation, definition, evaluation, and structuration; and 3) sustainability, technical specifications, ergonomics, and production by means of understanding the design's limits and effects.

One of the most critical objectives in buildings is their adaptation to environmental conditions in order to optimize energy performance as well as the thermal and visual comfort of the occupants. This issue is relevant, for example, in public buildings that incorporate large glazed surfaces, where overheating and a lack of thermal and visual comfort are common, especially in Mediterranean countries. The headmost cause of increased air conditioning loads is direct solar radiation on transparent surfaces. However, the significant losses through glazed surfaces in cold climates also cannot be ignored. Therefore, solar protections must be used, thus observing the harmful effects of the absence of solar radiation on energy loads in winter. This chapter aims to study and compare construction materials recently introduced to the market, such as dynamic glasses, with significantly higher thermal and optical performance than traditional glasses.

Latin America is one of the most urbanized regions. It implies a high demand for housing for the low-income population of this region in the face of climate change. The COVID-19 pandemic caused a delayed technology transition in the region. Although recently it has been observed that the construction industry and Latin American governments from the pandemic are promoting disruptive actions in this industry. The promotion of robotic construction in Latin America is an alternative to this problem. Construction robotics is an opportunity for sustainable development in Latin America. It is essential to update the construction sector by orienting it towards implementing this type of technology. Technology transfer plays a prominent role in the dissemination of this new paradigm. What is the dynamic of construction robotics to develop affordable housing in Latin American cities? This chapter aims to identify the dynamic of construction robotics to develop affordable housing in Latin American cities.

The present research is part of a wider qualitative study that aims to assist cities in the South Pacific developing countries with the recovery of natural areas through a green infrastructure-based approach, following a case study method. The overarching purpose of this study is to pinpoint relevant contributing elements for the successful implementation of the green infrastructure approach aiming at providing Peruvian coastal cities with novel sustainable environmental policies. To foster the conservation of the natural or semi-natural ecosystems converging with the cities, the following specific objectives were set:

1) to carry out a comprehensive physical and spatial analysis of the natural areas of Chimbote and Nuevo Chimbote; 2) to review the Peruvian regulatory framework, at the national and municipal levels, as well as international standards on the conservation of natural areas; and 3) to explore a scenario featuring the disappearance of these Peruvian coastal natural spaces and its associated consequences.

This study aims to analyze the implications of green knowledge and technology in organizational green innovation, urban green innovation, and green roofs. The analysis is supported by the assumption that green sharing knowledge and technology is basic to organizational green innovation and urban green innovation areas practices, operations, and activities. The methods employed are based on the analytical-reflective and descriptive supported with the review of theoretical and empirical literature. The analysis concludes that green knowledge sharing is relevant to create and develop the green technology with positive implications for organizational green innovation, urban green innovation areas, and green roofs.

Historically, water has played an essential role in choosing the location and settlement of a habitat and, consequently, in the configuration of the landscape, building, and urban context. The quality of the inhabitants' hygiene and health depends on water and sewage treatment. Water supply and sanitation are crucial to achieving end-user quality and enjoyment of the home and the city. This case study focuses on the city of Madrid, intrinsically related to water from its origin and name. Therefore, the "mayra" abounds in Madrid and is the "mother of water." The latter refers to Madrid's location surrounding a large fountain that produced a stream that flowed into the Manzanares River. It recovers photographs of the landscape found in unpublished historical and military archives with the intention of showing certain excavations that affected the El Pardo Woodlands, such as Janini's artesian wells and other missing projects in the Manzanares River Basin.

Residential buildings are the most common scenario of fatal fires in Spain, and over the past decade, there has been no significant decrease of fatalities in this kind of accident, even a slight increase in recent years. In this context, a systematic methodology to evaluate the risk level of residential buildings, previously grouped into building typologies, is presented. Its application to different housing typologies allows covering large urban ensembles, providing an important information to regulators and fire services. The factors that most increase the level of fire risk are identified, analyzing them both by their construction stage and by their morphology.

Infrastructure and industrial projects continue to expand, resulting in a steady rise in demand for cement. However, it is claimed that nature and the environment suffer significant harm throughout the cement manufacturing process due to the cement factory's greenhouse gas emissions. Additionally, economic difficulties have emerged as a result of energy usage during manufacturing. As a result, waste materials have begun to be utilized in geopolymer concrete in place of cement. Therefore, environmentally friendly materials with a low carbon footprint have been in high demand across the world's building sector. Activators are used with waste materials as binders of geopolymers without using cement. Thus, geopolymers have gained importance due to their environmental compatibility, sustainability, and durability. This chapter comprehensively reviews the environmental suitability of geopolymers produced using waste materials. In addition, the authors tried to collect most geopolymer physical, mechanical, microstructural, and durability properties.

We are living in rapidly changing times where the world is running low on its carbon budget in trying to rapidly transition to net zero. We are also living amidst transformations, innovation, and operational modernization to be able to provide for a Smart City, including climate health. Future smart cities demand a balanced blend of data and technology to create diversely inclusive and sustainable solutions. In this chapter, the author focuses on how smart buildings can be designed for a sustainable future, exploring the possibilities of converting existing buildings into interoperable spaces, powered by technology and data. While the author outlines the differences between brownfield and greenfield smart buildings, he makes a case for each and explains their potential in achieving productive sustainability. The chapter delineates some of the policy implementations required to align smart buildings with sustainable climate action and substantiates global practices with ROI data. The author concludes the chapter by offering the way forward in the urban sustainability journey.

Smart cities have so far focused on improving citizens' standards of living, with emphasis on better transportation, improved housing spaces and workplaces, and energy conservation. A pandemic-stricken world emerging to adopt a new normal has shifted the focus to health and hastened the need for leveraging technology, including but not limited to Blockchain, artificial intelligence (AI), and the internet of things

Preface

Global environmental challenges are increasing, such as climate change and rapid urbanisation, and human influence on the environment continues to grow. Many of the resulting risks lead to diseases and negative impacts on health and quality of life. It is now essential to develop more sustainable and healthy environments, with a greater focus on prevention through actions targeting the root causes of disease.

Urban communities play a critical role in achieving sustainability overall. They are where most of the world's population lives, and they comprise high concentrations of services, consumption, and waste. To date urban communities rely heavily on surrounding peri-urban and rural areas to sustain their highly concentrated levels of consumption and waste. This displacement of resourcing and sink functions represents an unsustainable pattern of urbanization that accelerates the decline of global ecosystems services rather than supporting them through the compensatory contributions of peri-urban and rural areas.

By focusing on reducing environmental and social risk factors, almost a quarter of the global burden of disease can be avoided through better health promotion strategies, improved prevention and hygiene measures and safer management of toxic substances in homes and workplaces.

In this context, the main objective of this book is to highlight the interdisciplinary connections between the environment and human health, focusing mainly on new ideas and suggestions for promoting both sustainable development and human health and well-being.

The overall aim is to create a new approach to the analysis of human impacts on the natural environment and, conversely, to determine how the environment can modulate human lifestyles and behaviours and thus human health.

Urban Sustainability and Energy Management of Cities for Improved Health and Well-Being explores opportunities and challenges urban communities face, as they seek to become sustainable systems embedded in their diverse and complex social and environmental contexts. We invite contributions that envision sustainable solutions for urban infrastructure, housing, food and water security, ecosystems services, quality of life, human livelihoods, governance, and policies that improve the sustainability and resilience of urban communities.

This research book will update the current state of knowledge on the subject, since the authors of the chapters will all be researchers and professors from prestigious universities. It will not be a theoretical book, but will be based on where current research is taking us in terms of urban sustainability and energy management.

The book has been divided into the following chapters:

CHAPTER 1: BIOCLIMATIC CHARACTERISATION METHODOLOGY OF A CITY – THE CASE OF MÁLAGA, SPAIN

There is a two-way interaction between the climate and the city. Significant thermal differences may appear between one area and another in big urban areas with diverse urban fabric. This research determines the bioclimatic characterisation of all city's neighbourhoods, focusing on the case study of the city of Málaga, (Spain) using the definition of Urban Bioclimatic Areas (UBA). The methodology used is based on selecting key factors (geographic and climatic) and in multi-criteria analysis, applied at territorial and urban scale, by Geographic Information Systems (GIS). The results obtained show the microclimatic thermal differences in the city that may help to develop specific urban proposals to achieve the Sustainable Development Goals: SDG.3-Health and well-being, SDG.11-Sustainable cities and communities and SDG.15-Life of terrestrial ecosystems, within the framework of action proposed by the New Urban Agenda 2030 and heathier neighbourhoods.

CHAPTER 2: CONCEPTUALIZATION OF AN AIR-EXTRACTION SYSTEM TO MITIGATE COVID-19 TRANSMISSIONS INSIDE PUBLIC TRANSPORTATION BUSES

Data collected by the United Nations (UN) suggest that infections arise after a close contact with infected people through nasal and oral secretions released when an infected person coughs, sneezes or talks. In public transportation buses an important amount of people gather every day for long periods of time, making the air pollution inside these transportation systems a major risk for transmissions. Therefore, the objective of this chapter is to know criteria and strategies for a conceptualization of an air-extraction system inside public transportation buses, based on the detailed study applicability of the: a) product-user interaction; technical data, dimensional relationship and evaluation; b) creative process, ideation, definition, evaluation and structuration; c) sustainability, technical specifications, ergonomics and production, by means of understanding the design's limits and effects.

CHAPTER 3: DYNAMIC GLAZING WITH HIGHER THERMAL AND OPTICAL PERFORMANCE FOR ZERO ENERGY BUILDING DESIGN

One of the most critical objectives in buildings is their adaptation to environmental conditions in order to optimize energy performance as well as the thermal and visual comfort of the occupants. This issue is relevant, for example, in public buildings that incorporate large glazed surfaces, where overheating and a lack of thermal and visual comfort are common, especially in Mediterranean countries. The headmost cause of increased air conditioning loads is direct solar radiation on transparent surfaces. However, the significant losses through glazed surfaces in cold climates also cannot be ignored. Therefore, solar protections must be used, thus observing the harmful effects of the absence of solar radiation on energy loads in winter. This chapter aims to study and compare construction materials recently introduced to the market, such as dynamic glasses, with significantly higher thermal and optical performance than traditional glasses.

CHAPTER 4: DYNAMIC OF CONSTRUCTION ROBOTICS TO DEVELOP AFFORDABLE HOUSING IN LATIN AMERICAN CITIES

Latin America is one of the most urbanized regions. It implies a high demand for housing for the low-income population of this region in the face of climate change. The COVID-19 pandemic caused a delayed technology transition in the region. Although recently it observed that the construction industry and Latin American governments from the pandemic promoting disruptive actions in this industry. The promotion of robotic construction in Latin America is an alternative to this problem. Construction Robotics is an opportunity for sustainable development in Latin America. It is essential to update the Construction Sector by orienting it towards implementing this type of technology. Technology transfer plays a prominent role in the dissemination of this new paradigm. What is the dynamic of construction robotics to develop affordable housing in Latin American cities? This chapter aims to identify the dynamic of construction robotics to develop affordable housing in Latin American cities.

CHAPTER 5: GREEN INFRASTRUCTURE AS A NATURE-BASED RECOVERY FOR NATURAL AREAS IN DESERT DEVELOPING COUNTRIES OF THE SOUTH PACIFIC COASTAL STRIP – THE PERUVIAN CASE STUDY OF CHIMBOTE AND NUEVO CHIMBOTE

The present research is part of a wider qualitative study that aims to assist cities in South-Pacific developing countries with the recovery of natural areas through a Green Infrastructure-based approach, following a case study method: Chimbote and Nuevo Chimbote, Peru. The overarching purpose of this study is to pinpoint relevant contributing elements for the successful implementation of the Green Infrastructure approach aiming at providing Peruvian coastal cities with novel sustainable environmental policies. To foster the conservation of the natural or semi-natural ecosystems converging with the cities, the following specific objectives were set: i. to carry out a comprehensive physical and spatial analysis of the natural areas of Chimbote and Nuevo Chimbote; ii. to review the Peruvian regulatory framework, at the national and municipal levels, as well as international standards on the conservation of natural areas; iii. to explore a scenario featuring the disappearance of these Peruvian coastal natural spaces and its associated consequences.

CHAPTER 6: GREEN KNOWLEDGE AND TECHNOLOGY AND ITS IMPLICATIONS IN ORGANIZATIONAL GREEN INNOVATION, URBAN GREEN INNOVATION SPACES, AND GREEN ROOFS

This study aims to analyze the implications of green knowledge and technology in organizational green innovation, urban green innovation, and green roofs. The analysis is supported by the assumption that green sharing knowledge and technology is basic to organizational green innovation and urban green innovation areas practices, operations, and activities. The methods employed are based on the analytical-reflective and descriptive supported with the review of theoretical and empirical literature. The analysis concludes that green knowledge sharing is relevant to create and develop the green technology with positive implications for organizational green innovation, urban green innovation areas and green roofs. Keywords: Organizational green innovation, green knowledge, green technology, green urban areas.

CHAPTER 7: LANDSCAPE AND WATER NETWORKS – IMPACT ON HEALTH FOR THE SMART CITY: CASE STUDY – EL PARDO AND THE MANZANARES RIVER BASIN

Historically, water has played an essential role in choosing the location and settlement of a habitat and, consequently, in the configuration of the landscape, building, and urban context. The quality of the inhabitants' hygiene and health depends on water and sewage treatment. Water supply and sanitation are crucial to achieving end-user quality and enjoyment of the home and the city. This case study focuses on the city of Madrid, intrinsically related to water from its origin and name. Therefore, the "mayra" abounds in Madrid and is the "mother of water." The latter refers to Madrid's location surrounding a large fountain that produced a stream that flowed into the Manzanares River. It recovers photographs of the landscape found in unpublished historical and military archives with the intention of showing certain excavations that affected the El Pardo Woodlands, such as Janini's artesian wells and other missing projects in the Manzanares River basin.

CHAPTER 8: METHODOLOGY FOR THE EVALUATION OF FIRE SAFETY IN EXISTING URBAN RESIDENTIAL AREAS IN SPAIN – THE CASE OF SOCIAL HOUSING IN PAMPLONA

María Fernández-Vigil, Carmen Corrales, Raquel Garín, Juan Bautista Echeverría

Residential buildings are the most common scenario of fatal fires in Spain, and over the past decade there has been no significant decrease of fatalities in this kind of accidents, even a slight increase in recent years. In this context, a systematic methodology to evaluate the risk level of residential buildings, previously grouped into building typologies, is presented. Its application to different housing typologies allows covering large urban ensembles, providing an important information to regulators and fire services. The factors that most increase the level of fire risk are identified, analyzing them both by their construction stage and by their morphology.

CHAPTER 9: RESEARCH OF ALTERNATIVE ECOLOGICAL WASTE MATERIALS USED IN GEOPOLYMERS FOR SUSTAINABLE BUILT ENVIRONMENT

Infrastructure and industrial projects continue to expand, resulting in a steady rise in demand for cement. However, it is claimed that nature and the environment suffer significant harm throughout the cement manufacturing process due to the cement factory's greenhouse gas emissions. Additionally, economic difficulties have emerged as a result of energy usage during manufacturing. As a result, waste materials have begun to be utilized in geopolymer concrete in place of cement. Therefore, environmentally friendly materials with a low carbon footprint have been in high demand across the world's building sector. Activators are used with waste materials as binders of geopolymers without using cement. Thus, geopolymers have gained importance due to their environmental compatibility, sustainability, and durability. This book chapter comprehensively reviews the Environmental suitability of geopolymer produced using waste materials. In addition, the authors tried to collect most geopolymer physical, mechanical, microstructural, and durability properties.

CHAPTER 10: SMART CITY BUILDINGS FOR A RESILIENT, SUSTAINABLE FUTURE

We are living in rapidly changing times where the world is running low on its carbon budget in trying to rapidly transition to net zero. We are also living amidst transformations, innovation and operational modernization to be able to provide for a Smart city. Including climate health. Future Smart cities demand a balanced blend of data and technology to create diversely inclusive and sustainable solutions. In this chapter, the author focuses on how Smart buildings can be designed for a sustainable future, exploring the possibilities of converting existing buildings into interoperable spaces, powered by technology and data. While the author outlines the differences between brownfield and greenfield Smart buildings, he makes a case for each and explains their potential in achieving productive sustainability. The chapter delineates some of the policy implementations required to align Smart buildings with sustainable climate action and substantiates global practices with ROI data. The author concludes the chapter by offering the way forward in the urban sustainability journey.

CHAPTER 11: SUSTAINABILITY AND HEALTH IN A SMART CITY – HEALTH AT THE HEART OF A SMART CITY

Smart Cities have so far focused on improving citizens' standard of living, with emphasis on better transportation, improved housing spaces and workplaces, and energy conservation. A pandemic-stricken world emerging to adopt a new normal has shifted the focus to health and hastened the need for leveraging technology, including but not limited to Blockchain, Artificial Intelligence (AI), and the Internet of Things (IoT) to enable Smart Cities to move to the next level of wellbeing. With public health and sustainability in the spotlight, civic agencies and government bodies are now alert to potential future contingencies that cannot be ruled out. This chapter emphasizes the need for a future-proofed healthcare system within sustainable Smart Cities, that not only foresees and contains public health emergencies but also provides for its citizens' general wellbeing.

CHAPTER 12: URBAN HEALTH IN VULNERABLE ENVIRONMENTS – WATER, SANITATION, AND HYGIENE IMPROVEMENT ACTIONS IN SCHOOL AREAS IN MAKENI, SIERRA LEONE

Adequate access to sanitation and safe water is a main challenge to improve urban health in low-income countries. Diseases derived of precarious hygiene conditions are a major burden in African rapidly growing urban areas. These challenges are embedded in complex stakeholder networks and juxtaposing processes of governance, management, and technical upgrading and need to be addressed through a holistic approach. It is argued that schools are a key objective for sanitation, water, and hygiene actions in the city. Concerns for improvement of hygiene conditions among school children have risen in the context of contemporary sanitary crisis. Furthermore, schools play a pivotal role between the public and the private realms and a potential to foster change. The study focuses on WASH actions implemented in schools of the city of Makeni, Sierra Leone. It seeks to define a set of context specific recommendations to improve the health-related conditions in the school environment.

Recent events have challenged the way the built environment is designed, managed and experienced. Droughts, wildfires and a global pandemic have undoubtedly forced the building industry to find ways to cope with unpredictable events and to respond to a broader and more complex set of requirements. Today's environment demands new design processes, construction techniques, occupancy practices, and management strategies to increase the resilience of the built environment to extreme, uncontrollable, and unpredictable events while providing healthy and sustainable environments for people. This Topic invites researchers that address this topic to reflect on what the new concept of sustainability for the built environment should be, and to guide new research directions.

In this book, we look for a range of innovative research and evidence-based policy ideas that demonstrate the importance of housing research and policy for health and wellbeing.

Editor

Roberto Alonso González-Lezcano
Universidad CEU San Pablo, Spain

Chapter 1
Bioclimatic Characterisation Methodology of a City:
The Case of Málaga, Spain

Emilia Román López
Universidad Politécnica de Madrid, Spain

Ester Higueras García
https://orcid.org/0000-0002-0182-8884
Universidad Politécnica de Madrid, Spain

Francisco Javier Neila González
Universidad Politécnica de Madrid, Spain

ABSTRACT

There is a two-way interaction between the climate and the city. Significant thermal differences may appear between one area and another in big urban areas with diverse urban fabric. This research determines the bioclimatic characterisation of all city neighbourhoods, focusing on the case study of the city of Málaga (Spain) using the definition of urban bioclimatic areas (UBA). The methodology used is based on selecting key factors (geographic and climatic) and in multi-criteria analysis, applied at territorial and urban scale, by geographic information systems (GIS). The results obtained show the microclimatic thermal differences in the city that may help to develop specific urban proposals to achieve the Sustainable Development Goals: SDG.3-Health and well-being, SDG.11-Sustainable cities and communities, and SDG.15-Life of terrestrial ecosystems, within the framework of action proposed by the New Urban Agenda 2030 and heathier neighbourhoods.

INTRODUCTION

Cities are located in places where the conditions of sun exposure, wind, surface or underground water courses and topography were decisive for their origin and foundation. However, with their growth,

DOI: 10.4018/978-1-6684-4030-8.ch001

evolution and transformation of urban uses and activities, the reverse has occurred, i.e., cities have conditioned the temperature, humidity and local wind system, establishing a specific urban climate called *urban microclimate* (Oke, 1987). The microclimate has been analysed by many theoreticians, given its importance in the presence of the urban heat island phenomenon, together with the conditions of comfort for citizens in urban public spaces (Neila et al, 2020), among other issues. This research is focused on studying the significant differences in the microclimate that appear in urban contexts, of medium and large extension, between some areas of the city and others, with the aim of establishing the bioclimatic characterisation of the different urban areas. For this purpose, a selection of a city's intrinsic and extrinsic conditions is undertaken, in order to determine urban areas with the same patterns of temperature, humidity and wind, establishing *Urban Bioclimatic Areas (UBA)*. The spatial definition of these areas in a city is very appropriate in order to propose differentiated environmental planning measures, specific bioclimatic ordinances, or singular urban design proposals by areas. These serve to increase the hours of thermal comfort in urban public spaces, regulating air temperature, increasing environmental humidity, enabling pleasant breezes in the summer and curbing cold winds in the winter. Improving the urban microclimate positively influences the air quality and temperature in cities and can even help reduce the hours used for heating and cooling inside buildings. It also has a direct influence on the prevention of diseases caused by the characteristics of the urban environment and on promoting people's health, due to reducing pollutants and improving outdoor conditions for outdoor activities (such as walking, sports, or, also, others mainly related to cultural, recreational and leisure events). All these measures make cities a driving force for action to achieve the Sustainable Development Goals, principally number 3-*Health and Well-being* and 11-*More Sustainable Communities*, within the action framework proposed by the New Urban Agenda 2030.

This paper, which is the result of a collaboration with the Urban Environment Observatory of Málaga (Observatorio de Medio Ambiente Urbano de Málaga -OMAU), aims to determine a multi-scale and spatial analysis methodology, which adequately integrates the extrinsic and intrinsic factors of a city to delimit differentiated *Urban Bioclimatic Areas (UBA)*, and from there to enable actions and strategies to improve the urban microclimate so that they can be included in specific plans, projects or programmes. Once the methodology has been defined, it is applied to the case study of the city of Málaga, a city located on the southern coast of Spain.

BACKGROUND

Cities are located in places with specific geographical and climatic conditions, which influence the thermal comfort of the people who live there. Likewise, and mainly in medium and large cities, there is an urban microclimate that is differentiated from its surroundings, characterised by a decrease in relative humidity, higher temperatures and a complex wind system in its analysis. (Oke, 1987)

In fact, the city's microclimate has been the subject of numerous theoretical and practical analyses since the 20th century, which have served to establish that the relief, geomorphology, presence of vegetation and surface watercourses are determinants of the microclimate for each location (WMO, 1976). According to Landsberg (Landsberg, 1947), the microclimate is a factor related to land use, since, among other aspects, temperature is affected by the presence of vegetation and water as well as by soil and subsoil conditions. Of particular mention are the theoretical and analytical studies of one of the first environmental ecologists, Ian McHarg (McHarg, 1960), who proposed an analysis method that superimposed sectoral layers to delimit territorial areas with intrinsic conditions and determinants that should be taken into account in environmental planning (Fanger, 1967). Also, on scales more related to architecture, the authors Maurice and Max, established principles of optimal bioclimatic design for heat loss and gain with respect to the comfort conditions of people and the climate of a place (Maurice & Max, 1990), which were already outlined in the four climatic regions proposed by the Olgyay brothers (Olgyay, 1963). However, there are fewer studies that establish the interaction between the city and its territory from the integration of external and internal variables to determine the urban microclimate, a factor that will be undertaken in this research.

The sensation of thermal comfort in a place is subjective and depends on the thermal conditions of the environment, solar radiation, air temperature, temperature of the surrounding surfaces, relative humidity and wind speed, among the most significant extrinsic conditioning factors (Neila, 1990). To the above must be added the person's age, activity they are engaged in, gender and clothes they are wearing (Santamouris, 2001). Taking these aspects into account, the local bioclimatic chart is shown in a graph, through the analysis of dry temperature in °C and relative humidity in %, thermal comfort conditions of 80% to 90% of people in a given place (Olgyay, 1963). The PET (Physiological Equivalent Temperature) index establishes comfort from the objective values (temperature, humidity) and subjective values of the subject (from their sensation of comfort). This index integrates external factors (Ali-Tourdert & Mayer, 2006) with the subjective factors of the individual, which are considered very appropriate for the research. This research is based on the Adapted Comfort Climograph-ACC (Neila, 1990) to determine thermal comfort, using data recorded at various weather stations in the city, in order to subsequently establish the needs and strategies, month by month, to achieve comfort conditions for citizens.

Addressing climate from a mesoscale is truly complex, due to the large number of interacting factors. In this sense, research conducted for Hong Kong Bay (Ng, E., et al., 2004) is of particular mention, which resulted in *The Urban Climatic Analysis Map (UC-AnMap)*, with significant contributions for the assessment and design with wind, beyond the previous assumptions (Garcia & Fuentes, 2005) and other previous manuals (Allard & Santamouris, 2003) whose approach was more generalist. However, the analysis unit selected for Hong Kong is not perfectly accurate for the detailed urban design, since the pixel size is excessively large, not coinciding with the morpho-typological characteristics of the city. Nor does it consider the thermal comfort of people in these areas. Both issues are addressed in the methodology proposed in this research.

Undoubtedly, the urban eco-systemic vision of the city in its territory has had important reflections that have been taken onto account for this research (Ferrão & Fernández., 2013) (Chrysoulakis et al., 2014) (Li et al.,2018) (CDai et al., 2001), where the framework for the understanding of the flows of matter and energy is established, with the support of GIS tools. The relationships between the two systems that affect numerous disciplines are very complex (Dijst et al., 2018) and many previous studies bet on the singularized calculation in a city, such as the Chinese ones (Huiyue et al., 2019) (Zhang et al., 2013) where city form characteristics are incorporated from intensity, efficiency and impact, or those of Russian cities (Arefiev et al., 2015), or African (Nyeko, 2012), or Italian (Pollo et al., 2020), or Spanish (Galan & Perrotti 2019).

On the other hand, remote sensing systems using satellites and, in general, Geographic Information Technologies have introduced a real revolution in the generation of maps and data with a level of updating and detail that was unthinkable two decades ago, with advantages and disadvantages depending on the study to be conducted (Mojtaba et al, 2018). The realisation of maps to visualize the dynamics of the Urban Heat Island (UHI) by remote sensing has validated these systems in the analysis of cities in the territory (Neila et al, 2020). This research is based on geospatial data from the Download Centre of the National Geographic Institute (Centro de Descargas del Instituto Geográfico Nacional -CNIG)[1] and the Urban Environment Observatory (Observatorio de Medio Ambiente Urbano - OMAU)[2] of Málaga to draw up both territorial and urban maps, with the aim of determining urban bioclimatic areas of the city.

In view of the advances and limitations of previous studies undertaken (Bourbia &Boucheriba 2010), (Geros et al., 2005), (Nikolopoulou & Lykoudis, 2007), the main objective of this research is to establish a methodology that integrates territorial and urban factors starting from the thermal comfort of people in hot and cold months. This is in order to delimit urban areas, using Geographic Information Systems, which have differentiated microclimatic conditions in accordance with their urban structure and morpho-typology, and that will be called: Urban Bioclimatic Areas (UBA), differentiating between hot and cold months. The city's characterisation in these UBAs will enable bioclimatic plans, programmes or projects to be drafted, which are differentiated by neighbourhoods or urban districts, specifically addressing the thermal and microclimatic conditioning factors that occur in each period.

METHODOLOGICAL STRUCTURE

The method proposed for this study consists of four distinct stages:

Stage One: Referring to the search and collection of data, where a series of extrinsic variables of a territorial nature were selected, such as climate, relief, slopes, surface water, solar radiation, wind and vegetation, and variables intrinsic to the city, such as density, urban structure, height of buildings, presence of trees in streets and green areas.

Stage Two: Local climate conditions are determined from people's sense of comfort, both in winter and summer, using tools such as climographs, charts and solar arcs. Climate data from several weather stations should be recorded if the city is large. This stage is undertaken thanks to the city's Adapted Comfort Climographs (ACC), conducted for cold and warm months.

Stage Three: Preparation of thematic cartographies, through Geographic Information Systems (GIS), to characterise the territorial and urban structure (neighbourhoods and large neighbourhoods), urban density, hypsometry, clinometry, green areas and ecosystems, solar radiation and wind in different periods (hot and cold), according to comfort conditions throughout the year obtained from climographs. From this information it has been possible to establish what the climatic, territorial and urban determinants of the city for bioclimatic characterisation are. The multi-criteria decisions of these spatial analyses have been contrasted with the contributions of specific studies (Malczewski 2010).

Stage Four: Stage related to the integration and synthesis of results obtained in the previous stages, through the reclassification of values and the application of algorithms using multi-criteria analysis and map algebra, which will allow the city's *Urban Bioclimatic Areas (UBA)* to be delimited. The following figure shows this methodological process

INTRINSIC AND EXTRINSIC CONDITIONING FACTORS OF A CITY, IN THEIR TERRITORIAL CONTEXT, FOR BIOCLIMATIC CHARACTERISATION

The main innovation of the methodology used is the integration of territorial and urban parameters analysed with those obtained with respect to the thermal comfort of people and their differentiated

Table 1. Methodological structure

<table>
<tr><th colspan="6">Stage one: Search and collection of data</th></tr>
<tr><th colspan="2">Climatic data</th><th colspan="2">Territorial data</th><th colspan="2">Urban data</th></tr>
<tr><td colspan="2">Temperature
Relative humidity
Solar radiation
Wind</td><td colspan="2">Clinometry
Hydrography
Vegetation
Soil uses
Ecosystems</td><td colspan="2">Administrative delimitation
Density
Building height
Urban green areas
Tree planting in streets</td></tr>
<tr><th colspan="6">Stage two: Determining local climate conditions from a comfort level</th></tr>
<tr><td colspan="6">Local bioclimatic needs according to Adapted Comfort Climographs (ACC)</td></tr>
<tr><th colspan="2">Solar radiation</th><th colspan="2">Wind</th><th colspan="2">Relative humidity</th></tr>
<tr><td>Cold months, winter</td><td>Hot months, summer</td><td>Cold months, winter</td><td>Hot months, summer</td><td>Cold months, winter</td><td>Hot months, summer</td></tr>
<tr><td>Radiation capture</td><td>Protection from radiation</td><td>Protection from the wind</td><td>Exposure to wind</td><td>Low humidity</td><td>High/moderate humidity</td></tr>
<tr><th colspan="6">Stage three: Urban-territorial analysis and drafting of thematic cartographies</th></tr>
<tr><th colspan="3">Territorial scale</th><th colspan="3">Urban scale</th></tr>
<tr><td colspan="3">Relief maps (clinometric)
Hydrology maps
Vegetation map
Solar radiation map (winter and summer)
Wind maps (winter and summer)</td><td colspan="3">Urban relief maps (clinometry)
Density map
Building height map
Map of urban green areas
Map of tree planting in streets</td></tr>
<tr><th colspan="6">Stage four: Integration and synthesis of results</th></tr>
<tr><td colspan="6">Selection and reclassification of factors and decision-making criteria</td></tr>
<tr><th colspan="3">Territorial scale</th><th colspan="3">Urban scale</th></tr>
<tr><td colspan="3">Selection and reclassification of values.
Period of cold months
Selection and reclassification of values.
Period of warm months</td><td colspan="3">Selection and reclassification of values.
Period of cold months
Selection and reclassification of values.
Period of warm months</td></tr>
<tr><th colspan="6">Map algebra / multi-criteria analysis</th></tr>
<tr><td colspan="6">Result: Bioclimatic characterisation of the city</td></tr>
<tr><td colspan="6">Urban Bioclimatic Areas (UBAs)
Data sheets by Major Neighbourhoods - Cold Months
Data sheets by Major Neighbourhoods - Hot Months</td></tr>
</table>

Source: Compiled by the authors

needs for summer and winter periods. A selection of climatic, urban and territorial factors that directly influence the determination of bioclimatic areas in a city was used for this, in order to analyse and represent them cartographically by creating sectoral maps (using Geographic Information Technologies). Subsequently, these values were reclassified in order to perform an integral analysis of all the variables, using the multi-criteria methodology, and differentiating between cold and warm periods, according to the results obtained by the Adapted Comfort Climograph (ACC).

The Adapted Comfort Climograph (ACC) creates an area of comfort based on the genetic fingerprint of the local people, who are adapted to their climate, the clothes they wear at different times of the year, the metabolic activity that may take place inside buildings or outdoors, and the internal temperature of

Figure 1. Adapted Climate Climograph (ACC) for Málaga, in winter and summer situation
Source: Compiled by the authors

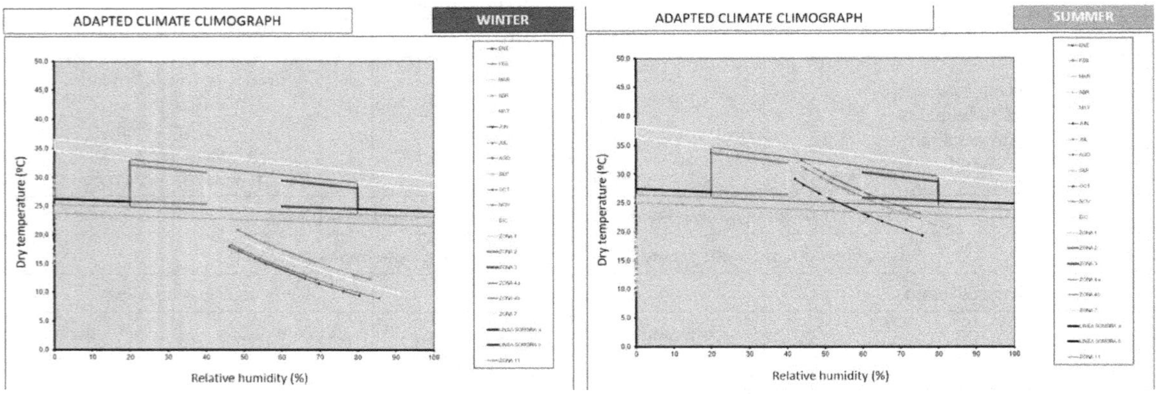

the building's surfaces, a consequence of the external conditions, temperature and radiation, the quality of the construction, mass and insulation, and the internal loads, equipment use, occupancy and activity.

The result is an area of relative temperatures and humidity, which provides conditions of comfort if the building or person is kept shaded and the air is calm, where there will be a maximum of 10% of unsatisfied people (PPI); there is a second complementary area with a maximum of 20% of unsatisfied people. The area varies in position according to the variables described above. Usually we work with only three different areas, linked to changes in clothing, one for summer conditions, when people wear light clothing; for the climographs we have used 0.4 clo, which is light summer clothing. For winter, a 0.9 clo clothing has been used, which is the classic clothing for an interior in slightly colder winter conditions. For transition periods, spring and autumn, an intermediate value of 0.7 clo has been used.

In the area where the comfort area is located, the local climate or microclimate is represented for the months in which the clothes used for that area are worn. If the whole day-types coincide within the area, it means that the climatic conditions are perfect and only the building has to be kept shaded, especially the openings, and the air is calm. If this is not the case, the climograph provides guidelines for designing bioclimatic strategies in outdoor and indoor urban spaces.

For example, it may show that there is a need to provide solar radiation, which will lead to a general design of public spaces, buildings or openings to facilitate this. In this case, the building envelope must be designed with capacity, thermal mass, to store the captured energy. If the need for capturing energy does not apply to the whole day, i.e. there are hours when solar energy must be collected and others when the openings must be shaded, then the climograph indicates the ranges of hours for each period.

The climograph can propose the use of ventilation, either permanent or occasional, in this case generally nocturnal. Ventilation may be thermal, to reduce the sensation of heat, or ventilation to remove excess humidity. If it is nocturnal, it will tell us that it is necessary to store that freshness in the building's mass. It will indicate when internal loads may be sufficient to heat the interior and may propose evaporative cooling or simple supply of humidity, and the use of thermal insulation.

The following is an explanation of the territorial and urban factors taken into account, which will spatially discriminate some areas from others based on their bioclimatic nature, based on numerous environmental analyses (Mc Harg, 1969; Higueras, 2006). Relief, surface hydrography, vegetation, sun

Table 2. Territorial bioclimatic conditions of clinometry

Bioclimatic conditions	Clinometry	
	Flat and moderate 0 -16% slope	**Steep > 16% slope**
Sun exposure on slopes	Few hours of sunshine	High solar radiation
Surface water courses	With plains, meanders, sedimentary areas with high permeability	Relief embedded in valleys and mountains, high run-off and low permeability
Wind	Little wind except for the area along the coastline	Anabatic and katabatic winds, day and night
Green infrastructure and vegetation	Type of species according to soil composition and slope	Type of species according to soil composition and slope

Source: Compiled by the authors

exposure and wind are considered in the territorial scale and density, building height, urban green areas, sun exposure and wind in the urban scale.

In the territorial scale, the topography can be assessed from hypsometry and clinometry. Hypsometry can condition the wind system, sun exposure of slopes, surface watercourses and vegetation growth. From a bioclimatic point of view, clinometry and orientation of the surface areas is an especially important variable to take into account because it conditions, among other things, the amount of radiation received by the surface areas, surface water courses, presence of vegetation and wind movement. The clinometric analysis conducted in this study establishes two major areas: those with flat, moderate slopes (0-4% / 4-8% / 8-6%) and those with steep slopes (> 16%). The following table summarises the bioclimatic conditions in flat, moderate and steep soils.

With respect to surface water courses, these can function as microclimatic regulators, since they are low areas where humidity and water accumulate (either run-off, seasonal or permanent). In these areas, vegetation tends to grow with greater vigour and growth, directly related to the reduction of ambient temperature and the increase in humidity due to the effect of evapotranspiration. Therefore, there are two main areas to consider, which are summarised in the following table.

Closely related to water, the territorial green structure is a major microclimatic regulator, as plants increase evapotranspiration, retain suspended particles, provide oxygen to the air; they are carbon sinks, shade the soil, retain water, slow down erosion, attenuate low-frequency noise and provide shelter and habitat for a large number of species (Aram et al., 2019). Discrimination into two groups has been conducted for this research, as follows:

Table 3. Territorial bioclimatic conditions of the blue infrastructure

Bioclimatic conditions	Blue infrastructure	
	Areas belonging to the surface water network and its area of influence	**Rest of Areas**
Relative humidity	High-Very high	Medium-Low
Temperature reduction	High	Low
Natural ventilation	Low	Medium-High if they are dividing

Source: Compiled by the authors

Table 4. Territorial bioclimatic conditions of the green infrastructure

Bioclimatic conditions	Green infrastructure	
	Areas with little vegetation, crops, bushes, shrubs	Areas with much vegetation, bushes and trees
Evapotranspiration.	Low-medium.	Medium-High
Oxygen supply and Carbon sequestration	Moderate	High
Solar filter	Low	Medium-High
Potential erosion brake	Medium-Low	High
Wildlife refuge	Medium	Medium-High

Source: Compiled by the authors

While the above sectoral analyses may be common in environmental and physical environment studies, sun exposure and natural ventilation studies, such as those presented below, are not as frequent. Wind conditions and incident solar radiation are absolutely key to delimit zones for the bioclimatic characterisation of an area. For this, it is necessary to previously consider the information obtained through the Adapted Comfort Climograph (ACC), mainly if there are significant differences in thermal comfort between the cold months (November, December, January, February and March) with respect to the hot months (June, July and August), as in the case study. If so, it is important to conduct differentiated wind and solar radiation studies for both periods. The sun's path in winter is shorter, with a smaller solar azimuth angle range than in summer and, above all, lower solar altitude. These two circumstances mean that radiation is reduced in the study area during the months of November, December, January, and February, with the minimum path being on 21 December (winter solstice). In addition, this period records the need to capture solar radiation, while in the hot period, which is when most solar radiation is received, there is a need for shading. As for ventilation, this will be necessary in the hottest months, but it will be necessary to protect against it in the coldest months, according to the results of the ACC. For the above reasons, it is necessary to have a detailed knowledge of these conditions, both at territorial and urban scales.

Urban morpho-typology has been analysed, with maps of building density and heights, and the green and blue infrastructure present in the city in order to study the intrinsic bioclimatic conditioning factors

Table 5. Radiation, ventilation and humidity requirements for comfort in cold and warm months

Variables	Season of the year	
	Cold months, winter situation	Hot months, Summer situation
Solar radiation	Capture of solar radiation	Protection from solar radiation, shading
Wind	Wind protection	Exposure to wind, breezes and mountain-valley winds
Relative humidity	Low relative humidity	High or moderate relative humidity if temperature is not excessively high

Source: Compiled by the authors

Table 6. Bioclimatic conditions at urban scale for cold and warm months

URBAN BIOCLIMATIC CONDITIONS	Season of the year	
	Cold months, winter situation	Hot months, summer situation
Density	Medium density zones that allow winter sunlight to enter the area	High density areas with shaded streets, but oriented according to prevailing winds
Height of the building	Urban canyon that allows sunlight to enter the city	Urban canyon to allow wind to enter
Urban green areas	Deciduous trees to allow sun exposure	Proximity to lush green areas within a proximity radius
Alignment of trees in streets	Deciduous trees in streets to allow for sun exposure	Street planting of trees to shade pavements
Solar radiation	Need for high solar radiation on streets and public spaces	Need for protection from direct solar radiation, shading of streets, etc
Wind	Need for wind protection	Need for exposure to mountain-valley winds and breezes

Source: Compiled by the authors

of the city, since both green areas and the presence of water are important regulators of the microclimate, with differentiated effects in winter and summer.

SUMMARY THROUGH MULTI-CRITERIA ANALYSIS: MAP ALGEBRA

For the preparation of environmental assessment studies, plans, programmes and urban planning and/ or territorial scale projects, it is necessary to seek compatibility between the pre-existing physical and natural environment with the different uses that are potentially to be implemented in a given location. The joint assessment of the territory through comprehensive studies using multi-criteria assessment is an effective tool, where the use of Geographic Information Systems (GIS) is an important technical support. According to Barredo (Barredo, 1996): *"Multi-criteria (and multi-objective) assessment is a set of techniques used in multidimensional decision and assessment models, within the field of decision making"*, offering the opportunity to obtain a balanced analysis of all facets of planning problems (Nijkamp &Van Delft, 1977).

GIS allows the combination of different criteria that can be assigned different degrees of importance or weights through multi-criteria analysis; however, this combination must be done in such a way that one criterion does not overshadow the others, which is a determining factor in achieving the desired characterisation. The relevance of the method considers the incorporation of determining factors of the relationships between the city and the context, with the criteria of integration, balance, and interaction between them. This requires the factors to be truly determinant, so that they are easily aggregated from both scales and are correctly weighted. To this end, we have followed the contributions of Ferrão & Fernández (2013) Chrysoulakis, et al., (2014) and Dijst et al., (2018) among others.

In order to integrate all the factors examined in this study, a reclassification was previously conducted, where the rating scale ranged from 1 (extremely low) to a maximum value of 5 (extremely high). This 5-value scale made it possible to differentiate between minimum and maximum values, with the existence of an intermediate value (3).

Table 7. Reclassification criteria

Scale for the valuation of bioclimatic variables				
Extremely low	**Low**	**Medium**	**High**	**Very High**
1	2	3	4	5

Source: Compiled by the authors

The final goal of this reclassification is to achieve the bioclimatic characterisation of Málaga's urban areas using map algebra. In each of the reclassification maps of the territorial and urban determinants, the valuations are represented chromatically, with the criterion that the most favourable values, with respect to the bioclimatic analysis and according to the period of the year analysed, are in green; intermediate scores are represented with yellow, and the least favourable valuations are in red.

The general criterion for reclassification is that if the determinant is favourable in the period taken into account and according to the needs and strategies detected in the climographs, higher values are assigned (green) and if it is less favourable the score will be lower (red). Therefore, the reclassification of the variables has been established according to the thermal needs detected for each period, so that for some of them this process has had to be conducted twice, and in a different way, according to these criteria. For example, areas highly exposed to wind in summer have been rated very positively compared to less exposed areas, according to the data provided by the ACC. However, areas with high exposure during the cold months were rated as very unfavourable compared to protected areas, also based on the ACC results.

Once the reclassification of the parameters taken into account has been conducted, the combination of factors can be undertaken using map algebra to subsequently establish the bioclimatic characterisation of the city and the delimitation of the *Urban Bioclimatic Areas* (UBA). As mentioned above, a distinction has been made in this analysis between hot and cold months, due to the different bioclimatic strategies required. The following is the combination of variables conducted for each period, which will allow us to obtain two delimitation maps of *Urban Bioclimatic Areas* (UBA)The selection of the determining factors at both territorial and urban scale is a decisive step in the characterisation pursued. For this purpose, the territorial scale factors are selected from the extrinsic geomorphological conditions surrounding the city, such as the relief, hydrographic basins, sunny slopes, and origin of the prevailing wind. The assessment ranges for each of these factors are derived from the climogram conditions of the site, where the needs for thermal comfort in both winter and summer of the city are determined. This

Table 8. Reclassification - Exposure to solar radiation in cold period

Parameter taken into account	Bioclimatic suitability	Value - reclassification
<250,000 Wh/m^2	Extremely low	1
250,000 Wh/m^2 to 350,000 Wh/m^2	Low	2
350,000 Wh/m^2 to 450,000 Wh/m^2	Medium	3
450,000 Wh/m^2 to 550,000 Wh/m^2	High	4
550,000 Wh/m^2 to 603,000 Wh/m^2	Extremely high	5

Source: Compiled by the authors

Figure 2. Fragment of sun exposure map and reclassification map in the cold months for the city of Málaga
Source: Prepared by the authors based on data from the National Geographic Institute (IGN).

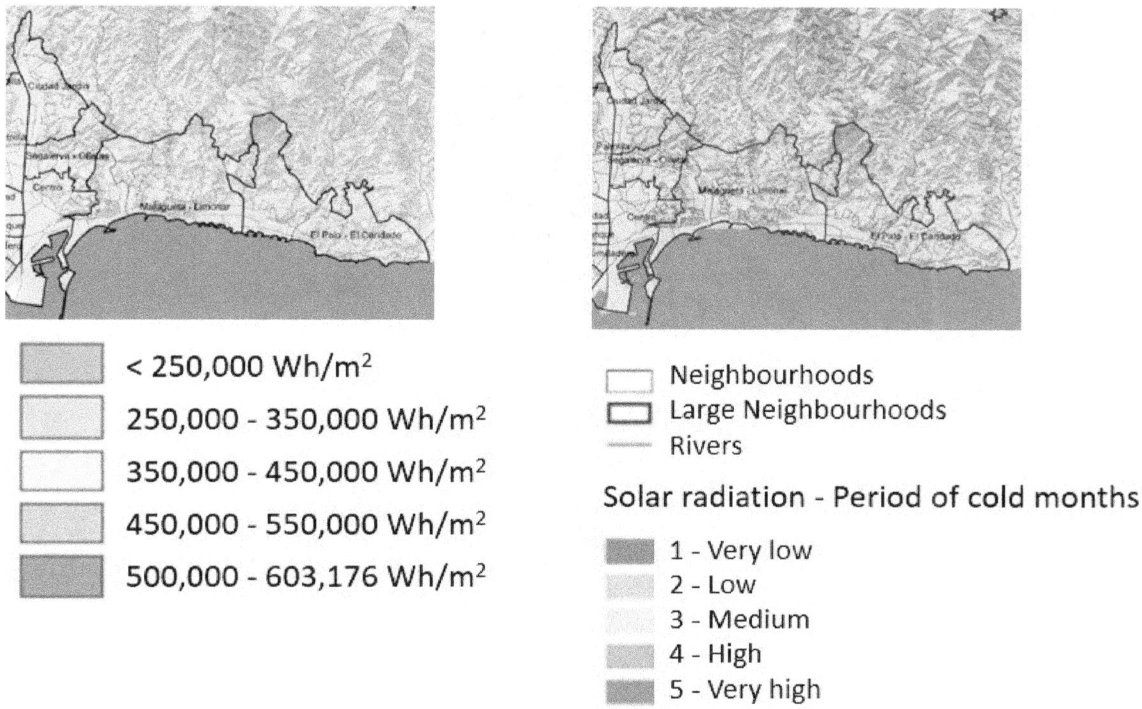

will vary from place to place, and will determine the fact that wind, for example, can have positive or negative values for summer or winter in different contexts.

THE CASE STUDY: THE CITY OF MÁLAGA (SPAIN)

Málaga is a city located in Andalusia, southern Spain. Its municipal area has a population of 578,460 inhabitants and an extension of 394.98 Km^2 (SIMA, 2021) and the urban fabric has a heterogeneous structure, with a variety of urban areas and a unique geographical situation, with the Mediterranean Sea

Table 9. Variables for the multi-criteria analysis by period of the year taken into account

Period of the year	Variables taken into account in the multi-criteria analysis
Period of cold months	Winds (cold months) + Solar radiation (cold months) + green areas and ecosystems (natural and artificial) + area of influence of tree-lined streets (equidistance of 15 m from street axis) + area of influence of parks (sup. > 20,000 m^2) + urban density
Period of hot months	Winds (hot months) + Solar radiation (hot months) + green areas and ecosystems (natural and artificial) + area of influence of tree-lined streets (equidistance of 15 m from street axis) + area of influence of parks (sup. > 20,000 m^2) + urban density

Source: Compiled by the authors

Figure 3. Hypsometry and spatial delimitation of the large neighbourhoods of the city of Málaga
Source: Prepared by the authors based on data from the National Geographic Institute (IGN in Spanish).

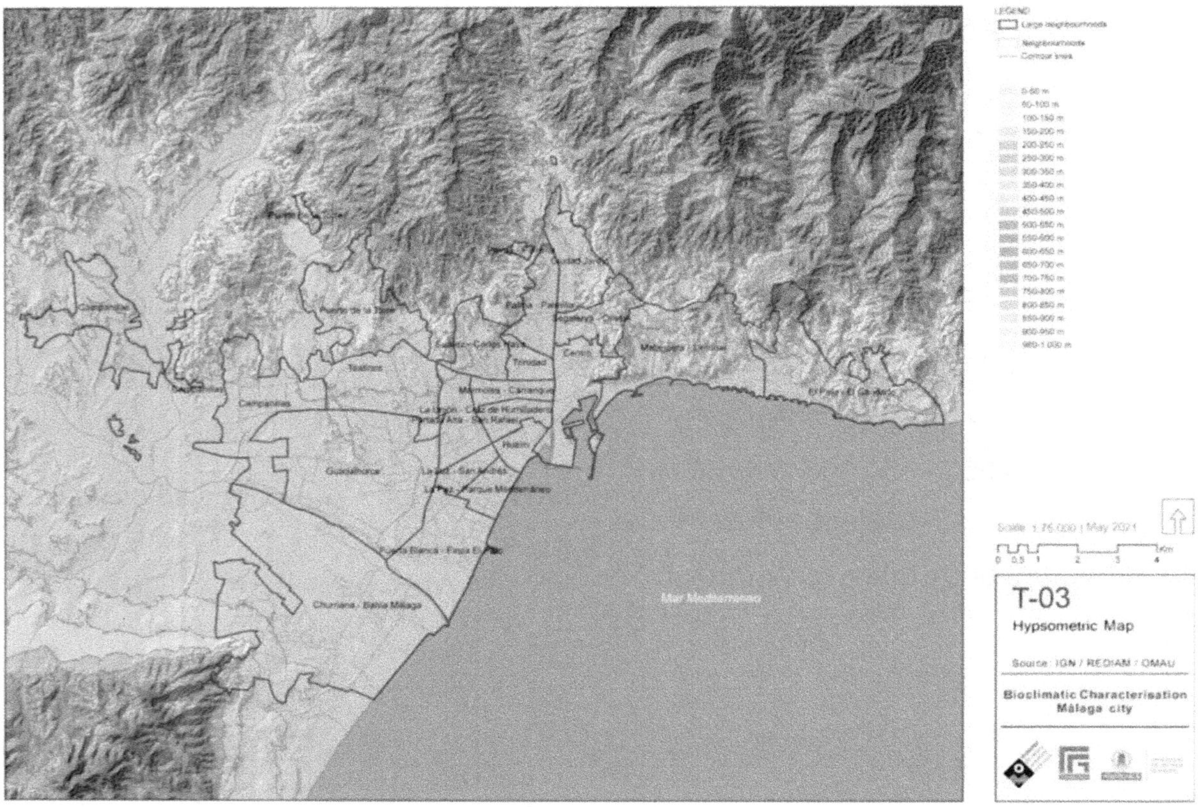

to the south, the Montes de Málaga five kilometres to the north-east (with the 1032 m high Cresta de la Reina) and the Montes de Málaga Natural Park; to the west it is crossed by the Guadalhorce River in an enclosed valley. The Guadalmedina rRiver flows from the Montes de Málaga, with five well-defined stream basins (Arroyo de las Vacas, Arroyo Chaperas, Humaina, Hondo and Arroyo de Los Frailes).

As mentioned, there are very heterogeneous urban fabrics, some very dense, others less so, areas with greater pollution due to anthropic activities, presence of urban parks of different sizes, practically flat areas and others with considerable slopes, etc. Administratively, the city is divided into neighbourhoods (297), which in turn are grouped, according to their characteristics, into Large Neighbourhoods (20). All these characteristics, typical of the urban structure and the territorial context in which the city is located, provide unique conditions for the urban microclimate and, therefore, for the bioclimatic characterisation of these areas, which is the purpose of the study.

The city also has a Climate Plan, which is the framework for this study. This Plan, called ALICIA, began to be drafted in December 2018 and has a direct association with the 2015 Málaga Urban Agenda (linked to the United Nations New Urban Agenda) and the Covenant of Mayors for Climate and Energy. It was approved in March 2020 and is structured around four main points: urban model and mobility, urban metabolism, biodiversity and social and economic cohesion. In strategic line 1 of the first point, dedicated to the urban model, its main aim is *"Adapting planning to the Urban Agenda and the Climate*

Plan, the compact, complex and proximity city" and one of the proposed actions is defined as follows: *Article 16: "To incorporate climate and energy efficiency criteria into urban planning. The Climate Plan requires that the tools for action on climate variables be taken into account in the city's urban plans, for which the results of this Plan associated with urbanism and planning should be included in the next reviews of the Plan"*

In this context, the study will provide information on the bioclimatic characteristics and conditions of the different neighbourhoods of the city, as issues to be taken into account and considered in the urban planning of the city and in other types of urban interventions, both for the existing city and for new developments.

DETERMINATION OF LOCAL CLIMATE CONDITIONS FROM A COMFORT LEVEL

As explained in the paper's methodology, the first step consists of collecting data and determining local climate conditions from the comfort point of view, through producing Adapted Climate Climographs (ACC) of the city of Málaga. Climatic data have been obtained from the State Meteorological Agency (Agencia Estatal de Meteorología - AEMET) in order to prepare these charts, specifically *Airport, Meteorological Centre and Port* and also from the Environmental Quality System stations of the Junta de Andalucía, specifically *El Atabal, Hilera and Martiricos.*

The following figure shows the isopleth diagrams, which show the months of the year on the abscissa axis and the hours of the day on the ordinate axis. Each cell appears in a different colour with the following legend: in dark blue, the need for solar radiation to achieve thermal comfort; in light blue, the need for internal loads; in grey, the cells showing the periods of thermal comfort for 80% of people (with 20% dissatisfied); in white, the cells with thermal comfort for 90% of the people (with only 10% dissatisfied); in yellow, the need for ventilation in order to feel comfortable; and finally in red, the hours of the day and months with excessive heat. Graphs corresponding to the climate change projections were also made, where the difference with respect to the previous graphs are crucially seen, with the presence of more hot hours than in the current situation.

The interpretation of the above diagrams shows three periods throughout the year: cold (winter), hot (summer) and intermediate (spring-autumn) months, with differentiated needs. They show the existence of thermal comfort in the middle hours of the day in the spring and summer months; they set out the need for solar radiation in the mornings in almost all months of the year, and the need for solar protection from the beginning of March to mid-September (south and west, very conflicting orientation). Natural ventilation would be required from mid-May to the end of August, from morning until dusk.

In addition to temperature and humidity, another determining factor for bioclimatic characterisation is wind. In this sense, the direction of the prevailing winds in the bay where the city of Málaga is located is shown, and in which the existence of a dominant direction throughout the year (WNW-SSE) can be appreciated, with two periods with different directions. In fact, in winter, winds predominate from inland to the coast, while in summer the direction is reversed, from the sea to inland. This circumstance is incredibly positive from a bioclimatic point of view for summer conditions, but negative during the colder months.

Finally, the study of solar radiation throughout the year will also be decisive, given that the sun's path is different in cold and warm periods, and that the needs to achieve thermal comfort are also different,

Figure 4. Isopleth plots according to climate change projections (RCP4.5 and RCP8.0) and data from the Martiricos weather station, Málaga
Source: Compiled by the authors with data from AEMET

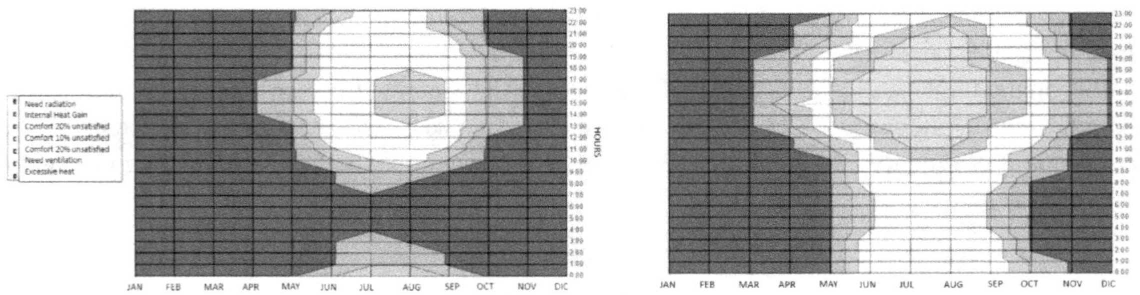

a relationship can be established between these variables to help the architect in making decisions on his projects, whether new construction or renovation, on the public space or on the building. For this purpose, solar arc studies or solar charts are usually conducted for each specific location.

The cylindrical solar chart represents the path of the sun corresponding to the latitude of each location. It is a chart that determines the two angles that define the sun's position on the celestial vault throughout the year: on the abscissa axis the solar azimuth angle (horizontal) and on the ordinate axis the solar altitude (vertical angle). The typical days of each month are represented by curves, where three singular points are highlighted: ortho (sunrise), zenith (maximum solar height, at 12 solar hours each day of the year) and sunset (dusk). The following figure shows how there is a need to capture solar radiation during the cold months (January, February, March), when the solar path is shorter, and the mornings of the intermediate months (October, April, May), together with a need for radiation protection in the warmer months, when the solar path is longer, mainly in the evening hours (June, July, August and September)

URBAN-TERRITORIAL ANALYSIS AND THEMATIC MAPPING

The next methodological stage consists of preparing all the sectoral analysis maps, from the territorial and urban scale, to then proceed to their reclassification from the point of view of their bioclimatic suitability (differentiated by periods). This analysis has been supported by Geographic Information Systems-GIS (ArcMap 10.8.1, ESRI) and comprehensive national databases. The complementarity and integrated analysis of these sectoral maps will help to understand the relationships with this scale, which condition Málaga's bioclimate, both in winter and summer.

Table 10. Wind direction by months in 16-heading Rose

Period	Cold							Hot				
Month	NOV	DEC	JAN	FEB	MAR	APR	MAY	JUN	JUL	AUG	SEP	OCT
Direction	WNW	WNW	WNW	WNW	WNW	SW	S	S	SSE	SSE	S	SW

Source: Compiled by the authors with data from AEMET

Figure 5. Cylindrical solar chart for Málaga (hot and cold period)
Source: Climate Consultant 6.

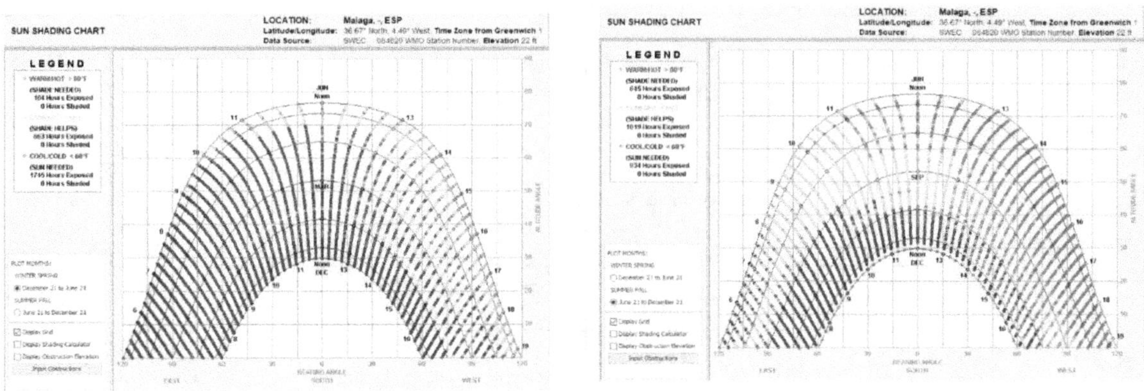

As can be seen in the following figure, a high percentage of Málaga's neighbourhoods are located on fairly flat terrain, with slopes ranging from 4-8%. However, there are certain urban areas that are located in areas with steep slopes (> 16%). Among the latter, we can highlight the following: El Palo-El Candado, Malagueta-Limonar, Segalerva-Olletas, Ciudad Jardín, Palama-palmilla, Suárez-Carlos Haya and Puerto de la Torre.

The research addresses a spatial approach of the qualitative aggregation of determinants at urban and territorial scale. Other authors have established comparative relationships via GIS for spatial analysis of territorial and urban data. Italian research by Riccardo Pollo, Matteo Trane, Matteo Giovanardi, (2020) and Galan J., Perrotti D., (2019); Chinese research by Huiyue W., Chen D., Duan H., Yin F., Niu Y., (2019) are valued for their methodology

The municipality is surrounded by areas of great eco-systemic and landscape value, such as the Montes de Málaga, to the north, or the mouth of the Guadalhorce River, to the south-west, where there is a small area of marshes. In the more urban area there is also a well-established structure of parks, as well as the presence, in some neighbourhoods, of abundant vegetation in private plots, streets and squares of the city. In this sense, the improvement of the microclimate is considered due to the presence of tree-lined streets, depending on the density of trees along the alignment, and of these large green spaces (surface area > 20,000m^2), additionally establishing an area of influence around them (150 m).

The analysis of incident solar radiation shows that, in general, this area receives a large amount of Wh/m^2 throughout the year, especially in the hottest months (when protection is needed as the main bioclimatic strategy). In the colder months, when there is a need for solar gain, south-facing slopes receive the most solar radiation. In this sense, the areas of El Palo-El Candado, Malagueta-Limonar, Puerto de la Torre and Segalerva-Olletas are the most benefited, due to their geographic configuration.

It should be noted that this is a general study of the entire municipality of Málaga. The analysis of solar radiation at a more urban level and concerning buildings would require specific studies, almost at plot level, where street orientations, street width-to-building height ratio, presence of trees, etc., are taken into account.

In this sense, urban density (inhabitants/hectare) has also been taken into account. The analysis shows that there are areas with different densities. The most populated neighbourhoods are generally located in more central positions with respect to the municipality and are as follows: Parque del Sur, 26

Table 11. Analysis of the variables of the urban-territorial environment

	Variable	Map	Objective	Ranges
ANALYSIS OF URBAN-TERRITORIAL SUPPORT	Administrative delimitation	Large Neighbourhoods and Neighbourhoods of the city of Málaga	Establish the spatial units of analysis	Delimited according to Málaga General Plan documents
	Neighbourhood population density	Density inhab/ha	Determine the densest areas from the least dense ones	< 50 inhab/ha 50-150 inhab/ha 150-250 inhab/ha 250-450 inhab/ha > 450 inhab/ha
	Relief_altitude	Hypsometric	Determine areas according to altitude	50 m intervals
	Relief_slopes	Clinometric	Determine flat, moderate and steep areas	0-4% slope 4% -8% slope 8-16% slope > 16% slope
	Green areas and ecosystems surrounding the city	Green areas and ecosystems Urban trees and area of influence of large parks	Delimit the green areas and natural, anthropic ecosystems on the perimeter and interior of the city	Natural, anthropic ecosystems, private green areas, tree-lined streets (tree density per street), urban green areas >20,000m^2 and their areas of influence (150 m)
	Radiation in hot months	Solar radiation on the substrate in the hot months established by the local climograph	Know the areas with the highest solar irradiance in the warm months	> 450,000 Wh/m^2 450,000-550,000 Wh/m^2 550,000-600,000 Wh/m^2 600,000-705,000 Wh/m^2
	Radiation in cold months	Solar radiation on the substrate in the cold period established by the local climograph	Know the areas with the highest solar irradiance in the cold months	> 250,000 Wh/m^2 250,000-350,000 Wh/m^2 350,000-450,000 Wh/m^2 450,000-550,000 Wh/m^2 550,000-603,176 Wh/m^2
	Wind in the cold months	Neighbourhoods in the cold period established by the local climograph	Determination of wind exposure, windward and leeward areas and wind channels. Cold months	Exposure to wind: Extremely low, low, medium, high, extremely high
	Wind in hot months	Neighbourhoods in the hot period established by the local climograph	Determination of wind exposure, windward and leeward areas and wind channels. Warm months	Exposure to wind: Extremely low, low, medium, high, extremely high

Source: Compiled by the authors

de Febrero, La Palmilla, Miraflores de los Ángeles, Suárez, La Breca, Camino de Suárez, Los millones, Nueva Málaga, 4 de Diciembre, Santa Marta, La Asunción, San Rafael, Alaska, Regio, San Carlos Condote, Haza Onda, Las Delicias, Nuevo San Andrés, Vistafranca, La Luz, Los Girasoles and La Paz.

Finally, as mentioned in previous sections, the wind analysis was conducted on a territorial scale, determining the areas most and least exposed to natural ventilation in the different periods considered. It has been found that there is a huge variation in wind exposure during the cold months, where the most affected areas are located in the Guadalhorce valley, compared to the warm months, where the most exposed areas are concentrated along the seafront.

Figure 6. Clinometric map, neighbourhoods and large districts of Málaga
Source: Prepared by the authors based on data from the National Geographic Institute (IGN in Spanish).

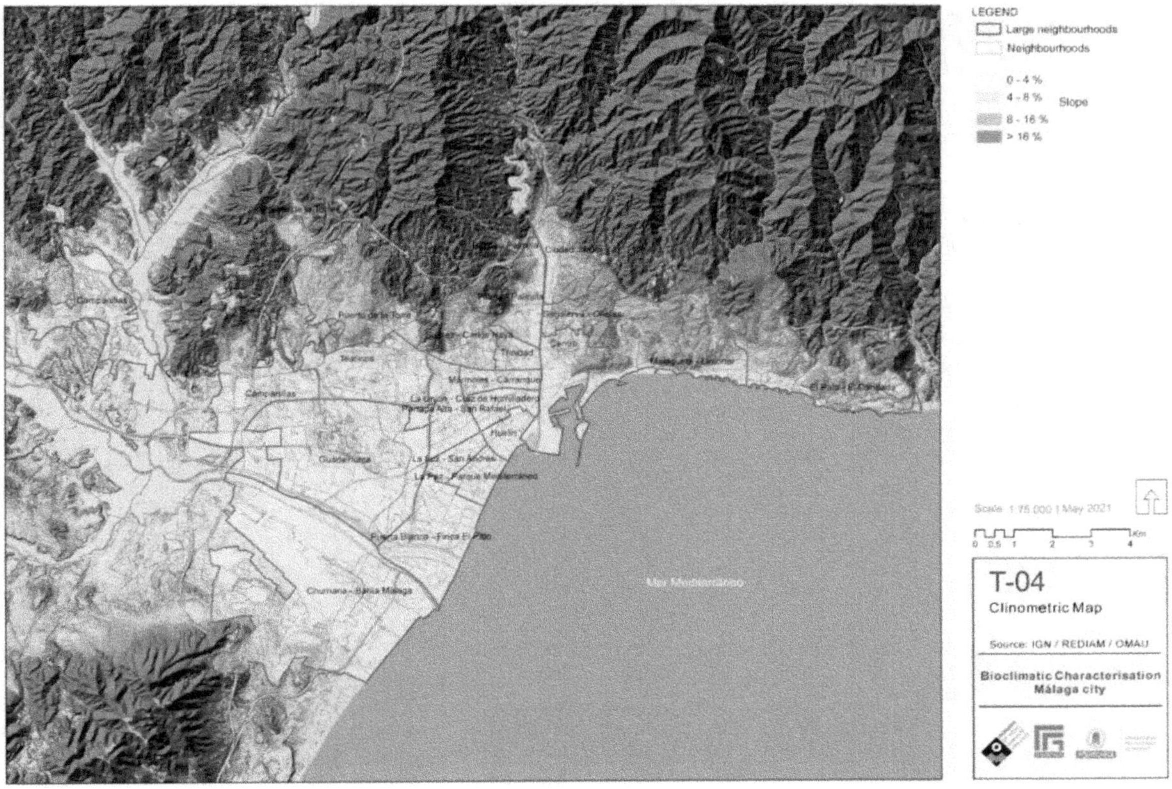

INTEGRATION AND SUMMARY OF RESULTS

As we have seen in previous points, the urban microclimate is unique and differentiated from that of the areas surrounding the city. For this study of bioclimatic characterisation of the city of Malaga, the analysis of the territorial scale has been considered relevant for the definition of urban bioclimatic area for the following reasons:

1. Málaga is a city by the sea. From a bioclimatic point of view, the sea is a thermal regulator that softens temperatures both in winter and summer, establishes the determination of sea breezes, and provides a complementary level of humidity to the environment by direct evaporation (more pronounced in the months of July and August)
2. The city is located on a fluvial plain of the Guadalhorce River and the Guadalmedina River valley, where there are conditions of higher humidity and channelling of the winds, which have a NE component in most months of the year, coinciding with the Guadalhorce valley
3. Málaga is located on the slopes of the Montes de Malaga, with some of its neighbourhoods in areas of moderate and high slope, which conditions sun exposure, as they have a south, south-east orientation of the slopes (sunny spot), which favours the capture of solar radiation.

Figure 7. Green areas, natural and anthropic ecosystems of Málaga
Source: Prepared by the authors based on data from the IGN and the Urban Environment Observatory

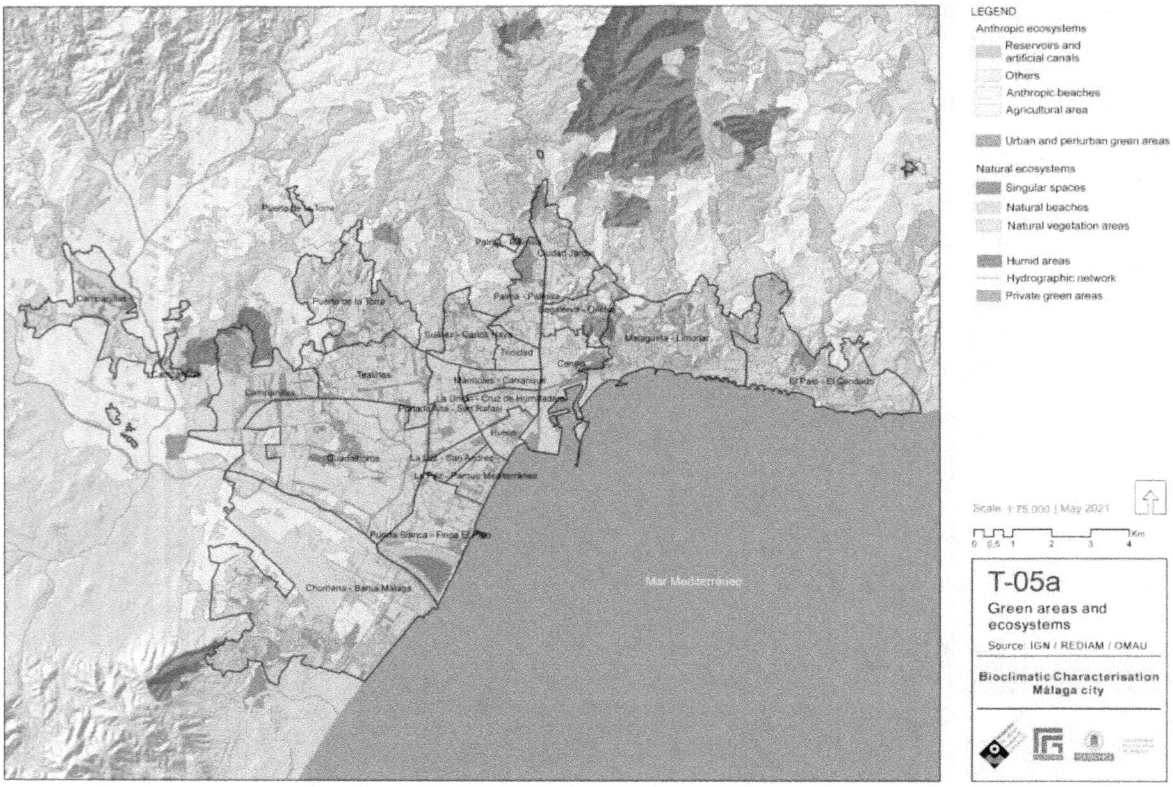

4. Important ecological areas are appearing around the city and many of them will become part of Málaga's Green Belt. These are areas with permeable, natural soil, and with the presence of vegetation (sometimes tall and sometimes short or crops), which function as thermal regulators of the whole and contribute both to the sequestration of carbon produced in the city and to the supply of oxygen, in an incredibly significant manner.

These four factors have a different influence on the hot and cold months and, therefore, will also result in different strategies in each case. However, it should not be forgotten that there are more cold months than warm ones, that almost the entire morning period is cool, cold or very cold, and that there are lengthy periods of thermal comfort in the midday hours; there are also periods of over-heating in the early afternoon in June, July, August and September, always with high ambient humidity values.

In terms of urban scale, as we have seen above, Málaga is made up of heterogeneous urban fabrics, the result of its growth over the centuries. This heterogeneity is reflected in the 11 ordinances of Málaga's General Urban Development Plan. The study of density, the location of urban green zones and areas, the degree of exposure to natural ventilation and solar radiation will also be decisive in the bioclimatic characterisation of the city.

In this methodological stage, the reclassification of the various aspects analysed is conducted, as explained in the methodology section, in order to subsequently conduct the multi-criteria analysis, sup-

Figure 8. Urban trees and areas of influence of Málaga's large parks
Source: Prepared by the authors based on data from the IGN and the Urban Environment Observatory

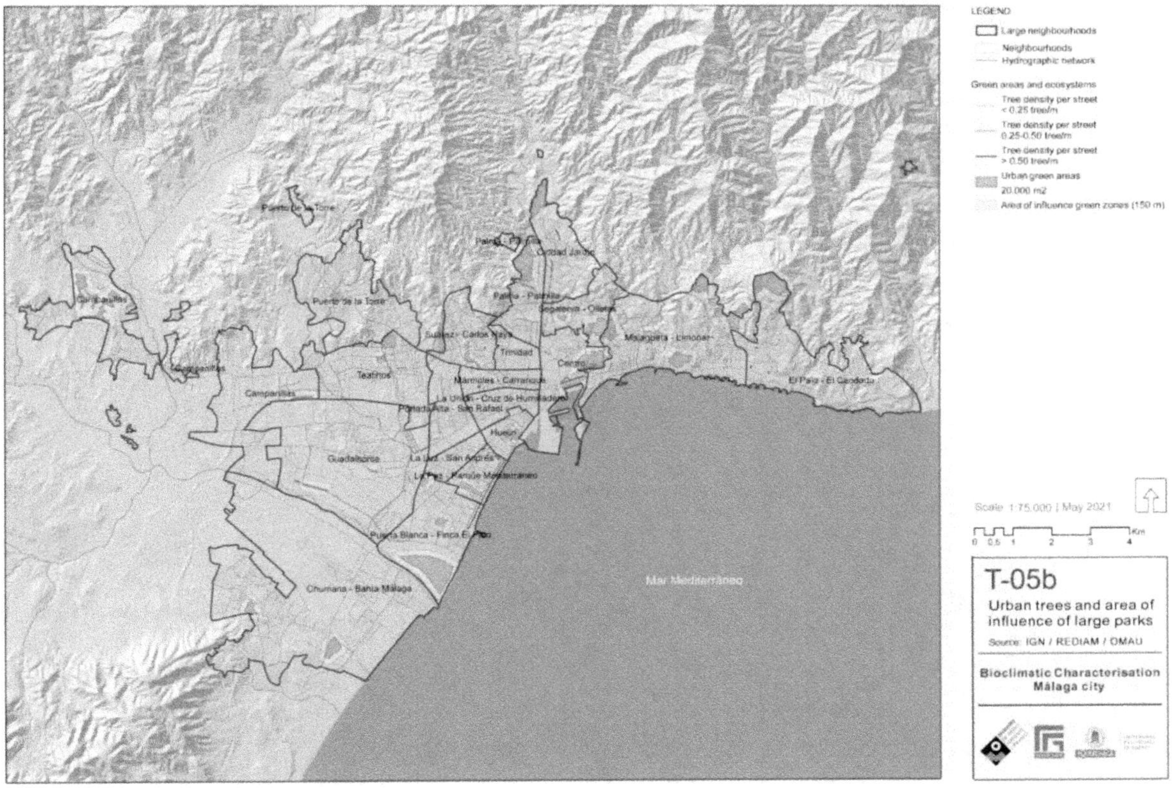

ported by map algebra (conducted with GIS). The following is an example of some of the results of this process. Firstly, those obtained with respect to wind assessment:

The differentiation of the assessment of the same aspect with respect to the period taken into account is emphasized. For example, for the ventilation analysis the resulting maps are as follows:

As for the valuation of green areas, they function as "refreshing" elements within the city, both in winter and summer. Through evapotranspiration, they improve the environment, trap CO_2, supply oxygen, lower the temperature and increase the humidity of the environment due to a photosynthesis process. In cities, vegetation can be found within the general system of public green spaces, in streets, and in the private landscaped areas of some plots. The presence and density of green areas is a determining factor for the city's microclimate. Its study has been detailed in two analyses, on the one hand, the one corresponding to the surface of green areas in urban and peri-urban areas and, on the other hand, the number of trees per linear metre present in streets, both with their respective areas of influence.

With regard to the justification of the categories and valuation adopted, the city of Málaga is developing the Green Infrastructure Plan in which the land outside the urban land is categorised into areas of green infrastructure, private green areas, artificial canals, beaches, agricultural land and others. To these are added the urban green areas, with the determinations of singular spaces, beaches, areas of natural vegetation and wetlands. For this study, the integration of all these aspects is understood to visualise the

Figure 9. Solar radiation at the winter solstice in Málaga
Source: Prepared by the authors based on data from the National Geographic Institute (Instituto Geográfico Nacional –IGN).

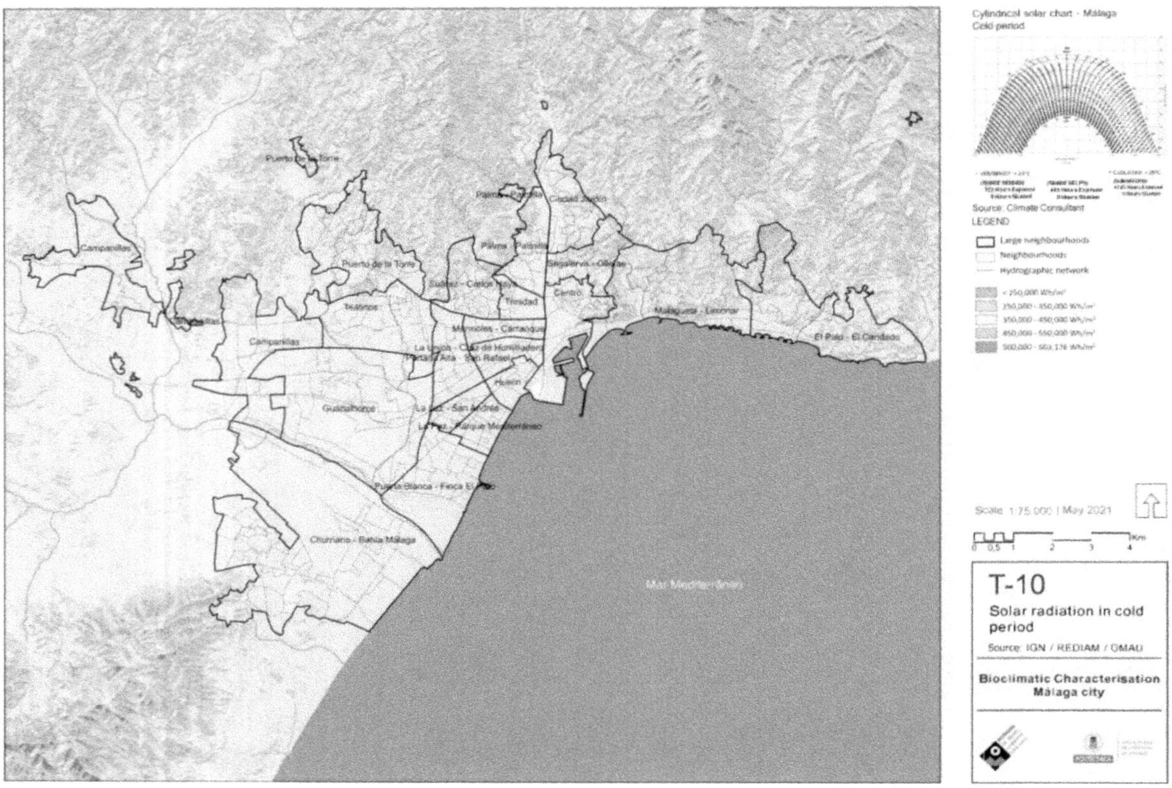

concept of green infrastructure and ecosystems, combining the territorial and urban scales in a spatial and environmental continuum.

Furthermore, it should be borne in mind that green areas with a larger surface area and density of trees have a positive bioclimatic effect on the urban fabric and people (parks larger than 20,000 m^2 were taken into account for this study). For this reason, these areas and their area of influence (150m from the perimeter of these areas) have been integrated in the final calculation, which has been valued with 1 extra point. The same assessment has been made with respect to the presence of private green areas.

Below are excerpts from some of the maps made for the characterisation and assessment of the aspects discussed above. In this case, we observe the presence of trees in streets.

Finally, once all the parameters considered have been reclassified, a map algebra is created that integrates them jointly, using Geographic Information Systems, and in which the Urban Bioclimatic Areas (UBA) are determined geographically, with different degrees of bioclimatic suitability of the municipality of Málaga, for hot and cold months.

In the case of the cold months, it is clearly observed how neighbourhoods more exposed to natural ventilation, located in the Guadalhorce valley, and with worse orientation with respect to solar gain have a lower valuation with respect to their bioclimatic characterisation. On the other hand, large neighbourhoods such as Malagueta-El Limonar or El Palo-El Candado have better ratings, among other aspects

Figure 10. Urban density of Málaga
Source: Prepared by the authors based on data from the IGN and the Urban Environment Observatory

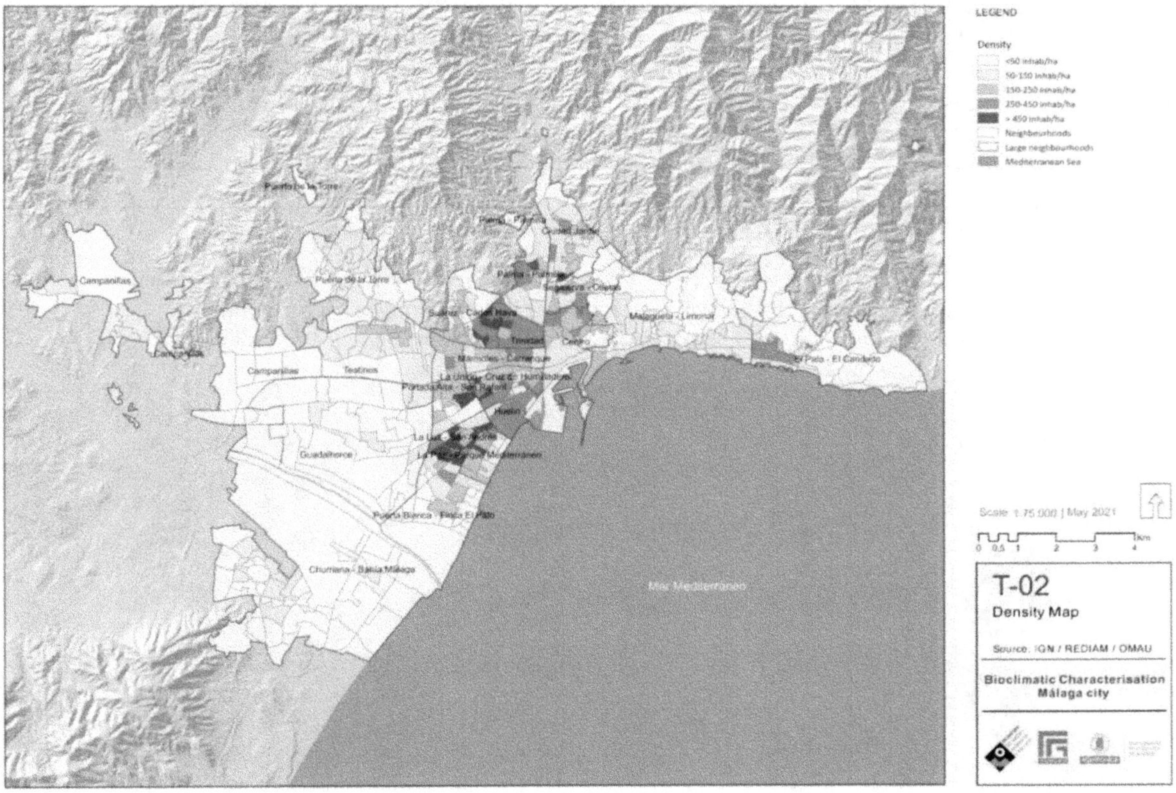

due to their good orientation (southern slope) and, therefore, better exposure to solar radiation during this period, as well as being protected from the prevailing cold winds.

During the warm months, the Urban Bioclimatic Areas (UBA) with the highest ratings do not coincide exactly with those of the cold period. In this case, the coastal front at the mouth of the Guadalhorce River is the most highly valued, due to its good exposure to the wind and the presence of vegetation. However, inland neighbourhoods, located in the central part of the municipality, are less well rated due to

Table 12. Reclassification - Wind exposure in hot period

Parameter taken into account	Bioclimatic suitability	Value - reclassification
Large neighbourhoods on the waterfront	Very High	5
2nd line from the waterfront	High	4
3rd line from the waterfront	Medium	3
4th line from the waterfront	Low	2
No breeze	Extremely low	1

Source: Compiled by the authors

Table 13. Reclassification - Wind exposure in cold period

Parameter taken into account	Bioclimatic suitability	Value - reclassification
Windward with respect to prevailing wind	Very High	1
2nd windward line	High	2
3rd windward line	Medium	3
Sheltered	Extremely low	4
Very sheltered	Extremely low	5

Source: Compiled by the authors

Figure 11. Reclassification for multi-criteria analysis. Wind exposure - cold months
Source: Prepared by the authors based on data from the IGN and the Urban Environment Observatory

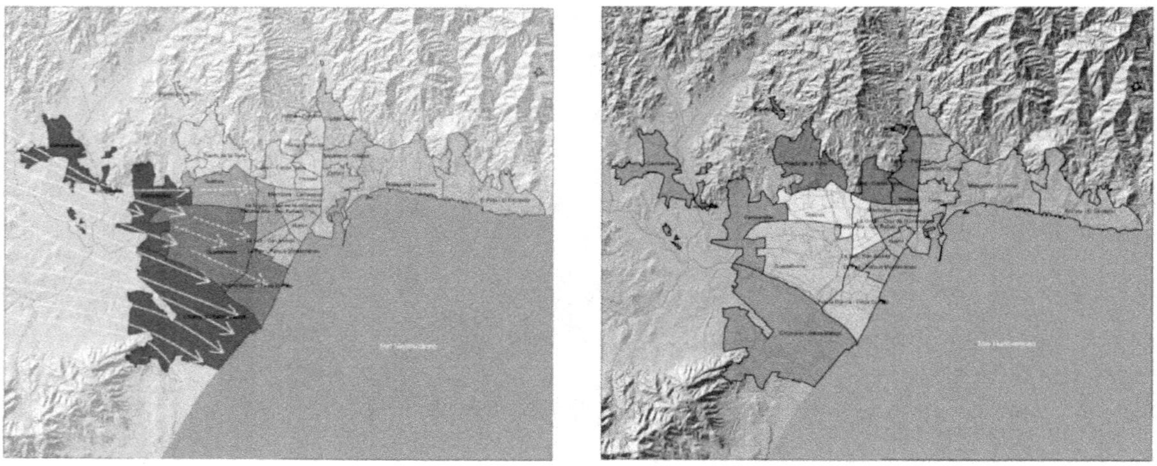

Figure 12. Reclassification for multi-criteria analysis. Wind exposure - warm months
Source: Prepared by the authors based on data from the IGN and the Urban Environment Observatory

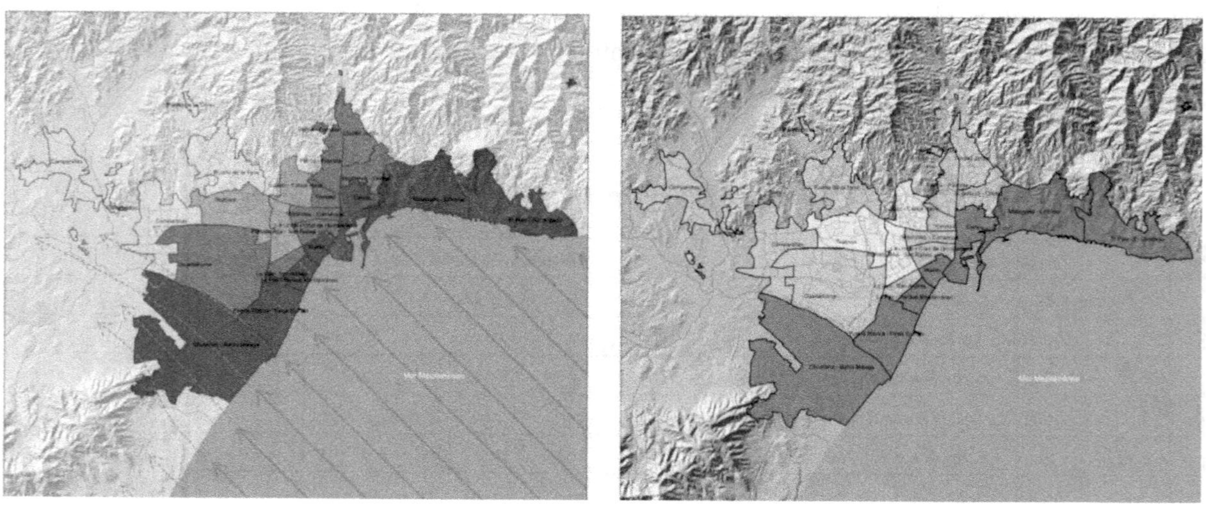

Table 14. Reclassification - Green areas and ecosystems

	Parameter	Aptitude	Reclassification value
Natural ecosystems	Unique spaces_	Very High	5
	Wetlands	High	4
	Beaches	High	2
Anthropic ecosystems	Urban and peri-urban green areas	Medium	3
	Agricultural area-others	Medium	3
	Beaches	Medium	2
Private green areas	Private green areas	Extra valuation	1
Large green areas	Urban green area > 20,000 m^2	Extra valuation	1
Influence Area	Area of influence of Large Urban Green Areas - 150 m	Extra valuation	1

Source: Compiled by the authors

their high density, low exposure to ventilation and lower presence of green areas and trees in the streets. This is the case, for example, of Trinidad, La Unión-Cruz del Humilladero or Portada Alta-San Rafael.

In other words, the city's Urban Bioclimatic Areas (UBA) vary throughout the year, depending on the needs that arise to achieve comfort according to the period of the year.

Figure 13. Reclassification - Area of influence of green areas and urban trees (map fragment)
Source: Prepared by the authors with data from IGN and OMAU

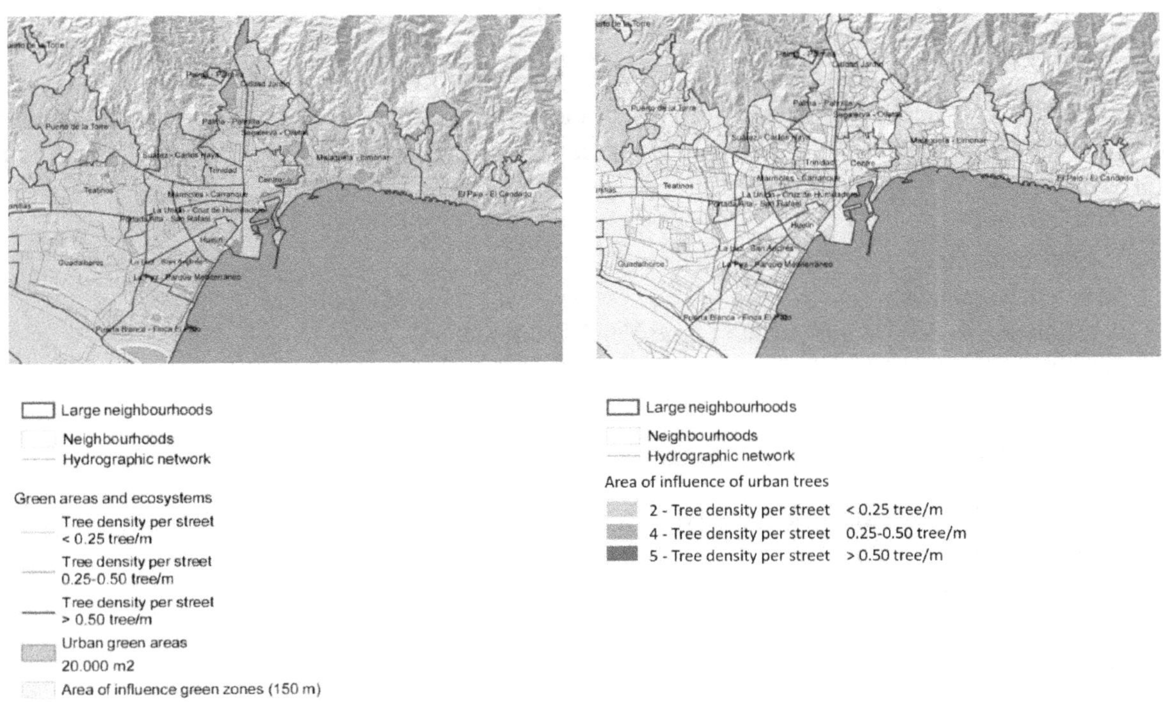

Figure 14. Bioclimatic characterisation in the period of cold months for Málaga
Source: Prepared by the authors with data from IGN

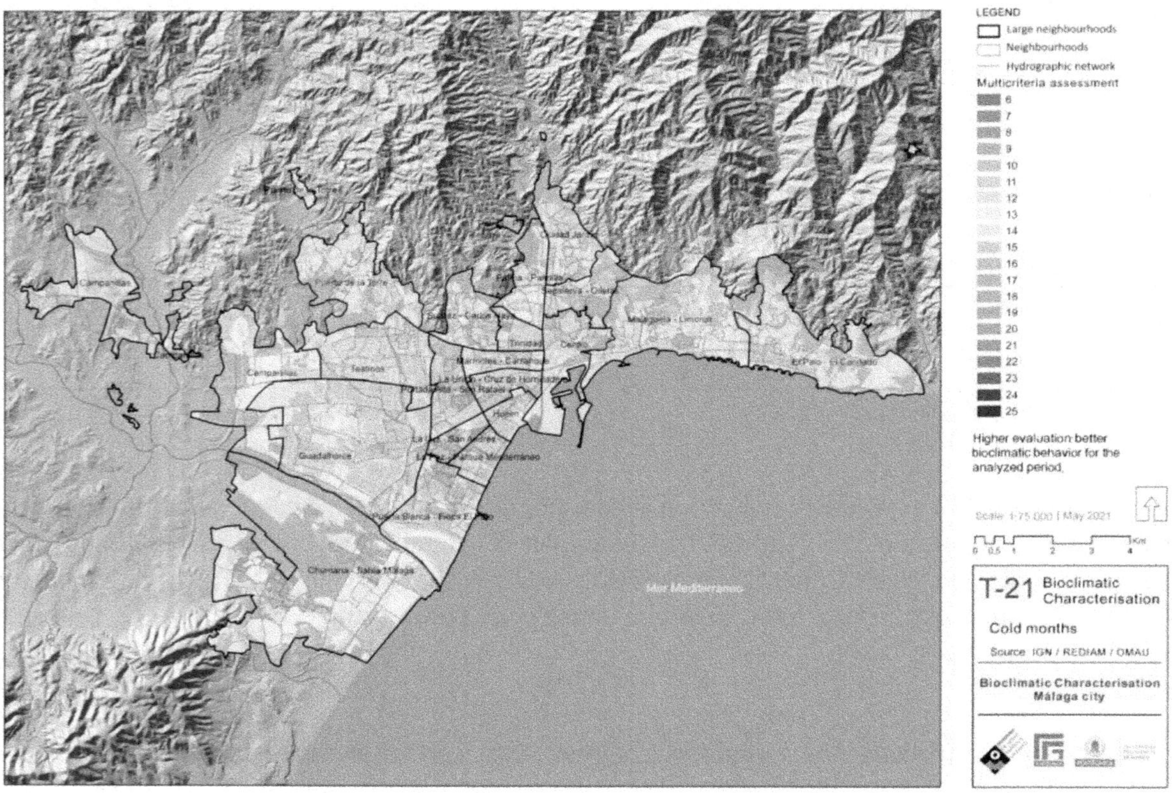

Finally, descriptive sheets are prepared for each of the 20 large neighbourhoods examined, in order to clearly and synthetically summarise, for each of them, the determinations on their territorial and urban conditioning factors, as well as their bioclimatic suitability. The data sheets contain the following information:

1. Urban structure
2. Urban determinants
   Administrative delimitation of the Large Neighbourhood
   Density
   Green areas and ecosystems
   Tree planting in streets
3. Climate determinants
   Solar Radiation in warm/cold months
   Wind in warm/cold months
4. Multi-criteria Assessment
5. Bioclimatic characterisation Cold Period
6. Bioclimatic characterisation Warm Period

Figure 15. Bioclimatic characterisation in the period of warm months for Málaga
Source: Prepared by the authors with data from IGN

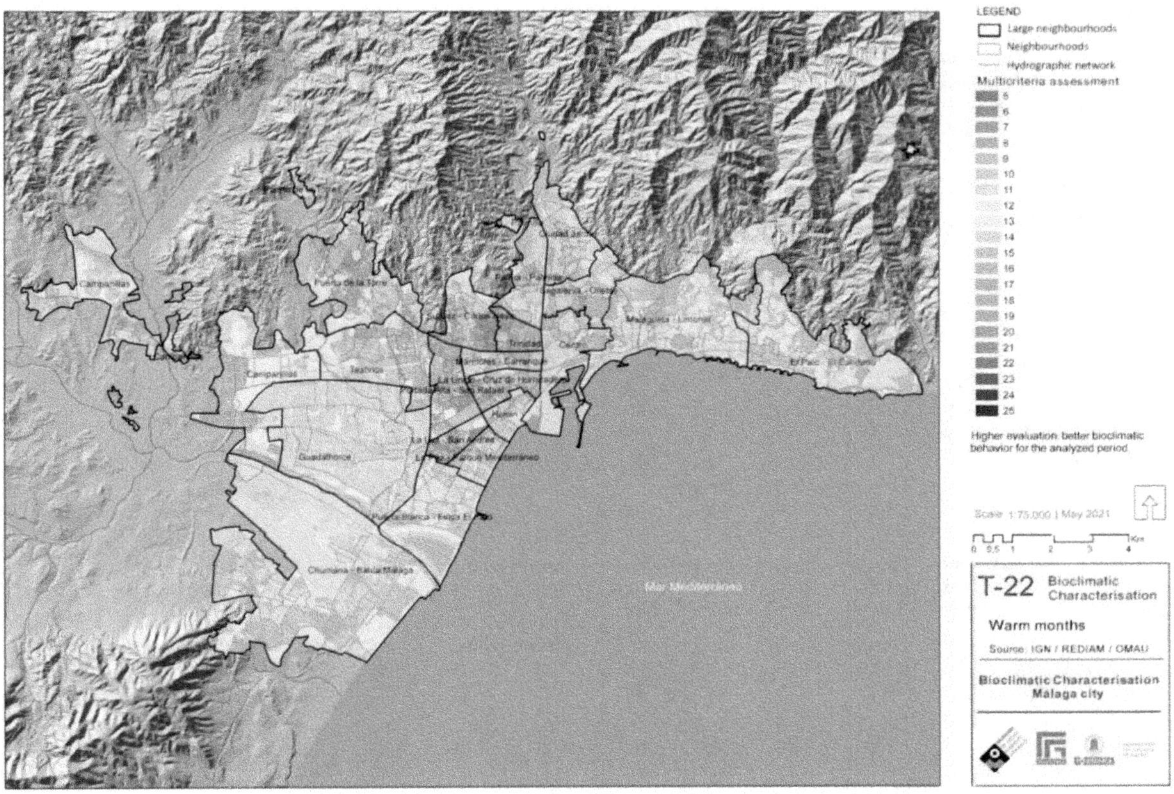

CONCLUSION

The degree of territorial suitability, according to certain activities, is different in each place. Their analysis and determination are key for harmonisation actions with the environment; they reduce impacts and improve the climatic conditions of each area, which means an increase in the health and quality of life of the people who live in these places. In this study, based on examining information in open databases and multi-criteria analysis using map algebra (GIS), specific cartographies have been obtained that visualize the different bioclimatic conditions throughout the year (hot and cold periods) in the different neighbourhoods of the city of Malaga. This study starts from the well-being of people and the extrinsic and intrinsic conditions of the place so as to make diagnostic decisions using geospatial data analysis. The results cannot be extrapolated to other contexts, but the methodology used will serve to adapt them to other urban situations.

The proposed methodology spatially integrates the conditioning factors of the physical environment outside and inside the city, understanding that there is an interaction between the two. The city is located in a climate, but activities and urban structure also condition the local microclimate. Both circumstances, related to geographical and climatic conditions, are analysed at the same level through the territorial and urban sectoral maps. In addition, discrimination of suitable areas is established based on the thermal

Figure 16. Multi-criteria assessment of the Large Neighbourhood Centre of Málaga
Source: Prepared by the author

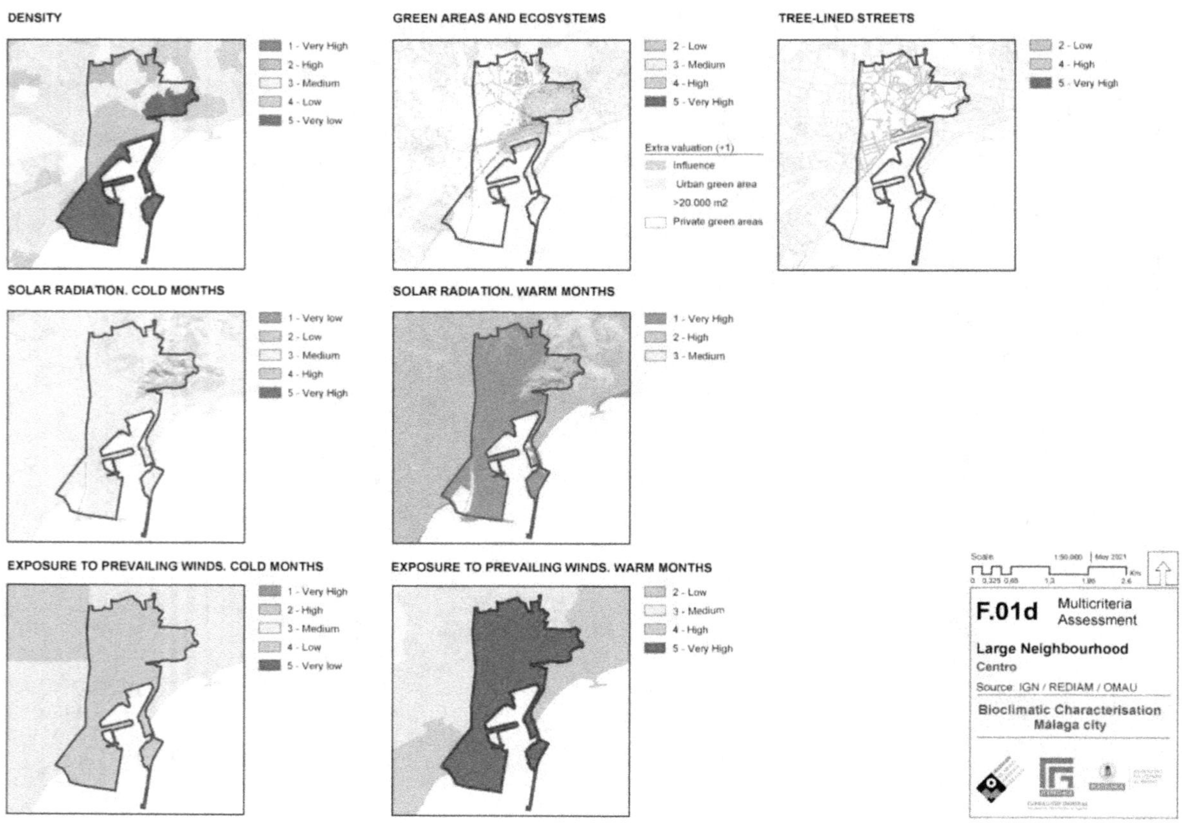

comfort conditions offered by the Adapted Comfort Climograph (ACC), with almost identical periods (equinoxes), and totally opposite periods, with cold (winter) and hot (summer) months. Depending on these climatic aspects, the sectoral maps are reclassified according to thermal comfort and the needs of sun exposure, wind or evaporation, which are the three main bioclimatic strategies established by Victor Olgyay (1963).

Due to the detailed information obtained in the existing databases (with high resolution digital models) it has been possible to make a spatial delimitation in accordance with the urban structure, starting from a morpho-typological analysis unit, which fits with the spatial relationship of the city. Although based on the principles of bioclimatic urban planning, it is not possible to establish final patterns that could be extrapolated from one place to another. The research proposes a general methodology that must be adapted and completed for each local context. The innovation of this research is to propose the weightings of the sectoral maps according to the thermal needs of people in each period of the year. Thus, in the case of the city of Malaga, it is necessary to think about the conditions during hot and cold months. In other Spanish territories, the needs will be different.

New research lines may continue with this research by establishing new projects that help to understand the intrinsic relationship of city factors in relation to a climate change scenario. It is also important to add other determinants of the urban heat island, such as urban surface finishes (mainly on pavements and façades).

The bioclimatic characterisation determined, with the delimitation of the *Urban Bioclimatic Areas* (UBA), will be key to draft a Local Bioclimatic Ordinance for the city of Málaga, with strategies for the bioclimatic design of urban public spaces to achieve more hours of thermal comfort; strategies to reduce the phenomenon of the urban heat island of the city; drafting of bioclimatic recommendations in urban areas with worse conditions in winter and summer; proposals for measures to improve the conditions of the urban environment, prevention of extreme heat stress and diseases the population suffer, related to the lack of comfort in buildings and public spaces. All this in order to achieve maximum adaptation of urban layouts and typologies to the microclimatic conditions of each area of the city, in order to establish the most appropriate guidelines and strategies in each of the locations and urban typologies.

The results obtained show the differences between some neighbourhoods and others in terms of making urban proposals that make it possible to achieve the Sustainable Development Goals: *SDG.3-Health and well-being, SDG.11-Sustainable cities and communities* and *SDG.15-Life of terrestrial ecosystems*, within the framework of action proposed by the New Urban Agenda 2030 and healthier urban communities (Higueras et al., 2021).

ACKNOWLEDGMENT

This work is part of a Technical Collaboration Agreement between the Urban Environment Observatory of Malaga (OMAU) and the Universidad Politécnica de Madrid with the ABIO and GIAU+S research groups.

REFERENCES

Akbari, H., Pomerantz, M., & Taha, H. (2001). Cool surfaces and shade trees to reduce energy use and improve air quality in urban areas. *Urban Environment*, *70*(3), 295–310. doi:10.1016/S0038-092X(00)00089-X

Ali-Tourdert, F., & Helmut, M. (2006). Numerical study on the effects of aspect ratio and orientation of an urban street canyon on outdoor thermal comfort and dry climate. Building and Environment, 41(2), 94-108.

Allard, F., & Santamouris, M. (2003). *Natural Ventilation in Buildings, a design handbook*. James and James.

Aram, F., Solgi, E., Higueras García, E., Mosavi, A. R., & Várkonyi-Kóczy, A. (2019). The Cooling Effect of Large-Scale Urban Parks on Surrounding Area Thermal Comfort. *Energies*, *12*, 3904. doi:10.3390/en12203904

Barredo Cano, J. I. (1996). *Sistemas de información geográfica y evaluación multicriterio en la ordenación del territorio*. Editorial RA-MA.

Bourbia, F., & Boucheriba, F. (2010). Impact of street design on urban microclimate for semiarid climate. *Renewable Energy, 35*(2), 343–347. doi:10.1016/j.renene.2009.07.017

Bröde, P., Krüger, E., Rossi, A., & Fiala, D. (2012). Predicting urban outdoor thermal comfort by the Universal Thermal Climate Index UTCI—A case study in Southern Brazil. *International Journal of Biometeorology, 56*(3), 471–480. doi:10.100700484-011-0452-3 PMID:21604151

Chen, L., & Ng, E. (2012). Outdoor thermal comfort and outdoor activities: A review of research in the past decade. *Cities (London, England), 29*(2), 118–125. doi:10.1016/j.cities.2011.08.006

Dai, F. C., Lee, C. F., & Zhang, X. H. (2001). GIS-based geo-environmental evaluation for urban land-use planning: A case study. *Engineering Geology, 61*(4), 257–271. doi:10.1016/S0013-7952(01)00028-X

Dijst, M., Worrell, E., Böcker, L., Lyons, G., Mokhtarian, P. L., Newman, P., & Perrels, A. (2018). Exploring urban metabolism—Towards an interdisciplinary perspective. *Resources, Conservation and Recycling, 1322*, 190–203. doi:10.1016/j.resconrec.2017.09.014

Fariña, J. (2000). *La ciudad y el medio natural*. AKAL.

Ferrão, P., & Fernández, J. E. (2013). *Sustainable Urban Metabolism*. MIT Press. doi:10.7551/mitpress/8617.001.0001

Fiala, D., Havenith, G., Bröde, P., Kampmann, B., & Jendritzky, G. (2012). UTCI-Fiala multi-node model of human heat transfer and temperature regulation. *International Journal of Biometeorology, 56*(3), 1–13. doi:10.100700484-011-0424-7 PMID:21503622

García Chávez, J. R., & Fuentes, V. (2005). *Viento y Arquitectura: El viento como factor de diseño arquitectónico* (3rd ed.). Trillas.

Geros, V., Santamouris, M., Karatasou, S., Tsangrassoulis, A., & Papanikolaou, N. (2005). On the cooling potential of night ventilation techniques in the urban environment. *Energy and Building, 37*(3), 243–257. doi:10.1016/j.enbuild.2004.06.024

Giridharan, R. S., Lau, S., Ganesan, C., & Givoni, B. (2007). Urban design factors influencing heat island intensity in high-rise high-density environments of Hong Kong. *Building and Environment, 42*(10), 3669–3684. doi:10.1016/j.buildenv.2006.09.011

Harg, I. M. (2000). *Proyectar con la naturaleza*. Gustavo Gili Barcelona.

Higueras, E. (2006). *Urbanismo Bioclimático*. Gustavo Gili.

Higueras, E., Roman, E., & Fariña, J. (2021). Guidelines for Healthier Public Spaces for the Elderly Population: Recommendations in the Spanish Context. Handbook of Quality of Life and Sustainability.

Hough, M. (1998). *Naturaleza y ciudad. Planificación urbana y procesos ecológicos*. Editorial Gustavo Gili. Barcelona Junta de Andalucía. *SIMA, Sistema de información Multiterritorial de Andalucía*. https://www.juntadeandalucia.es/institutodeestadisticaycartografia/sima/ficha.htm?mun=29067

Landsberg, H. E. (1947). Microclimatology. *The Architectural Forum, 86*(3), 114–119.

Li, H., & Kwan, M. (2018). Advancing analytical methods for urban metabolism studies. *Resources, Conservation and Recycling*, *132*, 239–245. doi:10.1016/j.resconrec.2017.07.005

Lynch, K. (1975). *Planificación del sitio*. Ed Gustavo Gili.

Malczewski, J. (2010). Multiple Criteria Decision Analysis and Geographic Information Systems. In M. Ehrgott, J. Figueira, & S. Greco (Eds.), *Trends in Multiple Criteria Decision Analysis. International Series in Operations Research & Management Science* (Vol. 142). Springer. doi:10.1007/978-1-4419-5904-1_13

Matzarakis, A. (2001). Assessing climate for tourism purposes: Existing methods and tools for the thermal complex. *Proceedings of the first international workshop on climate, tourism and recreation*, 101–112.

Matzarakis, A., Rutz, F., & Mayer, H. (2007). Modelling radiation fluxes in simple and complex environments—Application of the RayMan model. *International Journal of Biometeorology*, *51*(4), 323–334. doi:10.100700484-006-0061-8 PMID:17093907

Mirzaei, P. A., & Haghighat, F. (2010). Approaches to study Urban Heat Island – Abilities and limitations. *Building and Environment*, *45*(10), 2192–2201. doi:10.1016/j.buildenv.2010.04.001

Mojtaba, P., Mastani, M., Joybari, A., Parham, A., Mirzaei, J., & Haghighat, F. (2019). *Urban heat island, urban climate maps and urban development policies and action plans*. Environmental Technology & Innovation. doi:10.1016/j.eti.2019.100341

Neila, F., Sánchez-Guevara, C., Núñez, M., Román, E., Gómez, G., & De Luxán, M. (2020). *Urban heat island, urban climate and building energy consumption*. MODIFICA Project. García-Maroto Editores S.L.

Neila, J. (1990). *Arquitectura bioclimática*. Munilla Leira.

Ng, E., Tam, I., Ng, A., Givoni, B., Katzschner, L., Kwok, K., Murakami, S., Wong, N. H., Wong, K. S., Cheng, V., Davis, A., Tsou, J. Y., & Chow, B. (2004). *Final Report – Feasibility Study for Establishment of Air Ventilation Assessment System*. Technical Report for Planning Department HKSAR.

Nikolopoulou, M., Baker, N., & Steemers, K. (2001). Thermal comfort in outdoor urban spaces: Understanding the human parameter. *Solar Energy*, *70*(3), 227–235. doi:10.1016/S0038-092X(00)00093-1

Nikolopoulou, M., & Lykoudis, S. (2007). Use of outdoor spaces and microclimate in a Mediterranean urban area. *Building and Environment*, *42*(10), 3691–3707. doi:10.1016/j.buildenv.2006.09.008

Nyeko, M. (2012). GIS and Multi-Criteria Decision Analysis for Land Use Resource Planning. *Journal of Geographic Information System*, *4*(4), 341–348. Advance online publication. doi:10.4236/jgis.2012.44039

Oke, T. R. (1987). *Boundary Layer Climates*. Routledge.

Olgyay, V. (1963). *Design with Climate: Bioclimatic Approach to Architectural Regionalism, Manual de Arquitectura Bioclimática para arquitectos y urbanistas*. GG.

Pollo, R., Trane, M., & Giovanardi, M. (2020). Urban Metabolism, modelli interdisciplinari e progetto a scala microurbana. Firenze University Press. doi:10.36253/techne-9857

Sukopp, H., & Werner, P. (1989). *Naturaleza en las ciudades. Desarrollo de flora y fauna en áreas urbanas*. Ministerio de Obras Públicas y Urbanismo.

Zhang, Y., Liu, H., & Chen, B. (2013). Comprehensive evaluation of the structural characteristics of an urban metabolic system: Model development and a case study of Beijing. *Ecological Modelling, 252*, 106–113. doi:10.1016/j.ecolmodel.2012.08.017

ADDITIONAL READING

Arefiev, N., Terleev, V., & Badenko, V. (2015). GIS-based Fuzzy Method for Urban Planning. *Procedia Engineering, 117*, 39–44. doi:10.1016/j.proeng.2015.08.121

Ferrão, P., & Fernández, J. E. (2013). *Sustainable Urban Metabolism.* MIT Press. doi:10.7551/mitpress/8617.001.0001

Huiyue, W., Chen, D., Duan, H., Yin, F., & Niu, Y. (2019). Characterizing urban building metabolism with a 4D-GIS model: A case study in China. *Journal of Cleaner Production, 228*, 1446–1454. doi:10.1016/j.jclepro.2019.04.341

Santamouris, M., Asimakopoulos, D. N., Assimakopoulos, V. D., Chrisomallidou, N., Klitsikas, N., & Tsangrassoulis, A. (2001). *Energy and climate in the urban built environment* (M. Santamouris, Ed.). University of Athens.

KEY TERMS AND DEFINITIONS

Bioclimatic Urbanism: Planning by knowing the sun and wind conditions of a place in order to establish recommendations to improve the peoples´ thermal comfort in outdoor spaces.

Climate Change Scenario: Knowledge of the changes in temperature, winds, rainfall patterns and periods of drought that will occur in cities and take them into account in order to establish adaptation and mitigation strategies for urban areas.

Healthy Neighbourhoods: Addressing health problems that are concentrated in urban areas due to the presence of pollution, poor accessibility to green spaces and sedentary urban lifestyles, by urban regeneration

Spatial Algebra Analysis: Use of map algebra to establish a multi-criteria sectorial analysis to make an integrated synthesis map for complex decision making.

Sun Exposure: The relief conditions where a city is located and the solar radiation of the city at each time of the year. This can be used as a factor in urban design through the bioclimatic design of the city.

Urban Bioclimatic Areas: Are defined as homogeneous urban areas where the same urban microclimate conditions are found. These areas are affected by conditioning factors ranging from the territorial scale (geomorphology, wind, sun, and green structure) to urban conditioning factors (as density, street geometry, and presence of vegetation).

Urban Thermal Comfort: Thermal comfort of most people in a city. It is a subjective feeling, but it is possible to determine a degree of satisfaction of 80% of people in both hot and cold weather.

Wind: Relief conditions and the presence of the coastline determine the wind regime of a place. This circumstance can be used to improve the thermal comfort of cities through bioclimatic urban planning design.

ENDNOTES

1 Download Centre of the National Geographic Institute (Centro de Descargas del Instituto Geográfico Nacional -CNIG): https://centrodedescargas.cnig.es/CentroDescargas/index.jsp

2 Urban Environment Observatory (Observatorio de Medio Ambiente Urbano - OMAU): https://www.omau-malaga.com/

Chapter 2
Conceptualization of an Air-Extraction System to Mitigate COVID-19 Transmissions Inside Public Transportation Buses

Omar Eduardo Sánchez Estrada
https://orcid.org/0000-0002-0108-0642
Universidad Autónoma del Estado de México, Mexico

ABSTRACT

Data collected by the United Nations (UN) suggest that infections arise after a close contact with infected people through nasal and oral secretions released when an infected person coughs, sneezes, or talks. In public transportation buses, an important amount of people gather every day for long periods of time, making the air pollution inside these transportation systems a major risk for transmission. Therefore, the objective of this chapter is to know criteria and strategies for a conceptualization of an air-extraction system inside public transportation buses, based on the detailed study applicability of the 1) product-user interaction (technical data, dimensional relationship, and evaluation); 2) creative process, ideation, definition, evaluation, and structuration; and 3) sustainability, technical specifications, ergonomics, and production by means of understanding the design's limits and effects.

INTRODUCTION

The World Health Organization (WHO) in collaboration with the University of Bath (United Kingdom) have developed an air-quality model to collect data from satellite measurements monitored in cities and rural areas from over 3,000 locations. Such model confirms that 92% of the world's population inhabit in areas that exceed limits established by the organization (WHO, 2020). Exposure to polluted air is related to more than 3 million deaths per year. In addition, humanity suffers from a pandemic derived from a SARS-CoV-2 disease, which produces pneumonia and a severe breathing difficulty syndrome. The current pandemic has generated concern among the world community, Covid-19 is rapidly spread-

DOI: 10.4018/978-1-6684-4030-8.ch002

ing in many ways and one of them is ventilation and air conditioning systems, which work with air transporting a group of pollutants, virus and bacteria while moving. In the case of public transportation the same phenomenon occurs when people get exposed according to the place and time they use for transportation; if there is a Covid-19 infected person inside the transportation system, he or she will excrete drops of saliva when sneezing, coughing, breathing or talking out loud, and the extension size of such expelled particles is from 1 to 5 mm in an approximate space of 1 to 2 m (Yu. et al., 2004). Droplets excreted by the infected person will become bioaerosols (Adhikari et al., 2019), and the particles have an approximate diameter of 0.3 to 100 µg; in such a way that the breathable fraction of 1 a 10 µg is worrisome (Tindale et al., 2020).

Therefore, bad air quality increases death rates every year (European Environment Agency, 2005). Recent studies show that the atmospheric aerosol contains solid or liquid particles that are suspended in the air of relatively closed spaces. The United States Environmental Protection Agency (EPA) affirms that particle contamination is a combination of solid particles (dirt, dust and smoke) or liquid drops in the air; while human sight only sees a few, others can only be analyzed with an electronic microscope. PM10 particles are formed by thick mode particles (10 micrometers and smaller), while PM2.5 are formed by fine mode particles (2.5 micrometers) (EPA, 2006). According to the Agency, an acceptable PM standard in the atmosphere for short-term health is defined as: (daily average of 24 hours) 35 $\mu g/m^3$ for PM2.5 and 150 $\mu g/m^3$for PM10. The standards for the World Health Organization (WHO) are set as.5 25 $\mu g/m^3$ for PM2 and 150 $\mu g/m^3$ for PM10. However, they highlight the fact that even low standards could affect human health (WHO, 2016). This Organization's statistics show that 4.6 million people die each year due to bad air quality illnesses (Cohen et al., 2017). For ventilation and air conditioning systems in public transportation, the air filter is an important component in order to stop and possibly eliminate Covid-19 particles (Elsaid and Ahmed, 2021). Different public transportation systems address mobility needs; however, buses are the main base in different countries around the world, and along the use of fuel, air filtration of old buses and interior designs as well as ventilation, among others, make a vehicle a high risk place to get Covid-19 if used as a work place or a transportation system.

Nowadays, natural ventilation with fresh air can be controlled with different technical solutions, its efficiency and simplicity considerably improve energy savings and its limits depend on exterior air in great measure, such limits are found in an air temperature of 20-26 °C (Axley and Emmerich, 2002). For instance, current ventilation designs for new commercial and residential buildings are not enough to provide ventilation rates that prevent infections inside (Morawska et al., 2020). The Federation of European Heating, Ventilation and Air Conditioning Associations (REHVA) recommended an increase in air supply with a greater exchange rate of air and to avoid its recirculation; they have also published documents where the use of air systems is recommended in order to damp exterior air by means of open spaces disposition, considering 100% in order to eliminate recirculation (REHVA, 2020). Consequently, design as an environment-transforming discipline must be implemented as a group of creative activities that are managed with systematic methods and techniques, efficiently implementing hidden opportunities in everyday technology. This project does not only develop its possible implementation, it also proposes a practical exercise of interdisciplinary collaboration, a technical data analysis suitable for future projects and important information for designers, producers, teachers, users, etc.

BACKGROUND

Recent research makes evident that SARS-CoV-2 transmission is given by aerosols, some doctors and staff inside hospitals have used clear plastic barriers and plexiglass boxes to prevent solid and liquid particles from spreading; however, these actions limit patient access and medical activities. According to Matava and Denning (2020), an alternative to reduce aerosol spreading and to not affect procedures and personal follow-up of patients may be a high-flow air extractor, which is defined as a high-efficiency portable filtration unit that allows 235 L s^{-1} (500 $feet^3$ min^{-1}) and has the capacity to modify a conventional room into a negative pressure room. The extractor is composed by a HEPA (High Efficiency Particulate Air) filter that eliminates 99.97% of all 0,3 μm or larger pathogens in the air, then filtered air can be adapted to a purged system or direct it to the exterior. High flow directed extraction combines a high-efficiency particulate air (HEPA) filter and a high-flow suction system, this technique is characterized by the high-flow air extractor functioning. Scientists developed an experimental model to determine efficiency of particle elimination that is similar to human aerosols; they simulated aerosols by means of using two particles: ultrafine testing dust which varies in size from 1 to 20 μm and essential oils with a variation in size from 1nm to 1μm; they also worked with an essential oil diffusor as a continuous feeder of aerosol in order to simulate human breathing in a determined time period.

Other projects work to promote natural ventilation by using heating and radiative refrigeration, which is developed through infrared radiation exchange among people gathered in a determined place and the surrounding area; it works with minimum energy in air-based systems of intensive use in order to generate thermal comfort. Heat transference exchanged by radiation to surfaces and to air convection in environmental conditions with similar temperatures is nearly equivalent (Chowdhury et al., 2008), that is 50% of one's thermal perception is increased by the area's temperature. In that way, when manipulating the area's temperature going up or down, comfort can be kept in a larger air temperature range, which allows for natural ventilation hours to increase considerably (Aviv et al., 2021). It is important to highlight that efforts to mitigate the pandemic are permanent, such is the case of the United Christian Hospital in Hong Kong that modified its technical structure based on positive pressure into a negative pressure for patients with viral infections spread by air, and as a result propagation of infected people with the virus decreased (Cornelius et al., 2020). Likewise, in Singapur hospitals the use of high-technology particle air filters facilitated the use of negative pressure with exhaust fans connected to high-efficiency microfilters into windows, this proposal eliminated 99% of fine material, as the air exchange is set at a speed of 25 times per hour (Gralton et al., 2011; Phoon et al., 2019).

In the other hand, UV technology and air filters such as the HEPA and electrostatic ones work to improve air quality. For instance, ultraviolet radiation eliminates virus and bacteria, and ultraviolet rays disintegrate DNA in cooling files, burning external walls of virus and microbes to avoid their contamination (Elsaid and Ahmed, 2021). HEPA (High Efficiency Particulate Air) filters are used to stop contaminants from coming through and to destroy 99.97% of all air pathogens, 0,3 μm or larger; they are used in airspace, microelectronic, optics, food processing and chemical industries, among others (HEPA filter, 2021). Electrostatic filters adapt to different ventilation systems to purify air. The electrostatic term is related to incoming positive charge particles, which are attracted by surfaces in the opposite charge filters and vice versa; in this step all air articles are magnetically added to the filtering system and remain together until the system is cleaned out (Brinner, 2021).

Although aerosols exposure can be mitigated by means of social distancing and use of face masks, smaller volatile particles must be controlled with efficient ventilation systems design. In countries such as India, being exposed to PM related to public transport during short distance trips has proven effects in health (Knibbs et al., 2011), evidence shows air in the bus cabin may be contaminated by external air. The risk estimator due to COVID infection in the air, developed by Miller et al. (2020), helped estimate infection probabilities in taxis and buses, these operations considered an assumption of only one infected person, the latent period of infection is larger than the mathematical model length, then infectious aerosols move uniformly throughout the vehicle's interior volume and are destroyed by means of the vehicle's air filtration, viral inactivation, gravitational deposition and air change per hour, so the chance of being infected is related to the number of inhaled virus.

Therefore, it is important to recover using public transportation within the last stages of an extended pandemic, but always in a safe manner for users, and this will be possible if every country's authorities study the people's demand to mitigate the risks of getting infected. The WHO recommended the world governments and transportation providers to clean out any crowded area or objects one has contact with (WHO, 2020). It has been shown that in tropical countries such as Brazil, wind speed, and radiation and temperature increase are all very important weather factors in order to gradually reduce the pandemic effects (Rosario et al., 2020). Hence, avoiding infections inside public transportation such as buses is not restrictive to theory, technique, system, model or concept, the important part will be finding a formula to develop a system that avoids this killing virus from spreading.

METHOD

This research considers three stages that are related to the proposed objectives, and has a descriptive character, starting with the design's theory to identify the process' general needs, the research's current state and the practice in order to determine the design's criteria and strategies when conceptualizing an air extraction system in public transportation buses in order to avoid Covid-19 infections. This work takes as a reference the Metropolitan Area of Guadalajara (ZMG, acronym in Spanish), capital of the State of Jalisco in Mexico, as it has a type C mesothermal mild weather with some variations during dry and humid seasons.

First Stage

Applicability Study

According to Davydova and Frías (2016), for the ZMG minimum and maximum day temperatures and warm nights often increase during spring. Inside public transportation, thermal stress is present due to local climate change, therefore passengers' health is threaten by external factors, such as carbon dioxide or particulate matter. Also, a lack of ventilation along user-saturation, particularly those who are developing biological respiration processes, degrade interior air quality. From all auxiliary components in vehicles, air conditioning systems are the ones that use more energy, dynamic modelling of steam compression systems (white box models) are not adequate to develop air conditioning models for buses due to alterations caused by opening and closing doors, shades in streets and random number of possible passengers getting in and off the vehicle (Chen et al., 2021). These assessments are the beginning of a detailed analysis based on factors interconnecting ideas from a design's creative process. Determination of formal properties (functional and structural links) will allow developing a coherent unit from

Figure 1. Integrated thinking for project requirements
Source: Own elaboration

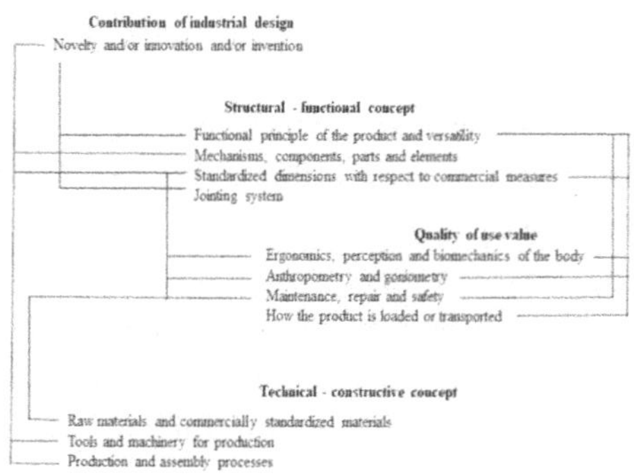

producer to user so as not to conceal structural weaknesses, but to define the object's properties. Inside design theory, research based on recognizing social research's reflexive character as part of a real event to build a concept must be gathered, and observations need to be orderly so as to build a comprehensive interpretation of the designing team, producers, suppliers, external actors, buyers, distributors, and users. It must be highlighted that design from its first stages may favor access control and passenger saturation for any transportation system, according to the needs and mobility demand of users. Figure 1 shows the integral thinking basic structure for a project's requirements, prior to defined first activities for developing a product or social service.

Second Stage

Ergonomic Variables for the System

The proposal includes making use of environmental factors and anthropometry as two relevant components of ergonomic variables, so as to move from the analysis to the design based on modelling data relying on the design's specific requirements. Understanding interactions among people using the bus for transportation and the system's elements are based on ergonomic intervention (man and built environment) stages; the best performance is sought, as well as the best way to adapt the air extractor's technical and operative characteristics when handling it, perceiving the object and communication, all based on behavior derived from social distancing protocols and interactions with different systems and subsystems in the built environment, which can be defined as the group of human being's concrete products (handrail, support straps, seats, control buttons, information dials, foldout chairs, etc.). Also, ecological and sociocultural factors condition interaction. Figure 2 shows such determinants.

As long as the system to be conceptualized (air extractor) has operative elements according to the user's interaction needs, the operator will also have handling ability for optimizing and improving the equipment's performance. Sensitivity, and physical and psychosocial motivation favor the interaction system. Therefore, in order to identify problems with interactions, it is necessary for the air extractor that will work as a simulator to be precise in its general structure, as the objective is to avoid mistakes already identified in previous projects, as well as to establish the moments when a functional alteration

Figure 2. Conditioning factors (interaction system)
Source: Own elaboration

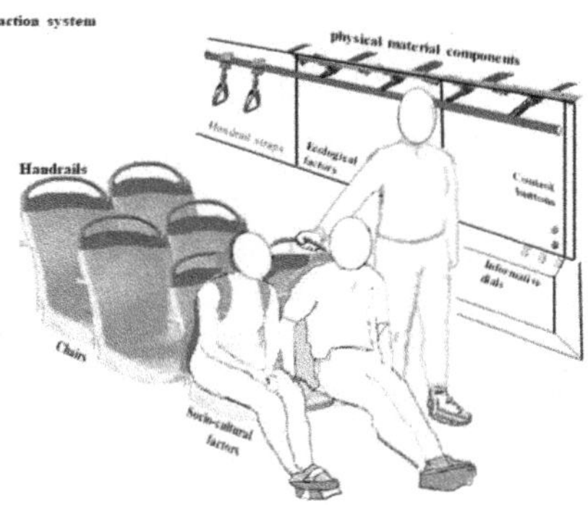

could happen, which provides with knowledge about the environment's components or elements that could affect the sequence of use.

Environmental Factors

Making temperature equal to the interaction mean, by yielding or accepting heat from convection, is a physiological quality of the human body, this depends on the difference or gradient of temperature between the skin and the air, or in the heat exchange due to radiation with the objects it interacts with directly or indirectly, it is also possible with heat transference due to evaporation or conduction. Crank operations or wheels are objects that promote conduction due to friction and the materials they are made of. In that way, the main physiological mechanism to dissipate heat is sweat; however, it is an indicator of discomfort that must be considered in conceptualization. Skin temperature fluctuates around 33°C and when environmental temperature is above, cold or heat sensations are produced. Comfort is disrupted with: a) air speed; b) relative humidity; c) radiant temperature; d) air temperature; and e) activity level. Every alteration gives place to different comfort sensations in equipment operators, transporters, distributors, technicians, buyers and users.

Anthropometry for Design

Precision in anthropometric aspects defines risks and promotes the system's stability. In order to define the components' place in an equipment, it is necessary to work with anthropometric studies in a determined population; the population is divided for implementing purposes from the smallest (in dimension) to the largest according to measurement type (height, weight, leg extension, etc.). Distribution curve of standard frequency is used to represent percentiles. Anthropometric characteristics of target population support conceptualization of the design by using the following lines: a) design for average population; b) design for the extreme, working with a variable's maximum and minimum of population that will be adjusted; c) design for rank, working with an heterogeneous mixed population, precise demand to define parameters to be used; and d) design for an adjustable range, it is possible to adjust dimensions of equipment's components according to the user's needs. Percentiles indicate people percentage among the population based on dimension, from the smallest to the largest according to a measurement type

Figure 3. Environmental factors
Source: Own elaboration

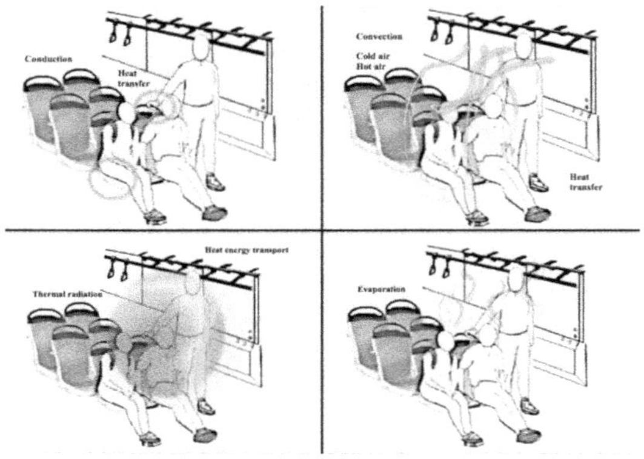

(height, weight, leg extension, etc.); Gaussian curve or distribution curve of standard frequency help represent percentiles. If a large population sample is taken and anthropometric parameters are evaluated, a statistically normal population distribution can be identified in a "Gaussian bell.

Figure 4. Anthropometric graph
Source: Own elaboration

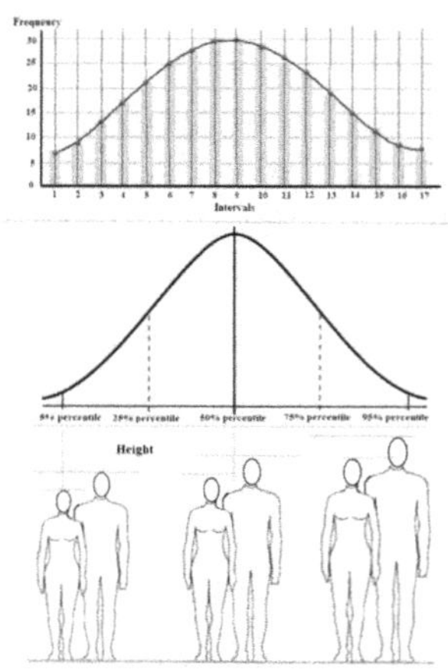

Third Stage

Design Proposal

Definition of a design project based on sustainability has become one of the most viable alternatives to transform exponential progress of environmental pollution, producers require to reassess their production lines, procedures, and human and environmental factors. Therefore, ISO14006 must be applied to developing, implementing and strengthening ecological methods of innovation and design, with special attention on the product's life cycle. It is required to know every life-cycle stage in order to avoid environmental impacts, whilst guaranteeing legal agreements fulfilment of every team member so as to strengthen innovation to promote permanent improvements in critical points that refer to taking environmental care. In view of the foregoing, conceptualization to design an air extraction system to mitigate Covid-19 infections or any of its variants in public transportation buses could be worked based on the following:

- Analysis stage
    - Closely look, analyze and synthetize everything related to extraction and filtration systems.
    - Identify technical, social and environmental problems and subproblems derived from the system's conceptualization.
    - Analyze historic, psychological and political circumstances to encompass knowledge on the subject.
    - Analysis of buses and their air conditioning systems.
- Creation stage
    - To determine shape, function, material and general structure to be used in the design's concept from different techniques so as to gather, improve and validate information.
    - To develop different conceptual alternatives with viable, desirable and feasible solution answers.
    - To evaluate and select the alternative that has previously been through the selection matrix, which has been approved and has the best score in different design requirements.

The following proposal has the intention to show an air extraction system alternative, so as to implement practical applications in public transportation buses, and somehow contribute in the efforts to reduce Covid-19 infections and to minimize death rates in places where it is applied. The intention is to use a high-flow air extractor that allows 235 L s^{-1} (500 $feet^3$ min^{-1}) and to have the capacity to modify the inside of a bus (passengers space) into a negative pressure space; such extractor may be composed of an HEPA (High Efficiency Particulate Air) filter, which can eliminate 99.97% of all air pathogens of 0,3 μm or larger. Air inside buses' cabin will be renewed every 5 minutes with exterior air that does not bring pathogens in. Therefore, recycled air in the cabin will go through an HEPA filter, guaranteeing pure air; its circulation will be from top to bottom coming in through the ceiling and going out through the floor; this allows to minimize longitudinal air flows throughout the cabin to avoid any type of infection. It is recommended to activate ventilation systems before passengers get on the bus and to activate the air extraction and depuration system. Figure 5 shows the proposal.

CONCLUSION

Existing technology must find different applications via designers, the current pandemic has significantly worried the whole scientific community, public transportation systems are modifying their safety and logistics protocols in general. This proposal is only one of many applications that can be promoted in order to decrease infections and deaths, among the most delicate problems. The project leaves an open possibility to adapt and utilize these systems in private cars, public transport collective systems, and any other space where people gather. Different ventilation, filtration and depuration systems such as the HEPA (High Efficiency Particulate Air) filter, electrostatic filters, air treatment units, air disinfection

Figure 5. Air extraction and purification system
Source: Own elaboration

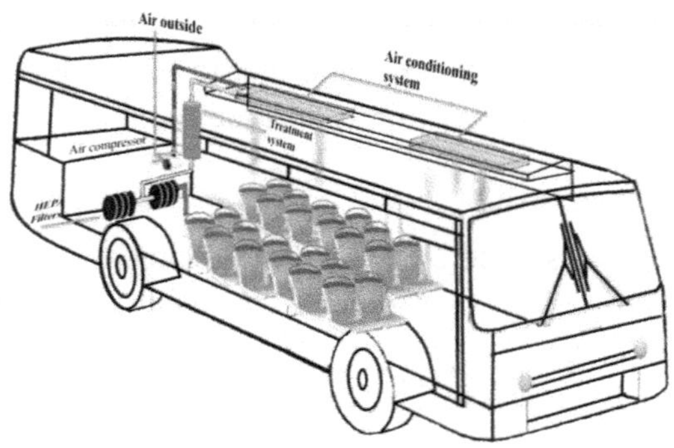

and purification systems, extraction units and ventilation units, must be put to the best use and be applied also in hospitals, educational and sports centers, shopping malls and possibly at home. Therefore, design as a creating activity that solves problems in an integral manner plays an important role in developing products that are manufactured with a specific goal and are usually multifunctional to increase people's life quality, and to keep a balanced utilization of natural, economic, cultural and political resources, while promoting equality and equity in perception, use, selling and utilization of any useful object. Therefore, it is necessary to work on a change in the personality of the designers and modify the basic principles of professional teaching in the area, based on a new dissemination of methodological proposals, adding the combination of technical criteria and quality of life.

It is important to mention that the presented theory along with the proposal's possible implementation in developing countries and first world countries is applied in a different manner; in the first case implementation possibility is conditioned to strong investments from the private sector, development and implementation of legislative and environmental measures, reconfiguration of objectives defined by entrepreneurs and collaborative work agreed upon achieving the purpose; in the second case it is required to update objectives permanently according to new technologies and to develop programs to consolidate and reach proposed long-term goals.

Figure 6. Air circulation
Source: Own elaboration

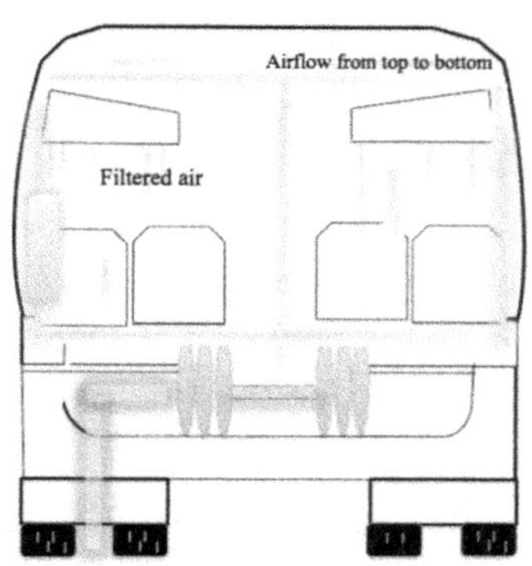

REFERENCES

AENOR Confía. (2022). *ISO 14006 Certification ecodesign management.* https://www.aenor.com/certification/environment/ecodesign

Adhikari, U., Chabrelie, A., Weir, M., Boehnke, K., McKenzie, E., Ikner, L., Wang, M., Wang, Q., Young, K., Haas, C. N., Rose, J., & Mitchell, J. (2019). A case study evaluating the risk of infection from Middle Eastern Respiratory Syndrome Coronavirus (MERS-CoV) in a Hospital Setting Through Bioaerosols. *Risk Analysis*, *39*(12), 2608–2624. doi:10.1111/risa.13389 PMID:31524301

Aviv, D., Chen, K. W., Teitelbaum, E., Sheppard, D., Pantelic, J., Rysanek, A., & Meggers, F. (2021). A fresh (air) look at ventilation for COVID-19: Estimating the global energy saving potential of coupling natural ventilation with new radiant cooling strategies. *Applied Energy*, *292*, 116848. doi:10.1016/j.apenergy.2021.116848 PMID:33776191

Axley, J. W., & Emmerich, S. J. (2002). A method for evaluating the suitabilityof climate for the natural ventilation of commercial buildings. Proceedings: Indoor Air.

BRINNER. 2021). *What is an electrostatic filter*. Retrieved from https://brinner.es/filtros-electrostaticos/

Chen, Z., Xiao, F., Shi, J., & Li, A. (2021). Dynamic Model Development for Vehicle Air Conditioners Based on Physics-guided Deep Learning. *International Journal of Refrigeration.*

Chowdhury, A. A., Rasul, M. G., & Khan, M. M. K. (2008). Thermal-comfort analysis and simulation for various low-energy cooling-technologies applied to an office building in a subtropical climate. *Applied Energy*, *85*(6), 449–462. doi:10.1016/j.apenergy.2007.10.001

Cohen, A. J., Brauer, M., Burnett, R., Anderson, H. R., Frostad, J., Estep, K., Balakrishnan, K., Brunekreef, B., Dandona, L., Dandona, R., Feigin, V., Freedman, G., Hubbell, B., Jobling, A., Kan, H., Knibbs, L., Liu, Y., Martin, R., Morawska, L., ... Forouzanfar, M. H. (2017). Estimates and 25-year trends of the global burden of disease attributable to ambient air pollution: An analysis of data from the Global Burden of Diseases Study 2015. *Lancet*, *389*(10082), 1907–1918. doi:10.1016/S0140-6736(17)30505-6 PMID:28408086

Cornelius, B., Cornelius, A., Crisafi, L., Collins, C., McCarthy, S., Foster, C., Shannon, H., Bennett, R., Brown, S., Rodriguez, K., & Bachini, S. (2020). Mass air medical repatriation of coronavirus disease 2019 patients. *Air Medical Journal*, *39*(4), 251–256. doi:10.1016/j.amj.2020.04.005 PMID:32690299

Davydova, B. V., & Frías, U. H. G. (2016). Detección de señales de cambio 22. climático en Jalisco, México. Memorías en Extenso de AIDIS, 2.

Elsaid, A. M., & Ahmed, M. S. (2021). Indoor Air Quality Strategies for Air-Conditioning and Ventilation Systems with the Spread of the Global Coronavirus (COVID-19) Epidemic: Improvements and Recommendations. *Environmental Research*, *199*, 111314. doi:10.1016/j.envres.2021.111314 PMID:34048748

EPA. (2006). *Particulate Matter Basics*. Available from. https://www.epa. gov/pm-pollution/particulate-matter-pm-basics#PM

European Environment Agency. (2005). *Communication from the Commission to the Council and the European Parliament—Thematic Strategy on Air Pollution* (COM (2005) 0446 Final).

Gralton, J., Tovey, E., McLaws, M. L., & Rawlinson, W. D. (2011). The role of particle size in aerosolised pathogen transmission: A review. *The Journal of Infection*, *62*(1), 1–13. doi:10.1016/j.jinf.2010.11.010 PMID:21094184

HEPA Filter. (2021). *HEPA filter applications*. Retrieved from https://filtroshepa.com.mx/default.html

Knibbs, L. D., Cole-Hunter, T., & Morawska, L. (2011). A review of commuter exposure to ultrafine particles and its health effects. *Atmospheric Environment*, *45*(16), 2611–2622. doi:10.1016/j.atmosenv.2011.02.065

Matava, C. T., Yu, J., & Denning, S. (2020). Clear plastic drapes may be effective at limiting aerosolization and droplet spray during extubation: implications for COVID-19. *Canadian Journal of Anesthesia/Journal canadien d'anesthésie, 67*(7), 902-904.

Miller, S. L., Nazaroff, W. W., Jimenez, J. L., Boerstra, A., Buonanno, G., Dancer, S. J., Kurnitski, J., Marr, L. C., Morawska, L., & Noakes, C. (2021). Transmission of SARS-CoV-2 by inhalation of respiratory aerosol in the Skagit Valley Chorale superspreading event. *Indoor Air*, *31*(2), 314–323. doi:10.1111/ina.12751 PMID:32979298

Morawska, L., Tang, J. W., Bahnfleth, W., Bluyssen, P. M., Boerstra, A., Buonanno, G., Cao, J., Dancer, S., Floto, A., Franchimon, F., Haworth, C., Hogeling, J., Isaxon, C., Jimenez, J. L., Kurnitski, J., Li, Y., Loomans, M., Marks, G., Marr, L. C., ... Yao, M. (2020). How can airborne transmission of COVID-19 indoors be minimised? *Environment International*, *142*, 105832. doi:10.1016/j.envint.2020.105832 PMID:32521345

Phoon, P. H., MacLaren, G., Ti, L. K., Tan, J. S., & Hwang, N. C. (2019). History and current status of cardiac anesthesia in Singapore. *Journal of Cardiothoracic and Vascular Anesthesia*, *33*(12), 3394–3401. doi:10.1053/j.jvca.2018.07.018 PMID:30131218

REHVA. (2020). COVID, R. (19). guidance document, April 3, 2020. REHVA, Federation of European Heating, Ventilation and Air Conditioning Association, Brussels, Belgium.

Rosario, D. K., Mutz, Y. S., Bernardes, P. C., & Conte-Junior, C. A. (2020). Relationship between COVID-19 and weather: Case study in a tropical country. *International Journal of Hygiene and Environmental Health*, *229*, 113587. doi:10.1016/j.ijheh.2020.113587 PMID:32917371

Tindale, L., Coombe, M., Stockdale, J. E., Garlock, E., Lau, W. Y. V., Saraswat, M., . . . Colijn, C. (2020). Transmission interval estimates suggest pre-symptomatic spread of COVID-19. MedRxiv. doi:10.1101/2020.03.03.20029983

WHO. (2016). *Ambient (outdoor) air quality and health*. https://www.who.int/es/news-room/fact-sheets/detail/ambient-(outdoor)-air-quality-and-health

World Health Organization. (2020). *Supporting healthy urban transport and mobility in the context of COVID-19*. Author.

Yu, I. T., Li, Y., Wong, T. W., Tam, W., Chan, A. T., Lee, J. H., Leung, D. Y. C., & Ho, T. (2004). Evidence of airborne transmission of the severe acute respiratory syndrome virus. *The New England Journal of Medicine*, *350*(17), 1731–1739. doi:10.1056/NEJMoa032867 PMID:15102999

Confía, A. E. N. O. R. (2022). ISO 14006 Certification ecodesign management. https://www.aenor.com/certification/environment/ecodesign

Davydova, B.V., & Frías, U.H.G. (2016). Detección de señales de cambio 22. climático en Jalisco, México. *Memorías en Extenso de AIDIS, 2*.

KEY TERMS AND DEFINITIONS

Anthropometry: Scientific tool that sizes the different segments of the human body for the design and adaptation of the physical environment that surrounds man's activity and with which he interacts directly.

Circumference: It is a closed measurement that follows the contour of the body. Therefore, this measure is not necessarily circular.

Curvature: It is a measurement from one point to another, following a contour, which is not usually closed or circular.

Distance: A straight line that measures from one point to another, between two marks on the body.

HVAC: Heating, ventilation, and air conditioning.

REHVA: Federation of European heating, ventilation, and air conditioning associations.

Scope: It is a measure from point to point, following along the axis of the arm or leg.

Thickness: Straight line that measures from one point to another horizontally, from front to back of the body.

Width: Straight line that measures from one point to another horizontally, crossing the body or a segment of it from side to side.

Chapter 3
Dynamic Glazing With Higher Thermal and Optical Performance for Zero Energy Building Design

Fernando del Ama Gonzalo
Keene State College, USA

Allison L. Zannoni
Keene State College, USA

Belen Moreno Santamaria
Universidad Politecnica de Madrid, Spain

Juan Antonio H. Hernandez Ramos
Universidad Politecnica de Madrid, Spain

ABSTRACT

One of the most critical objectives in buildings is their adaptation to environmental conditions in order to optimize energy performance as well as the thermal and visual comfort of the occupants. This issue is relevant, for example, in public buildings that incorporate large glazed surfaces, where overheating and a lack of thermal and visual comfort are common, especially in Mediterranean countries. The headmost cause of increased air conditioning loads is direct solar radiation on transparent surfaces. However, the significant losses through glazed surfaces in cold climates also cannot be ignored. Therefore, solar protections must be used, thus observing the harmful effects of the absence of solar radiation on energy loads in winter. This chapter aims to study and compare construction materials recently introduced to the market, such as dynamic glasses, with significantly higher thermal and optical performance than traditional glasses.

DOI: 10.4018/978-1-6684-4030-8.ch003

INTRODUCTION

Optimizing the energy performance and the thermal and visual comfort are the most important goals for new buildings design. This issue is relevant, for example, in public buildings, generally in those that incorporate large glazed surfaces, because of overheating problems and lack of thermal and visual comfort, especially in Mediterranean latitudes, such as in Spain. As mentioned, the most crucial cause of the increase in air conditioning loads, among others, is direct solar radiation on transparent surfaces. Accordingly, solar shielding elements are recommended, even though the absence of solar radiation can cause adverse effects on energy loads.

The first point that must be addressed in dealing with innovative transparent materials is to establish which ones are considered dynamic or static glasses. On the other hand, there are categories of insulating or low-emissive materials that have already been in the market for years and have significantly higher optical performance than traditional glasses. This consideration makes it necessary to include all the transparent elements whose optical and thermal characteristics differ from single glass or standard double glazing in this chapter. Therefore, the apparent differences found in these new products compared to simple glass or that with an air chamber are due to:

- A higher thickness.
- Deposition of a thin film (of various natures) on the flat glass.
- Filling the double-glazed chamber (with insulating materials, vacuum, or low emissive inert gases).

In the background section of this chapter, the double-skin facades will first be studied, then moving on to the study of glass, distinguishing between them in static and dynamic, focusing on Water Flow Glazing facades. Next, different solutions, including glass types and energy systems, will be tested through theoretical models to study the dynamic Water Flow Glazing's capabilities to improve their environments' energy saving and comfort. Finally, an economic analysis will show the Return of Investment (ROI) of different scenarios to explain the balance between cost and efficiency.

BACKGROUND

The Energy Performance Buildings Directive (EPBD) encourages procedures and materials to design energy-efficient and decarbonized facilities by 2050 (European Union, 2018). Zero-energy buildings generate as much energy from renewable technologies as their energy consumption (Sudhakar et al, 2019). Therefore, developing passive strategies is the first step to accomplishing Net Zero Energy goals. Further action comprises the integration of technologies for energy production and management (Frattolillo et al, 2020). It is a fundamental aspect of this chapter to differentiate between static glasses, those that do not change their properties over time and dynamic ones, those that are able to modify them, either by the external climatic conditions or by the user (Casini, 2015).

Static Glasses

Within static glasses, Allen et al. (2017) showed that the construction industry had several glazed products that improve the energy performance of single pane glass by addressing its weaknesses, such as thermal losses or low insulation in cold weather and thermal gains from warm weather. Solar Control Glasses have been the first improvement over traditional clear glasses that reduce the thermal load and excess lighting due to incoming solar radiation, either with the use of colored glasses in their mass or by utilizing surface chemical deposits in which the radiation was reflected directly to the outside (Hermanns et al., 2012). These glasses had advantages such as reducing the solar heat gain coefficient, aesthetic qualities due to the chromatic effects, and allowing views of the exterior space during the day from the inside, but not vice versa. The reduction of the solar heat gain coefficient was not always a positive fact during the year, especially in winter time.

A low-e coating worked as a trap for radiation, while simultaneously remaining transparent to daylight, and in some cases, to incoming solar radiation (Arici et al., 2015). Compared to the spectral characteristics of single glass, the transmittance was significantly reduced, and reflectance was up to 75% higher in the far-infrared (Cui and Mizutani 2016). Placing the film on face 2 (inner face of outer glass) reduces the U-factor and solar factor, as radiation was absorbed in the proximity of the external surface. However, suppose the film was placed on face 3 (outer face of the inner glass). In that case, the radiation passing through the external glass would hit the internal glass. Due to the presence of the external glass, it cannot return to the internal one, operating the window as a solar capture.

Aerogel is a material made up of silica particles and approximately between 90 - 99.8% air. It is characterized by a transparent and low-density open porous structure, giving it optimal optical and energy performance (Thapliyal and Singh 2014). However, so far, monolithic aerogel has been expensive, brittle and its transparency is not perfect (Smirnova and Gurikov 2018). The granule aerogel, called a nanogel, turns the glazing translucent, losing transparency and views. Therefore, these glasses provide high insulation values in exchange for sacrificing the views.

The fundamental principle of vacuum glass is to remove any gas in the chamber, thus improving its thermal insulation by dispensing the conductivity and convection of the gas present in the chamber (Lyu et al., 2019). The main advantage is the independence of the insulating properties of the glass concerning the thickness of the chamber, and the use of these low-e glasses, to reduce transmission. On the other hand, its main drawback is that the external and interior pressures generate tensions compressing glass panes towards the inside, with the risk that they will touch in the center.

A double-skin façade is a constructive solution that improves the envelope's energy efficiency, incorporating different types of glasses, both static and dynamic. The use of double-skin facades has been gradually imposed. Thus, there is a first glass surface (which forms the proper exterior façade) separated from a second glass surface (which constitutes the limit of the living space of the building), allowing, with restrictions, the movement of air in the chamber that define both surfaces (Van Roosmalen et al., 2021). Optimization of the double-skin façade would incorporate dynamic glass to it, and therefore capable of reducing the energy loads inside the house. The concept of the operation of a double skin of air is the circulation of a gas or a fluid, which after pre-heating or pre-cooling, is able to heat or cool the house, by introducing that air or by heat transmission in the case of glass with water (Norouzi and Motalebzade 2018). Double-skin facades enable the use of large glass enclosures. The negative impact on the energy behavior of the building is minimized through the study of the air chamber and the analysis of the thermo-fluid-dynamic problem (air circulation) to obtain energy savings in the house's air conditioning using the preheated air. An optimization of the double skin could be the incorporation of dynamic glass, as it would reduce the use of solar protection and even more the energy loads of heating and air conditioning.

In addition, by adding a third chamber, glass with water circulation could be incorporated into the study. However, if we compare both systems, as indicated by Lyu et al. (2018), the heat extraction capacity of glass with circulating water is much better than a double skin with air.

Dynamic Glasses

In this section, different dynamic glasses have been studied. From chromogenic glasses to circulating fluid glazing, both technologies have been in the process of research and commercialization over the last few years. Chromogenic materials can provide the building with an improvement in energy performance and avoid the lack of comfort produced by solar protections by changing its optical and solar properties (luminous transmittance and solar factor). However, its behavior would not be as effective, in principle, to avoid losses. Therefore, studying the characteristics of these types of glass and the analysis of others within the group of dynamics will allow us to find out which type of glass is more efficient against heat losses and gains.

Chromogenic glasses change their color-opacity either depending on the external conditions they are exposed to or by applying a slight electrical tension. The advantage of using these glasses is that they can work in hot and cold climates, both in winter and summer. In winter periods or cold temperatures, the glass can remain clear to allow the maximum entry of solar radiation, reducing the cost of heating and energy consumption by eliminating the use of blinds or solar protections. On the other hand, in hot periods or summer periods, radiation can be absorbed or reflected by the change of color or transparency of the window, reducing the interior temperature of the environment, with the consequent energy savings in air conditioning. According to Casini (2015) these glasses can be classified into two types, which we will describe and study separately.

Thermochromic glasses change their transparency in response to temperature, which is induced by a chemical reaction. The materials are polymer hydrogels, which vary from a transparent state (when the temperature drops) to a color diffusion, and white (when the temperature rises) which reflects light. In its activated state, the vision through it is dull. Thermochromics are essentially designed for skylights because, in their opaque state, they do not interfere with views as much as in a standard window.

Photochromic glasses change their transparency in response to light intensity, darkening as a response to radiation, typically ultraviolet, between wavelengths ranging from 300 to 400 nm and returning to their original properties in the dark. The phenomenon is the reversible change of a chemical reaction between two energy states with distinct absorption spectra. This change of state is usually induced by electromagnetic radiation.

Dispersed Liquid Crystals are an intermediate phase between crystalline solids and isotropic liquids. They are ordered fluids with anisotropic properties (Abdulmohsin et al., 2020). A variety of physical phenomenon makes them one of the most exciting subjects of modern science. Its unique properties due to their anisotropic optics and high sensitivity to external electric fields allow numerous practical applications. However, although these materials have reached a reasonably high maturation in the market, they are not very interesting from the point of view of energy efficiency for various reasons. First, in the transparent state, they must be continuously fed with energy consumption (20 W/m2), in addition to the difference in global transmittance between the transparent and opaque state (between 20 and 30% respectively in the visible and solar spectrum) with even more minor variations of the solar factor parameter.

Suspended Particle Devices (SPD) are electrically controlled glass which uses a thin layer of liquid, in which numerous microscopic particles have been suspended. The particles are messy and partially block solar transmission and sight in a dull state. Transparent electrical conductors allow the application of an electric field in the film of dispersed particles, aligning them and thus increasing the transmittance (Ghosh et al., 2015). The ignition requires about 100 volts, from the off state (colored, cobalt blue) to the on (almost transparent), and can be modulated to any intermediate state. The power required is 0.05W, both to turn it on and keep it in a constant transmission. Recent studies have proposed new models, which require a voltage below 35 volts, and new suspensions that would give different colors (green, red, and violet). These laminated glasses can be manufactured today in sheets of dimensions of 1.2x2.4 m, both flat and curved. However, the durability and solar optical properties have yet to be verified. These products are currently entering the market but at a high cost (Ghosh et al., 2016)

Electrochromic glazing (EC), a set of thin films applied to glass or plastic, can change its appearance from a clear yellow to a dark blue when a small voltage is applied, always allowing vision through it. A reversible electrochemical action causes the appearance of tinted glass. When the glass is not activated, the transmission is about 85%. The window allows a perfect view without distortions or fog in any state. The visible transmittance (T_v) and solar heat gain coefficient (SHGC) vary depending on the composition of the material. The U factor is not affected by dyeing, always remaining unchanged. Generally, the range between on and off varies according to the outdoor environmental conditions. The electrochromic window is set on a double insulating glass panel, where the EC layer is applied to the inner surface of the external glass. The ignition range is not only determined by the EC layer but also by the insulating glass panels. The external glass can be tinted or clear, and the internal glass can be of any type. The data of an EC glass is exposed when both outer glasses are updated laminated glasses. Macêdo et al. (1992) provided the following visible properties: visible transmittance ranges from 0.50 to 0.60 for the off state and 0.10 to 0.25 for the fully colored. The solar heat gain coefficient (SHGC) ranges from 0.48 to 0.09, and the U factor from 1.59 to 1.87 W/(m^2°C). Finally, the Shadow Factor varies from 0.67 to 0.60 for the off state and 0.30 to 0.18 for the fully colored. For EC windows with intermediate state control, one of the benefits would be a significant increase in views to the outside compared to conventional windows with sun protection. In critical situations, these windows would have the blinds lowered. Still, the visual characteristics are improved in the EC glass, in its colored state, concerning the comfort needs of the person in relation to the vision towards the outside. The reduction of solar heat gains can lead to significant savings in electrical energy in hot climates and when EC glass facades are located to the South, East, and West. On the other hand, EC glazing can cause glare since sometimes the EC is insufficient to control the luminosity. In addition, they cannot provide complete control of direct radiation, causing effects such as glare. Finally, long ignition time and non-uniform final color are visual problems that will have to be addressed in the future.

Water Flow Glazing's fundamental feature is a circulating water chamber between two laminated glass panels, providing the thermal load's control by absorbing infrared radiation in the water chamber. The circulating water, connected to a circuit driven by a pump, recirculates the fluid in a closed-loop (Gutai et al., 2020). This enclosure optimizes solar collection in winter and neutralizes thermal losses by circulating warm water through it. Overheating in summer is neutralized by circulating cold water through the chamber.

A typical U value of a double glass with an air chamber is 2.7 W/m²K. The stagnant water without circulation would achieve only the absorption of infrared radiation, which would radiate it inside once heated. The absence of the air chamber makes the U value rise to reach 5 W/m²K. When the water circulates, it absorbs all the excess heat, thus reducing the energy consumption inside the house. The Water Flow Glazing system can be applied to curtain walls, interior partitions, or glazed roofs. The system consists of two close-loops. The primary circuit consists of a circulation pump and an energy source for heating or cooling water. In contrast, the secondary distributes the water to the interior partitions or the glass facade and uses a mass flow rate of 0.1 l/s (Moreno et al., 2020). The energy between closed loops is transferred through a heat exchanger. The energy source can be a boiler, a heat pump, or solar thermal panels. Using a geothermal exchanger in winter and summer allows the water to be cooled or heated when it runs through the buried pipes. Gutai and Khebary (2021) introduced the Smart Water-filled Glass (SWFG), which includes the modification of transparency of the façade component by dying the fluid that circulates through the glazing. The findings confirm that the SWFG can save from 0% to 5.28% for climates with cooling demand and from 0.25 to 3.39% for cold climates, compared to WFG.

MATERIALS AND METHODS

In this section, the authors intended to explain the simulated facility and the different scenarios that has been considered, from the existing single glazing to the Water Flow Glazing coupled with a geothermal heat exchanger. Scenarios using conventional double glass has been simulated using Design Builder software with the Energy+ engine. The Water Flow Glazing scenario has been simulated with the tool validated in Moreno et al. (2020).

Description of the Prototype

The studied prototype was an entirely detached building of lightweight construction and a floor area of 41 m², raised 0.5 meters from ground level. Figure 1 shows the main volume of the prototype and the proposed geothermal heat exchanger coupled with a water-to-air heat pump. The heat exchanger was made up of 400 meters of pipes buried at a depth of 5 meters. The external diameter of the pipes was 32 mm. A soil density of 2600 kg/m³ and a 2.6 W/(mK) thermal conductivity have been considered to simulate the performance of water-to-air heat pumps (Torregrosa-Jaime et al., 2019).

The composition of the opaque enclosures was a sandwich panel with interior and exterior 4mm aluminum layer and a 4cm polyurethane foam. The thermal transmittance of these enclosures was 0.48 W/(m²K) for walls and roofs and 2.45 for the floor. The relationship between the glazing and the opaque enclosures in which they were located (*Window to Wall Ratio* or WWR) was 20% in the Northern façade, 40% in Southern and Western facades, and 85% in the Eastern elevation. The Window-to-Floor Area (WFA) was 30%. In addition, an aluminum frame with thermal break and a thermal transmittance U = 4 W/(m²K) has been considered according to UNE-EN ISO 10077-1 (2010), representing 10% of the glazing area. In the different simulated scenarios, infiltrations equivalent to 0.27 per hour have been considered, corresponding to the air permeability of non-practicable window carpentry per the UNE-EN 12207 (2000) windows and doors standard. The opening hours were from 8:00 a.m. to 10:00 p.m. with internal loads of 2 people sitting, according to ISO 7730 (2005). When it came to internal loads, computer equipment of 230 W has been considered. The properties of glazing are reflected in Table 1. The

Figure 1. Description of the prototype. 1. simulated prototype, 2. geothermal heat exchanger

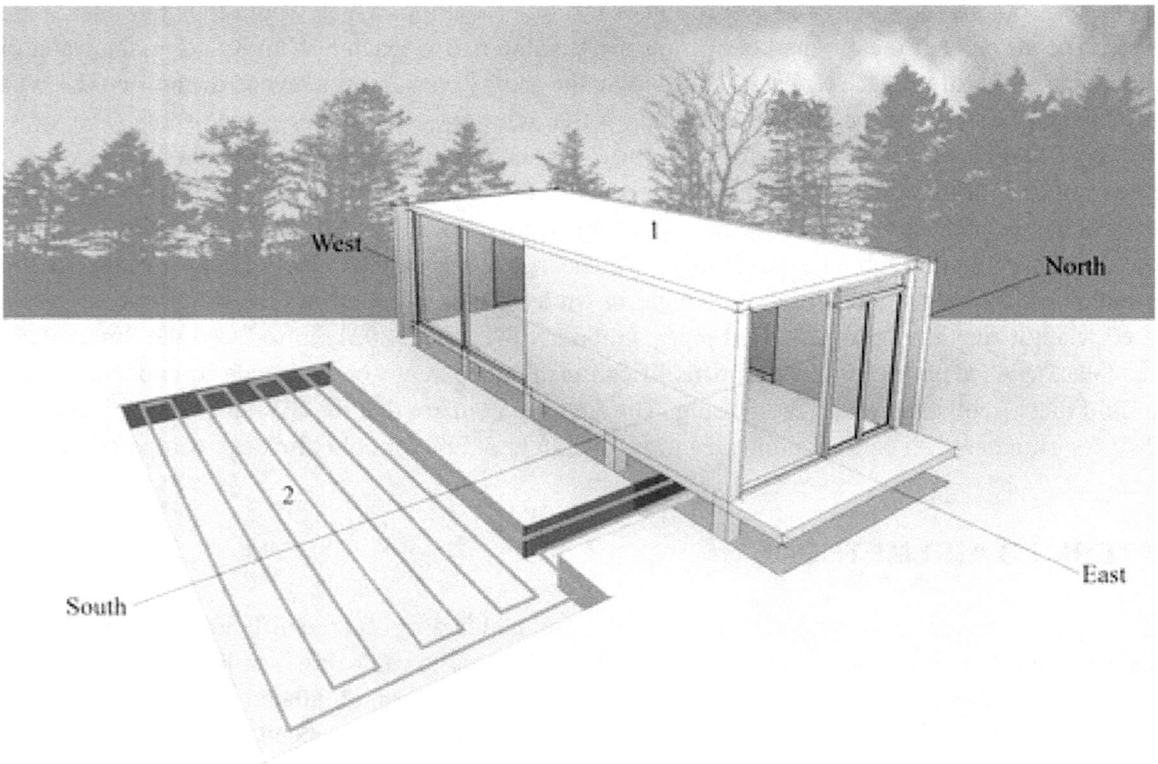

existing glazing was a single SGG Stadip Clear 44.1 8 mm. Scenario1 was simulated with clear double glazing (6mm/air cavity 8mm/6mm). Scenario 2 included double glazing with solar control. Finally, Scenario 3 considered WFG with laminated glass panes and a 12 mm water chamber.

Figure 2 explains the thermal behavior of water flow glazing. θ_{IN} represents the inlet temperature of the circulating water, while θ_{OUT} represents the outlet temperature. *R* and *T* are the variables that measure the reflectance and transmittance of the glazing, and i_0 represents the total incoming solar radiation. The temperature of the glass panels is considered steady over time. Finally, the spectral properties of the glazing assembly are dependent upon the variable incoming solar radiation, i_0.

Table 1. Thermal properties of the simulated glazing

	ṁ=0 Liter/ (min m²)		ṁ=1 Liter/ (min m²)	
Glazing	*U (W/m²K)*	*g*	*U (W/m²K)*	*g*
Single Glazing Clear 8 mm	5.42	0.75		
Double Glazing clear (6/8/6)	3.21	0.72	-	-
Double Glazing Solar control (6/8/6)	2.72	0.35	-	-
Water Flow Glazing (6+6/12/6+6)	4.79	0.45	0.56	0.22

Figure 2. Description of water flow glazing with dynamic U and g values. Solar radiation (i_0) on the glazing is divided into the reflectance (R), the energy transmittance (T)

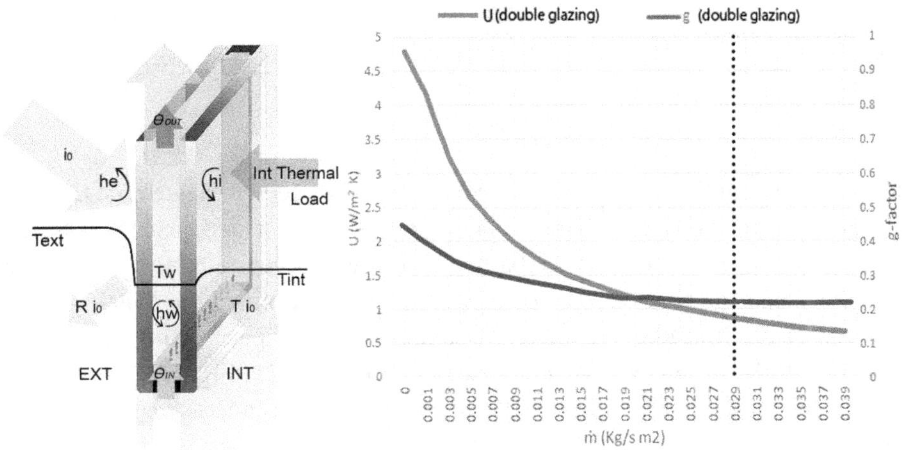

Equation (1) shows that U value in WFG depends on several variables.

$$U = \frac{U_i U_e}{\dot{m}c + U_e + U_i}, \tag{1}$$

where U_i measures the heat transfer between the water chambers and indoors, while U_e measures the heat transfer between the water chamber and outdoors, $\dot{m}$ is the mass flow rate, and c is the specific heat capacity of the fluid. Equation (2) shows the expression for the g factor for WFG taken from Sierra and Hernandez (2017).

$$g = \left(\frac{U_i}{\dot{m}c + U_e + U_i}\right) A_v + A_i + T, \tag{2}$$

where T is the glazing transmittance, A_i, the constant part of the internal heat transfer factor, and A_v, the net absorptance of the water.

RESULTS

This section shows the simulation conditions that have been considered, the energy demands of the analyzed prototype, and the consumption resulting from simulations. The original case study considered clear double glazing and air-to-air heat pumps. The rest of the scenarios considered a water-to-air heat pump fed by a horizontal geothermal exchanger. Table 2 explains the main features of air-to-air and water-to-air heat pumps, according to the values taken from Priarone et al. (2020).

Table 2. Specifications of the simulated heat pumps

	Heating Power (kW)	COP	Cooling Power (kW)	EER
Air-to-air heat pump*	11.60	3.6	11.40	3.2
Water-to-air heat pump**	13.17	4.8	10.32	4.6

* Original case study. ** Scenarios 1, 2, 3

The first step in analyzing the contribution of active glass has been to determine the energy demands of the prototype in its current situation. The existing glazing of the tested space was SGG Stadip Clear 44.1 8 mm laminar glass units from Saint Gobain. Therefore, it constituted the primary reference with which the rest of the scenarios proposed were compared and represented the existing state of the tested prototype. The thermal conditioning was made of an air-to-air heat pump with a Coefficient of Performance (COP) of 3.61 and an Energy Efficiency Ratio (EER) of 3.21. Figure 3 shows the monthly and annual energy demands the diagram of the heating, cooling and ventilation system (HVAC) corresponding to the original prototype. It shows that the maximum heating demand is 1177.9 kWh in December, followed by that obtained for January with 1109.4 kWh. The maximum cooling demand turned out to be 1082.3 kWh, and it occurred in July.

Figure 3. Energy demand of the case study. 1. clear 44.1 8 mm laminar glass, 2. air-to-air heat pump

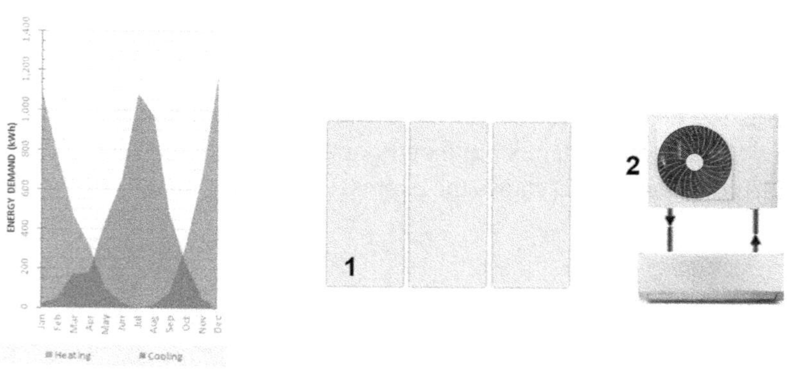

Scenario 1: Prototype Equipped with Geothermal Powered Water-to-Air Heat Pump and Clear Glass with Air Chamber

In this scenario, the energy consumption of a water-to-air heat pump was compared to the air-air heat pump with which the prototype was equipped in the first analysis. Its performance data was based on the temperature of the water and air. In scenario 1, insulating glass units with an air chamber in a 6/8/6 configuration had been used. The glass thermal transmittance (U) was 3.21 W/(m^{2}K) and the solar heat gain coefficient (g) was 0.72. The conditioning system was based on a water-to-air geothermal heat pump. Figure 4 shows the schematic view of the prototype's HVAC system used to maintain the interior conditions of the prototype within the control limits required by the RITE (2016). The blue and red lines represent the flow of water between the geothermal exchanger and the heat pump. It shows the heating, cooling, ventilation and pumping consumption. The maximum demand in winter was 301.9 kWh in December, whereas the maximum demand in summer turned out to be 389.6 kWh, and it occurred in July. The total yearly demand was 3406.74 kWh.

Figure 4. Scenario 1: consumption of the prototype and schematics. 1. clear double glazing, 2. geothermal heat exchanger, 3. water-to-air heat pump

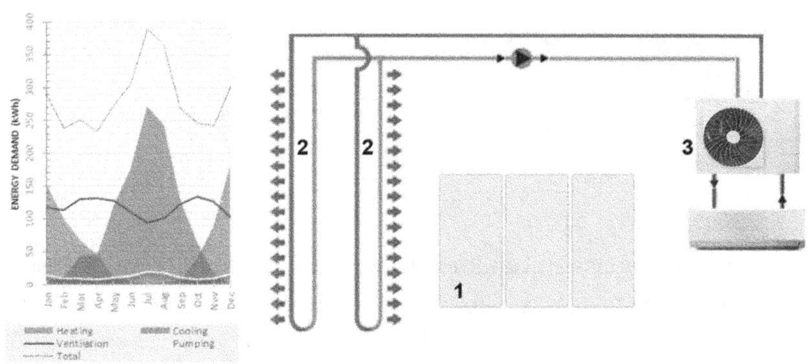

Scenario 2: Prototype Equipped with Geothermal Powered Water-to-Air Heat Pump and Solar Control Glass with Chamber

In this new scenario, the prototype was equipped with static solar control glass with a configuration 6/8/6. The glass thermal transmittance and solar heat gain coefficient were U = 2.72 W/(m^2K) and g= 0.41, respectively. Regarding the conditioning system, the water-to-air heat pump equipment already indicated has been used. Figure 5 shows the schematic view of the prototype's HVAC system. The only difference from scenario 1 was the solar control glazing used in the simulation. It shows the heating, cooling, ventilation and pumping consumption. The maximum demand in winter, including heating, cooling, ventilation and pumping energy, was 317.9 kWh in December, whereas the maximum demand in summer was 343.87 kWh in July. The total yearly demand was 3203.48 kWh.

Scenario 3: Prototype Equipped with Geothermal Powered Water-Air Heat Pump and Glass with Water Chamber in Continuous Operation Associated with the Conditioning System

The dynamic glass performed a dual function simultaneously. On the one hand, it reduced the solar loads that reached the interior of the building. However, at the same time, it acted as a solar collector, generating a certain amount of water at a temperature higher than the environment. In the simulation of dynamic glass, the reducing effect of solar charges had been achieved by identifying an equivalent static glass shown in Section 6.4 Simulation of glass with circulating water chamber. Therefore, from using this equivalent glass, it would be possible to simulate its influence on the interior environment of a specific room. Thus, once the states of dynamic glass had been characterized with water in circulation using different static glasses, evaluating the behavior of a system equipped with a glass of this type

Figure 5. Scenario 2: consumption of the prototype and schematics. 1. double glazing with solar control, 2. geothermal heat exchanger, 3. water-to-air heat pump

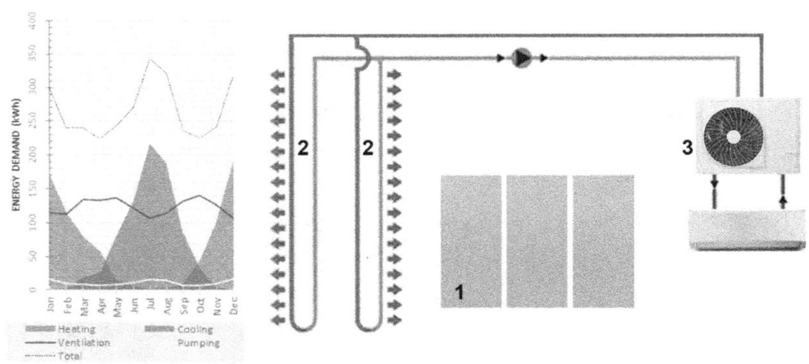

required alternating between one glass and another within the same simulation. Subsequently, utilizing a control element of the simulation program, the selection between one or the other would be carried out, emulating the activation of the water circulation, depending on the interior temperature of the module, as it occurred in the experimental installation. Equations (1) and (2) have shown how the values of the parameters *U* and *g* change with the mass flow rate. The role of Water Flow Glazing as a solar collector had been simulated considering two laminated glass panes (6+6mm) separated by a 12 mm water chamber. The water temperature leaving the chamber was given by equation (3).

$$\theta_{OUT} = \frac{i_0 A_v + U_i \theta_i + U_e \theta_e + \dot{m} c \theta_{IN}}{\dot{m} c + U_e + U_i}. \quad (3)$$

It was necessary to determine the values of the absorbance of the layers that made up the glazing. In this case, A_v was calculated from the absorbance for the external glass (A_1), the water chamber (A_w), and the internal glass (A_2). These values, shown in Table 3, had been obtained from the Window v7.4 program and Sierra and Hernandez (2017).

Table 3. Thermal and spectral parameters of Water Flow Glazing

	A_1 *	A_2 *	A_w **	A_v **
Water Flow Glazing	0.585	0.037	0.514	0.412

* values taken from Window v7.4 ** values taken from Sierra and Hernandez (2017)

The implementation of the model allowed the use of water heated by the glass to feed an air-water heat pump so that the thermal jump of the fluid that fed the heat pump was reduced, thus improving its efficiency. In addition, in an actual building, this water could also be used for the preheating of DHW or the thermal regeneration of the land in contact with the borehole heat exchanger if there was no demand for heating. Figure 6 illustrates that the maximum demand in winter, including heating, cooling, ventilation and pumping energy, was 267.9 kWh in December, whereas the maximum demand in summer was 183.87 kWh in July. The total yearly demand was 2160.26 kWh.

Figure 6. Scenario 3: consumption of the prototype and schematics. 1. heat plate exchanger, 2. water pump, 3. water flow glazing, 4. geothermal heat exchanger, 5. water-to-air heat pump

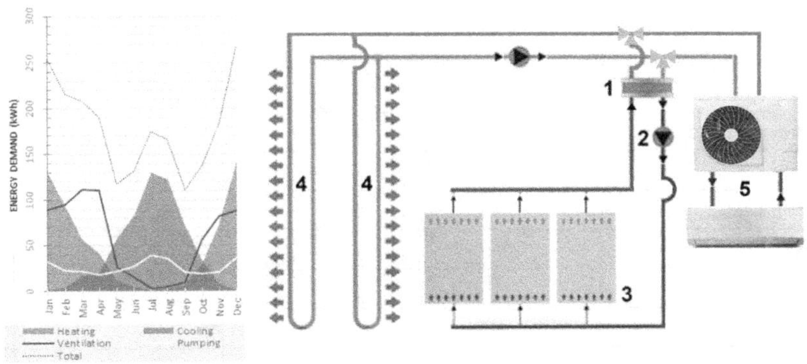

ECONOMIC ASSESSMENT OF SCENARIOS

The measures to improve their energy performance must consider profitability in terms of cost-effectiveness, among other factors. Thus, an economic evaluation of the simulated scenarios was presented, according to the Construction Price Generator of CYPE Engineers in Spain. This evaluation consisted of determining the Net Present Value (NPV) of the investment necessary to rehabilitate the facility assuming 30 years, understood as the difference between the current value of the action of the change of glass and facilities of the prototype, and the cost of carrying them out. In addition, the internal rate of return (IRR) and the return of investment (ROI) were calculated as well as the cash flows for each of the scenarios are included. Equation (4) shows the expression for Net Present Value, where R_t is the net cash inflow-outflows at year *t*, *i* is the discount rate, and *t* is a period of 30 years.

$$NPV = \sum_{t=1}^{n} \frac{R_t}{\left(1+i\right)^t}. \tag{4}$$

The electricity price was 0.25 €/kWh, and the discount rate was 2.5%, with a yearly increase of energy price of 10%. Equation (5) shows the expression for the return of investment.

$$ROI = \frac{Current\,Value\,of\,Investment - Cost\,of\,Investment}{Cost\,of\,Investment}. \tag{5}$$

Economic Assessment of Scenario 1

In scenario 1, there was a double-glazing envelope and a water-to-air heat pump with buried horizontal heat exchangers. As a result, Table 4 shows the budget of materials and equipment that made the glazing and heating and cooling system. The Net Present Value (NPV) turned out to be 16241.08€. In addition, the IRR was 6%, although the return of investment (ROI) was 19.7 years.

Table 4. Material cost of scenario 1

Materials and units	Amount	Cost (€/unit)	Total Cost (€)
Double glazing (m^2)	30.17	64**	1930.88
Water-to-Air Heat Pump	1	8200**	8200.00
Horizontal heat exchanger	1	2560**	2560.00
Control system	1	1975**	1975.00
PE pipes (m)	20	78.76**	1575.20
Total			16241.08

** Values taken from Construction Price Generator of CYPE Engineers

Economic Assessment of Scenario 2

Scenario 2 differed from scenario 1 in replacing the clear double glazing with solar control double glazing while maintaining the same configuration. Regarding the conditioning system, the water-to-air heat pump was maintained. Solar control glasses had a higher cost than clear glasses, so the renovation budget increased to 18956.38€, as shown in Table 5, and the prototype's cost per unit of area amounts to 517.02 (€/m^2). In this case, the NPV obtained was 35121.07€, while the IRR reached 8%, and the estimated return of investment (ROI) was 16.7 years.

Table 5. Material cost of scenario 2

Materials and units	Amount	Cost (€/unit)	Total Cost (€)
Double glazing with solar control coating (m^2)	30.17	154**	4646.18
Water-to-Air Heat Pump	1	8200**	8200.00
Control system	1	1975**	1975.00
Horizontal heat exchanger	1	2560**	2560.00
PE pipes (m)	20	78.76**	1575.20
Total			18956.38

** Values taken from Construction Price Generator of CYPE Engineers

Economic Assessment of Scenario 3

In scenario 3, passive glasses were replaced by active glasses with a circulating water chamber. Its operation, in this case, was associated with the conditioning system. Table 6 shows that this scenario implied an increase in the complexity of the facility, which directly affected the budget that rose to 24293.70€. The NPV was 23.076,80 €, the IRR was 5%, and the return of investment (ROI) was 20.5 years.

Table 6. Material cost of scenario 3

Materials and units	Amount	Cost (€/unit)	Total Cost (€)
Water Flow Galzing (m^2)	25.50	225*	5737.50
Double glazing with solar control coating (m^2)	6.30	154**	970.20
Circulating system (ud)	1	1200*	1200.00
PE pipes (m)	40	78.76**	3150.40
Other	-	-	550.60
Water-to-Water Heat Pump	1	8200**	8200.00
Horizontal heat exchanger	1	2560**	2560.00
Control system	1	1975**	1975.00
Total			24293.70

* Values taken from Santamaria et al. (2021); ** Values taken from Construction Price Generator of CYPE Engineers

Table 7 summarizes the performance of the prototypes considering the three different scenarios introduced in the mathematical model. As expected, scenario 3 showed the best energy performance, although it had the highest cost due to the price of the Water Flow Glazing components. Scenario 2 showed a good balance between the energy savings and the Return of Investment (ROI).

Table 7. Energy and economic assessment of the three scenarios

	Energy consumption (kWh/year)	Total Cost (€)	IRR (%)	ROI (years)
Scenario 01	3406.74	16241	6	19.7
Scenario 02	3203.48	18956	8	16.7
Scenario 03	2160.26	24293	5	20.5

FUTURE RESEARCH DIRECTIONS

Overall, this book chapter has shown that a set of algebraic equations from a simplified mathematical model helps architects and engineers understand the complex behavior of dynamic WFG envelopes at an early stage of the project. Future research could develop a complete mathematical model and its integration in commercial software to validate the hypothesis in a transient state over the years.

CONCLUSION

Static glazing can play a passive role in managing the interior comfort conditions by utilizing glazing coatings and gas cavities. On the other hand, dynamic glazing acts as an active facade component to handle the incoming solar radiation and keep indoor comfort. The most relevant finding of this research verified the dynamic condition of WFG through variable *U* value and *g* factor. The *U* value ranged from 5 to 0.66 W/(m^{2}K), and the *g*-factor ranged from 0.45 to 0.24 when varying the mass flow rate from 0 Liter/(m^{2}min) to 2 L/(m^{2}min). Furthermore, WFG technology could be coupled to water-to-water, or water-to-air heat pumps, increasing the Coefficient of Performance (COP).

The main conclusion of this chapter is that WFG contributed to energy efficiency, reducing CO_2 emissions and energy consumption.

The scenario with the lowest total energy consumption through the year was Scenario 3, with 2160.26 kWh. Its energy consumption was 32.5% lower than Scenario 2 and 36.6% lower than Scenario 1 (the highest energy consumption 3406.74 kWh).

However, it was pertinent to consider the economic factors to analyze the scenarios properly. In this case, Scenario 3 had the highest initial total cost of 24,293.70€, and Scenario 1 had the lowest initial total cost of 16,241.08€. Therefore, scenario 3's initial cost was 28.2% higher than Scenario 2 (being 18,956.38€) and 49.6% higher than that of Scenario 1.

Scenario 3, having such a higher cost than the others, was due to the fact that it had a significant increase in the complexity of the building and required 8 initial purchases whereas scenarios 1 and 2 only required 5. Moreover, the only difference in cost between Scenarios 1 and 2 was that of the double glazing (in scenario 1) and double glazing with solar control coating (in scenario 2). However, in scenario 3, while some of the things purchased were the same prices as scenarios 1 and 2, it also required additional, costly materials such as its water flow glazing, circulating system, and other, which added a total of 7,488.10€ to the initial total cost than that which was necessary in either scenarios 1 or 2.

Provided that Scenario 3 had the lowest energy consumption while simultaneously also having the highest total cost, it is of interest to analyze the ROI, NPV, and IRR of each of these scenarios to find the scenario most beneficial in terms of investments.

The lowest ROI between the scenarios was Scenario 2 with a ROI of 16.7 years. This was 18.0% less than Scenario 1 (which had the second lowest ROI of 19.7 years) and 32.3% less than Scenario 3 (which had the highest ROI of 20.5 years). Scenario 2 also obtained the highest NPV of 35,121.07€. Lastly, Scenario 2 had the highest IRR of 8% compared that of Scenario 1 (being 6%) and that of Scenario 3 (being 5%).

In conclusion, Scenario 2 was the most beneficial in terms of long-term investments while also having a total energy consumption 6.3% lower than that of the highest (Scenario 1) and an initial total cost which was 28.2% lower than that of its highest (Scenario 3).

ACKNOWLEDGMENT

This research has been supported by a KSC Faculty Development Grant (Keene State College, New Hampshire, USA).

REFERENCES

Abdulmohsin, H., Aritra, G., Senthilarasu, S., & Tapas, K. M. (2020). Evaluation of thermal performance for a smart switchable adaptive polymer dispersed liquid crystal (PDLC) glazing. *Solar Energy*, *195*, 185–193. doi:10.1016/j.solener.2019.11.024

Allen, K., Connelly, K., Rutherford, P., & Wu, Y. (2017). Smart windows—Dynamic control of building energy performance. *Energy and Building*, *139*, 535–546. doi:10.1016/j.enbuild.2016.12.093

Arici, M., Karabay, H., & Kan, M. (2015). Flow and heat transfer in double, triple and quadruple pane windows. *Energy and Building*, *86*, 394–402. doi:10.1016/j.enbuild.2014.10.043

Casini, M. (2015). Smart windows for energy efficiency of buildings. *International Journal of Civil and Structural Engineering*, *2*, 230–238.

Cui, Z., & Mizutani, A. (2016). Research on the reduction effect of transparent glass on cooling power energy consumption research on the reduction of cooling and heating loads by transparent solar heat absorbing glass panels (Part 2). *Journal of Asian Architecture and Building Engineering*, *15*(3), 651–658. doi:10.3130/jaabe.15.651

European Union. (2018). *Directive (EU) 2018/844 of the European Parliament and of the Council of 30 May 2018. Amending Directive 2010/31/EU on the Energy Performance of Buildings and Directive 2012/27/EU on Energy Efficiency* https://eur-lex.europa.eu/legal-content/EN/TXT/PDF/?uri=CELEX:32018L0844&from=EN

Frattolillo, A., Canale, L., Ficco, G., Mastino, C. C., & Dell'Isola, M. (2020). Potential for Building Façade-Integrated Solar Thermal Collectors in a Highly Urbanized Context. *Energies*, *13*(21), 5801. doi:10.3390/en13215801

Ghosh, A., Norton, B., & Duffy, A. (2015). Measured overall heat transfer coefficient of a suspended particle device switchable glazing. *Applied Energy*, *159*, 362–369. doi:10.1016/j.apenergy.2015.09.019

Ghosh, A., Norton, B., & Duffy, A. (2015). Measured overall heat transfer coefficient of a suspended particle device switchable glazing. *Applied Energy*, *159*, 362–369. doi:10.1016/j.apenergy.2015.09.019

Ghosh, A., Norton, B., & Duffy, A. (2016). Behaviour of a SPD switchable glazing in an outdoor test cell with heat removal under varying weather conditions. *Applied Energy*, *180*, 695–706. doi:10.1016/j.apenergy.2016.08.029

Ghosh, A., Norton, B., & Duffy, A. (2016). Behaviour of a SPD switchable glazing in an outdoor test cell with heat removal under varying weather conditions. *Applied Energy*, *180*, 695–706. doi:10.1016/j.apenergy.2016.08.029

Gil-Lopez, T., & Gimenez-Molina, C. (2013). Influence of double glazing with a circulating water chamber on the thermal energy savings in buildings. *Energy and Building*, *56*, 56–65. doi:10.1016/j.enbuild.2012.10.008

Gutai, M., & Kheybari, A. G. (2021). Energy consumption of hybrid smart water-filled glass (SWFG) building envelope. *Energy and Building*, *230*, 110508. doi:10.1016/j.enbuild.2020.110508

Gutai, M., Lee, S. T., Magori, B., Morishita, Y., Kheybari, A. G., & Spencer, J. (2020). Construction Aspects of Hybrid Water-Filled Building Envelopes. *Journal of Facade Design and Engineering.*, *8*, 127–152. doi:10.7480/jfde.2020.2.4784

Hermanns, M., del Ama, F., & Hernández, J. A. (2012). Analytical solution to the one-dimensional non-uniform absorption of solar radiation in uncoated and coated single glass panes. *Energy and Building*, *47*, 561–571. doi:10.1016/j.enbuild.2011.12.034

ISO 7730 (2005) *Ergonomics Of The Thermal Environment - Analytical Determination And Interpretation Of Thermal Comfort Using Calculation Of The PMV And PPD Indices And Local Thermal Comfort Criteria.* ISO: Geneva, Switzerland, 2005.

Lyu, Y. L., Chow, T. T., & Wang, J. L. (2018). Numerical prediction of thermal performance of liquid-flow window in different climates with anti-freeze. *Energy*, *157*, 412–423. doi:10.1016/j.energy.2018.05.140

Lyu, Y. L., Liu, W. J., Su, H., & Wu, X. (2019). Numerical analysis on the advantages of evacuated gap insulation of vacuum-water flow window in building energy saving under various climates. *Energy*, *175*, 353–364. doi:10.1016/j.energy.2019.03.101

Macêdo, M. A., Dall'Antonia, L. H., Valla, B., & Aegerter, M. A. (1992). Electrochromic smart windows. *Journal of Non-Crystalline Solids*, *147–148*, 792–798. doi:10.1016/S0022-3093(05)80718-5

Moreno, B., & Hernández, J. A. (2018). Analytical solutions to evaluate solar radiation overheating in simplified glazed rooms. *Building and Environment*, *140*, 162–172. doi:10.1016/j.buildenv.2018.05.037

Moreno Santamaria, B., del Ama Gonzalo, F., Pinette, D., Gonzalez-Lezcano, R.-A., Lauret Aguirregabiria, B., & Hernandez Ramos, J. A. (2020). Application and Validation of a Dynamic Energy Simulation Tool: A Case Study with Water Flow Glazing Envelope. *Energies*, *13*(12), 3203. doi:10.3390/en13123203

Norouzi, R., & Motalebzade, R. (2018). Effect of Double-Skin Façade on Thermal Energy Losses in Buildings: A Case Study in Tabriz. In *Energetic and Environmental Dimensions*. Academic Press. . doi:10.1016/B978-0-12-813734-5.00012-3

Priarone, A., Silenzi, F., & Fossa, M. (2020). Modelling Heat Pumps with Variable EER and COP in EnergyPlus: A Case Study Applied to Ground Source and Heat Recovery Heat Pump Systems. *Energies*, *13*(4), 794. doi:10.3390/en13040794

Santamaria, B. M., Gonzalo, F. A., Griffin, M., Aguirregabiria, B. L., & Ramos, J. A. H. (2021). Life Cycle Assessment of Dynamic Water Flow Glazing Envelopes: A Case Study with Real Test Facilities. *Energies*, *13*(8), 2195. doi:10.3390/en14082195

Sierra, P., & Hernandez, J. A. (2017). Solar heat gain coefficient of water flow glazing. *Energy and Building*, *139*, 133–145. doi:10.1016/j.enbuild.2017.01.032

Smirnova, I., & Gurikov, P. (2018). Aerogel production: Current status, research directions, and future opportunities. *The Journal of Supercritical Fluids*, *134*, 228–233. doi:10.1016/j.supflu.2017.12.037

Spanish Regulation of Thermal Installations in Buildings RITE. (2016). Factores de Emisión de CO2 y Coeficientes de paso a Energía Primaria de Diferentes Fuentes de Energía Final Consumidas en el Sector de Edificios en España. Instituto para la Diversificación y Ahorro de la Energía (IDAE).

Sudhakar, K., Winderl, M., & Shanmuga Priya, S. (2019). Net-zero building designs in hot and humid climates: A state-of-art. *Case Studies in Thermal Engineering, 13*, 100400. doi:10.1016/j.csite.2019.100400

Thapliyal, P. C., & Singh, K. (2014). Aerogels as Promising Thermal Insulating Materials: An Overview. *Journal of Materials, 2014*, 127049. Advance online publication. doi:10.1155/2014/127049

Torregrosa-Jaime, B., Martínez, P. J., González, B., & Payá-Ballester, G. (2019). Modelling of a variable refrigerant flow system in EnergyPlus for building energy simulation in an Open Building Information modelling environment. *Energies, 12*(1), 22. doi:10.3390/en12010022

UNE-EN 12207 (2000) *Windows and doors - air permeability - classification.* ISO: Geneva, Switzerland, 2010.

UNE-EN ISO 10077-1 (2010) T*hermal performance of windows, doors and shutters - calculation of thermal transmittance - part 1: general.* ISO: Geneva, Switzerland, 2010.

Van Roosmalen, M., Herrmann, A., & Kumar, A. (2021). A review of prefabricated self-sufficient facades with integrated decentralised HVAC and renewable energy generation and storage. *Energy and Buildings, 248*. doi:10.1016/j.enbuild.2021.111107

KEY TERMS AND DEFINITIONS

Building Envelope: The integrated elements of a building that separate its interior from the outdoor environment.

Circulating System: A group of devices that allows the water flow through the Water Flow Glazing and it is made of at least a water pump, a heat plate exchanger, and a precision flow meter and a thermometer for monitoring the inlet and outlet temperature.

Dynamic G-Factor: The coefficient used to measure the glazing's ability to transmit solar energy can change in response to an environmental, temperature, or electrical control. Water Flow Glazing can vary its dynamic g-factor by changing the mass flow rate.

Dynamic U-Value: A measure of heat loss in an element of the building envelope depending on the variable mass flow rate through the water flow glazing. It is indicated in units of Watts per meter squared per Kelvin $W/(m^2K)$.

Mass Flow Rate: The mass of a fluid which passes through the water flow glazing cavity per unit of time. It is indicated in units of kilogram per second.

Water Flow Glazing: A double or triple glazing with a water chamber connected to a circulating device allowing the flow of water through the glass panel in a closed-circuit exchanging heat with the environment.

Chapter 4
Dynamic of Construction Robotics to Develop Affordable Housing in Latin American Cities

Rolando-Arturo Cubillos-González
https://orcid.org/0000-0002-9019-961X
Universidad Pontificia Bolivariana, Colombia

ABSTRACT

Latin America is one of the most urbanized regions. It implies a high demand for housing for the low-income population of this region in the face of climate change. The COVID-19 pandemic caused a delayed technology transition in the region. Although recently it has been observed that the construction industry and Latin American governments from the pandemic are promoting disruptive actions in this industry. The promotion of robotic construction in Latin America is an alternative to this problem. Construction robotics is an opportunity for sustainable development in Latin America. It is essential to update the construction sector by orienting it towards implementing this type of technology. Technology transfer plays a prominent role in the dissemination of this new paradigm. What is the dynamic of construction robotics to develop affordable housing in Latin American cities? This chapter aims to identify the dynamic of construction robotics to develop affordable housing in Latin American cities.

INTRODUCTION

We are living in a crucial moment in Latin America. Significant world changes affect the future development of the region. One of them is the fourth industrial revolution. This new industrial revolution disruptively transforms the different structures of world society. But the fourth industrial revolution is not only the implementation of digital technologies. It represents considerable changes within the region companies that will impact their ability to survive locally and internationally (Rossato, 2018). That is why

DOI: 10.4018/978-1-6684-4030-8.ch004

the impacts of this revolution in Latin American countries will probably be more disruptive than in other parts of the world. The challenge for the regional industry is enormous to adapt to this new revolution.

Many jobs are at risk in Latin America. It estimates that 27% of formal employees are doing routine tasks. These jobs could replace by robots. In this area, Latin America has 31.6% of the companies looking for qualified workers, which is higher than the world average of 21.2% (Framil & Ovanessoff, 2018). If robots take over routine tasks in Latin America, regional companies can unleash innovation. Also, they can invest in jobs that emphasize human skills, and they will appreciate these experiences because they are essential for change. With machines that do heavy technical work, it is possible to drive multidisciplinary environments with higher productivity.

The fourth industrial revolution is an opportunity to close the gap with advanced economies. The introduction of this type of technology may favour the region. Latin America needs to design strategies that are relevant to its economies. It means understanding the strategic industrial areas like the construction sector that will be growth leaders in the region (R. A. Cubillos-González & Cardoso, 2020). The population growth of the region demands a high production of housing. Therefore, Latin American governments must turn the fourth industrial revolution into an opportunity (R.-A. Cubillos-González, 2021).

The Covid-19 pandemic can reconsider the relationship between architecture and medicine. In other words, between the daily human habitat and health. To promote Health environments. That will be a Latin American opportunity for the construction sector to develop new technologies that give better environments to the city (López-Escamilla et al., 2020). So, the fourth revolution could offer new tools to develop an innovative process in the Latin American construction sector. For example, the introduction of construction robotics in housing development in Latin America could change productivity in the region.

Besides, the COVID-19 pandemic consequences pose a restructuring of the labour market and a change in the urban structure in the long term (Kang et al., 2020). On the other hand, the high demand for housing implies thinking about the better development of Latin American cities. If human beings want to be healthy, we need a healthy environment (Ng, 2020). What commits governments and companies to meet the sustainable development goals (SDG). The resilient design in buildings increases well-being and reduces vulnerability to new pandemics. Construction robotics is an opportunity to add resilient and sustainable strategies in the design of housing in Latin American cities.

Construction robotics adoption improves productivity in the construction industry. This process confronts two main challenges (Yahya et al., 2019): first, the high cost to maintain the robotic technologies. Second, constructions firm could have a strategic partnership with high technology. Because this is the best method to improve robotics implementation. The technological development of construction robots is underway globally. Technology scenarios are crucial for workers' skills to use construction robots' technology (Pan et al., 2020). The integrated scenario approach is the axis for developing the construction robotics industry in Latin America.

Automation technologies are implementing in the housing industry around the world. Those technologies help form strategies this industry requires to meet sustainability criteria (Bannova & Kristiansen, 2014). Construction robotics without human supervision has crucial economic benefits (Melenbrink et al., 2017). So, the integrated scenarios approach robotic assembly considers affordances of on-site construction in the housing production as the main benefit to reduce cost in the Latin American construction industry.

Construction Robotics focuses on the implementation of automation to renew the construction industry and to stop its current low productivity (McKinsey Global Institute, 2017). Currently, the implementation of new technologies is lagging because of the lack of digital technology in construction. Robotic industrialization refers to the transformation of low-level components into higher-level components robot-supported industrial (Thomas Bock & Linner, 2015). Conventional construction reached its limits. Because automation in construction could achieve what other manufacturing industries have already successfully shown. Then world construction industry would elevate the sector to the level of the rest of the manufacturing fields (T Bock, 2014; Thomas Bock, 2007; Thomas Bock & Linner, 2015).

On another hand, the world's population is ageing at an unprecedented pace, and it is not only a severe crisis in the developed world. So, it is needed to develop smart housing and cities that integrate Automatic assistive technologies developed in a better sustainable environment for people. Robotic automatization could provide an innovative affordable housing solution to mitigate the impact of an ageing society.

This chapter will answer the question What is the dynamic of construction robotics to develop affordable housing in Latin American cities? The chapter aims to identify the dynamic of construction robotics to develop affordable housing in Latin American cities. It describes the barriers and opportunities to impulse innovative technologies. It explains the new construction systems opportunities in affordable housing in the region. After, it shows the role of innovation in Latin American affordable housing in cities. Finally, it presents the conclusion.

BARRIERS AND OPPORTUNITIES TO IMPULSE INNOVATIVE TECHNOLOGIES IN LATIN AMERICAN CITIES

The World Economic Forum (WEF) proposes six main challenges for the construction sector at the international level (Castagnino et al., 2018): 1) Urbanization and housing crisis. 2) Energy consumption and climate change. 3) Great infrastructure gap worldwide. 4) The construction industry is a global consumer of resources. 5) Great demand from the construction sector in emerging countries. 6) Aging workforce. These six challenges imply notable challenges for the construction sector worldwide, as this industry must face crucial disruptive technological changes (see Figure 1).

These challenges directly affect Latin America. The construction sector faces barriers and opportunities to respond to the need for technological change. Planners will have analysed the urbanization process of Latin American cities. The increase in demand for housing implies an unusual emphasis on the appropriate model of sustainable buildings in the region. On the other hand, the region faces changes in a short time in the industrialization paradigms. Latin America has not consolidated the processes before the fourth industrial revolution. At the same time, the Latin American construction industry needs to strengthen its capacities more quickly.

The Latin American construction industry requires the strengthening of four factors (Castagnino et al., 2018): High Levels of Productivity, Efficiency, Innovation, Technological Digitization. The strengthening of capacities will allow the construction sector in Latin America to develop solutions in virtual environments. Then, the construction of smart cities will allow the consolidation of sustainable buildings in the region. It will let improvements in the productivity levels of the sector. So, Construction firms have to identify the speed of digital technology adoption (García de Soto et al., 2019). Likewise, it is crucial to differentiate the requirements of each of these four levels. Each construction firm's capabilities are different, that advisable to evaluate differentiated strategies for each construction firm.

However, the Latin American construction sector faces barriers to robotization adoption in Latin America, such as the climate change effects in the region. Lack of responses to rapid urbanization in Latin America. Depletion of resources in the sector. Loss of competitiveness in the construction sector in the face of New Global Changes (see figure 2). Another factor is the technological obsolescence of the construction industry in the region.

Figure 1. Challenges of the construction industry worldwide
Source: Own elaboration (2021).

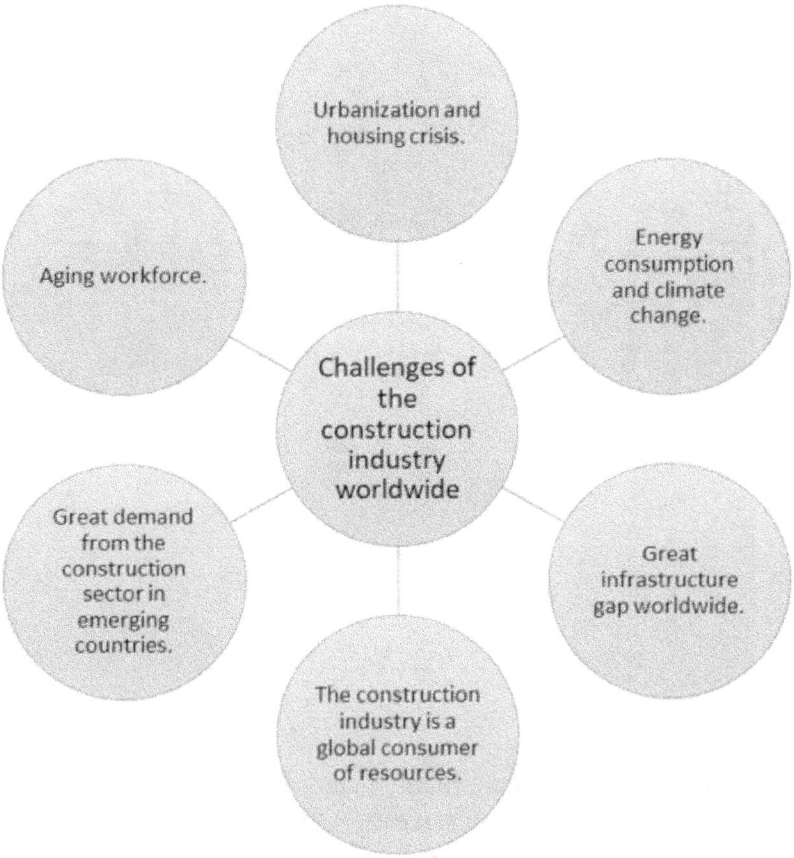

Indeed, it seems a technology difference between Latin American construction firms. They generated a technological exclusion between these actors. That is one of the main barriers to overcome. In other words, there is a need for digital technology transfer between the different constructions firms that build the Latin American city. Which should establish a dialogue between them that allows the change that the region requires.

We should carry out an adequate technology transfer to the construction sector in Latin America. Since we run the risk of technological exclusion for the Latin American city.

How to overcome the barriers of technological exclusion? The answer is to strengthen the industry through technological innovation that provides quality to inhabitants of Latin American cities. Latin America is full of opportunities for the technological development of the construction sector. Figure 3 shows the four opportunities that should focus on the region.

Figure 2. Barriers to the implementation of construction robotization in Latin America
Source: Own elaboration (2021).

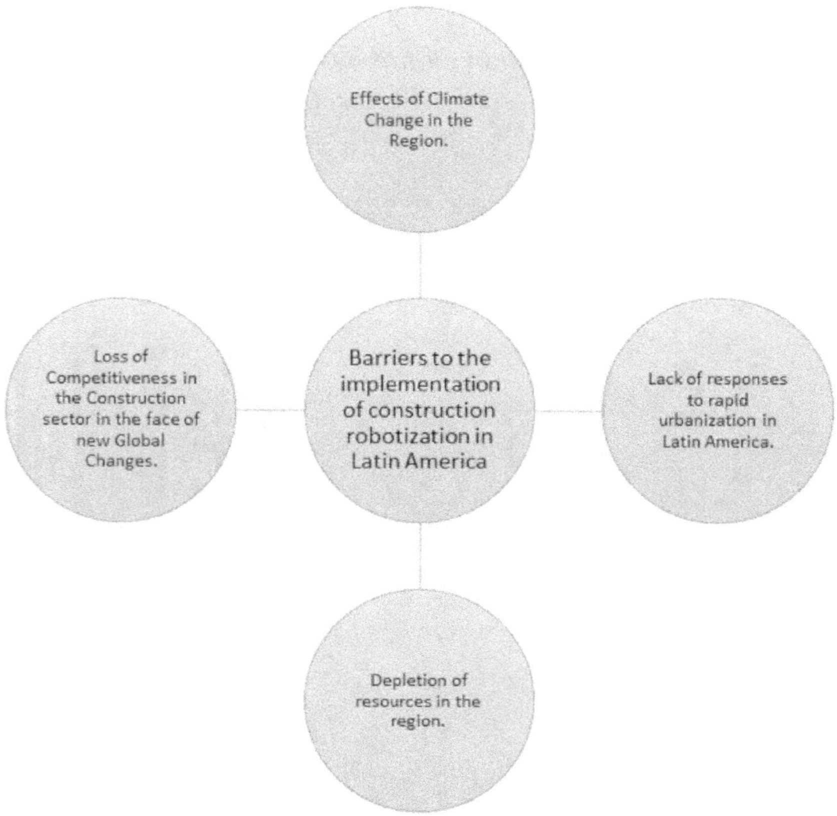

Promoting construction robotics is an opportunity to reduce the poverty gaps that still exist in Latin America. Robotization is not an issue alien to the region (Müller et al., 2019). The participation of Latin American countries is not insignificant for the region. Mexico and Brazil have consolidated internationalized experiences with the robotization of their industries (Lee, 2019). It can transfer and adapted to other Latin American countries.

Latin American Universities plays an important role in this process. The academy must take an active position to develop the Construction robotics process for the production sector in each country of Latin America. Strengthening the current innovation model of the three propellers, to then migrate from the linear models of technology transfer to the new parallel non-linear models of technology transfer (Hilkevics & Hilkevics, 2017).

The exponential development of the adaptation of construction robotics will allow the implementation of physical-digital collaboration environments that will permit the region to take a great leap in its technological development processes. It leads us to strengthen Collaborative processes in the construction industry. It will introduce Man-Machine Environments, which let us build an integrated city balanced with Nature. It will admit a significant reduction in environmental impacts in the region to a high degree.

Figure 3. opportunities for the implementation of construction robotization in Latin America
Source: Own elaboration (2021).

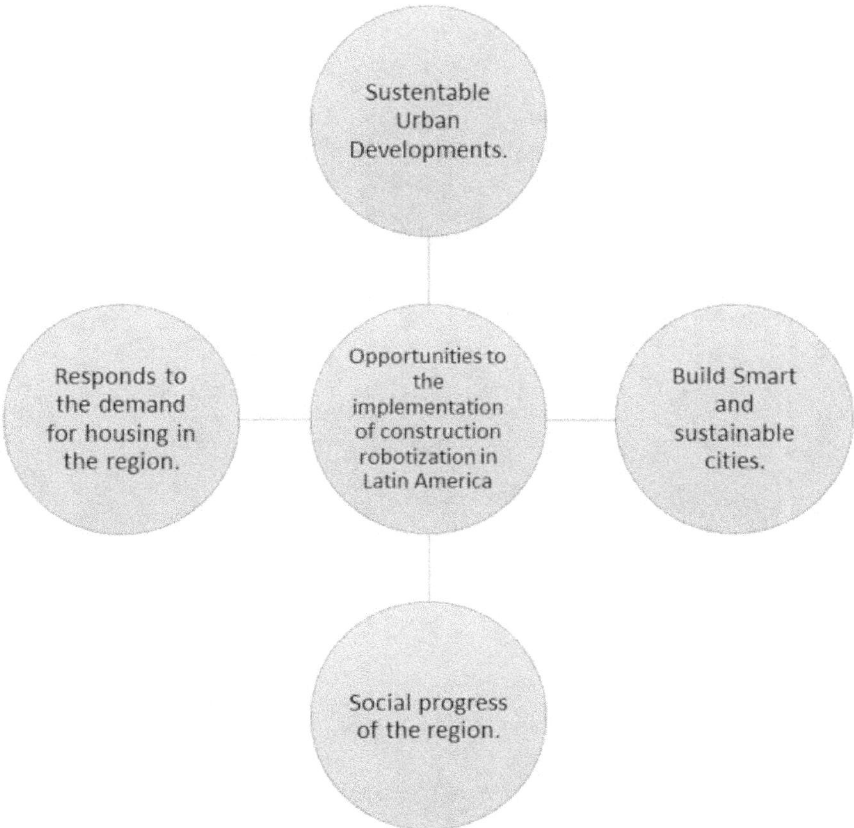

Some research works propose a concept of dynamic vertical urbanism (Hu et al., 2020), featuring constant vertical urban transformation by applying construction robotics technologies. This vertical city concept integrates functions with the help of construction robotics technologies, open building principles, and the BIM process (see figure 4). The concept of Robot-Oriented Design emphasizes the idea that all parameters shall have been considered at the earlier design and production stages (T.-A. Bock, 2017). The authors explain that this concept can evolve following social, economic, and environmental. This concept could provide different actors with a new design tool for future Latin American cities.

Other authors (Hazem et al., 2021) explain that construction faces problems after the pandemic construction firms need to provide a safe environment for working. So, using robots to mitigate the pandemic in different construction tasks is a better idea. Also, this solves congestion on construction sites which is necessary for the pandemic constraints. It is an opportunity to give better work conditions to the construction workers in Latin America. Because construction firms can develop their worker's skills and give a high level of training.

Figure 4. Dynamic vertical urbanism
Source: Own elaboration (2021) adapted from (Hu et al., 2020, 37).

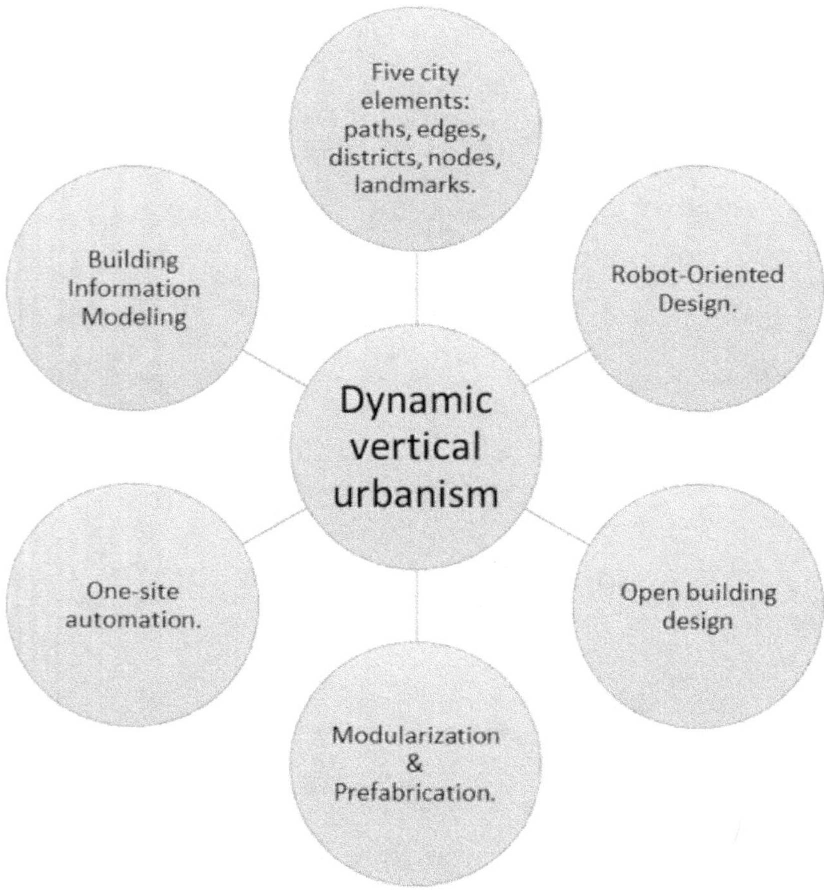

Despite the described above, there is a crucial opportunity in Latin America. Some studies (Biörck et al., 2020) explain the short and long-term trends of how the construction sector can emerge after the pandemic. So, this region is full of opportunities that offer the construction sector a significant change based on adequate management of its resources. These opportunities will allow the region to focus its efforts on the following changes:

Short terms

- **Increased digitization**. The construction industry is shifting to remote ways of working. For example, it will be a big drive to use building information modelling (BIM).
- **Supply chains resilient.** Contractors are building chains resilient and identifying alternative suppliers to mitigate future extreme events.

Long terms

- **Augmented consolidation.** The industry is consolidated to economies of scale and support investment in R&D, and technology.
- **Vertical integration.** The industry is starting to vertically integrate to increase efficiency and as a route to standardization and control of design and execution.
- **Further investments in digital technology and building systems innovation.** Construction robotics is proven to increase productivity will become even stronger. It will see an increase in R&D spending to develop new standardized building automation systems.
- **Increase in off-site construction.** It expects to see the industry gradually push fabrication off-site and manufacturers to expand their range of prefabricated subassemblies.
- **The acceleration toward sustainability.** Governments will encourage measures to meet carbon reduction and demand toward more sustainable buildings that promote healthier lifestyles.

NEW CONSTRUCTION SYSTEMS OPPORTUNITIES IN THE AFFORDABLE HOUSING IN THE LATIN AMERICAN CITIES

Construction robotics can increase production rates in the construction industry. It leads to significant time-saving in the construction process. Construction robotics allow a tenfold reduction in construction time. For example, these innovations have developed the concrete printing process in the industry (Soto & Skibniewski, 2020). There are a big number of constructions robotics systems, but it will explain only three mains of them that will have a crucial drive-in housing production, particularly to affordable housing production in Latin America.

- **Parallel Kinematic Manipulation (PKM).** The PKMs is used in-situ and for prefabrication processes. (Klöckner et al., 2020). These robots operate in different machining tasks. Wall masonry is adapted to the PKM structure. The robot is placed parallel to the wall it wants to build at a distance from the wall. The robot will be used for the pick-up, place of blocks, and building of the wall.
- **Single-task construction robots (STCR).** STCRs execute a single task on-site. This robot does not try to automate the main sections in the construction site. So it avoids large investments (Linner et al., 2020).
- **Robot Oriented Design (ROD).** ROD is the management of robotic technology that generate life cycle oriented in the construction context. It was developed for improving the construction processes and component design (T. Bock, 2015).

According to an international report (OECD et al, 2020), digital transformation can help Latin America to recover more quickly from the COVID-19 crisis. The construction sector in Latin America got an additional challenge the climate change. so digital transformation will be a tool to give answers to this difficulty.

It is crucial to increase the innovation processes that allow the execution of actions that respond to the effects of climate change. Indeed, promoting robotization in the construction industry could increase productivity by 50 to 60 per cent for companies in the region oriented to innovation (McKinsey Global Institute, 2017).

This type of industry needs to change to take actions that interrupt the climate change effects. So, the construction sector has an important role in these actions (Castagnino et al., 2018). Therefore, the construction sector needs to develop collaborative management models from robotization (see figure 5). That would allow a quick technology innovation in the construction sector. This is particularly true about affordable housing demand.

Figure 5. Collaborative management models from robotization
Source: Own elaboration (Rolando Cubillos, 2021)

To Latin America, the concept of affordable housing goes beyond the cheap housing concept. If people cannot have cities services next to them, it cannot be said to be affordable. The reasons for a lack of affordability vary from country to country. But commonly include incoming costs and changes in family composition. Latin American construction sector must consider different forms of tenure. They can offer more options to city residents' demands, for example, they include build-to-rent or co-housing (W.E.Forum, 2019).

In affordable housing production land and construction cost is often the biggest barriers in developing housing. To reduce these costs down it has to include (Darko et al., 2017; W.E.Forum, 2019; Witthoeft & Kosta, 2017):

- Minimizing bureaucracy can add significantly to project costs. The technology to provide solutions here.
- Emerging construction digital technologies like construction robotics could bring down operational costs.
- Drive green technologies to support developing alternative construction materials.
- Developing collaborative public-private partnerships on training to address construction sector skills.

THE ROLE OF INNOVATION IN THE LATIN AMERICAN AFFORDABLE HOUSING IN THE LATIN AMERICAN CITIES

Latin American cities governments must define long-term plans for increasing affordable housing production to the increased population demand. They need to minimize urban sprawl and evaluate their future densities. So, it is crucial to reduce political considerations that could stop the development of new affordable housing in their cities. Innovations in affordable housing in Latin America should include the following technology drives: 1) 3D-printed housing. 2) Design for Manufacture and Assembly. 3) Construction Robotics. 4) Prefabricated Housing.

Affordable housing in Latin America requires to drive technology that can be housing built faster and cheaper to respond to current demand. Construction robotics provides quality and speed of production. It is a positive impact on the environment and city communities. This digital technology should develop large-scale affordable housing and reduce emissions in the city.

The COVID-19 pandemic has highlighted the shortcomings of current urbanization patterns in Latin America. The informal and formal developments require special attention on quality and affordability for low-income families (Cities Alliance, 2020). For that reason, the construction sector must focus its efforts on innovative processes that allow the production of affordable housing with quality and low cost (Jennifer Graeff, 2015).

The industry must identify the low-income families' needs demand in Latin American cities to help for reducing social inequity and drive economic growth. Besides, the Construction industry, the government, and academia must work together to design public policies strategies to develop better cities in the region.

SOME INNOVATION CASE OF STUDY IN LATIN AMERICA

Innovation in Latin America has been assumed as a strategy to improve the competitiveness of companies and products. The literature review identified a prevalence of studies in Brazil, Mexico and Colombia (Quintero Sepúlveda et al., 2021) This study shows that innovation has been the right path for the region to promote a better standard of living and enhance the functioning of institutions in the region.

The relationship between the variables that make up the innovation index and innovation capabilities is observed as a priority in the Latin American context. The results of this study indicate that technical efficiency indices are similar in Latin American countries. Therefore, the greater the innovative capacity, the higher the rate of innovation in the region.

Regarding the market outlook (Triton-Market-Research, 2021) the industrial robots market in Latin America will increase at an annual growth rate of 9.21% during the years evaluated from 2021 to 2028.

Brazil, Mexico, and the rest of Latin America make up the overall market in the study region. For example, according to the report, the Argentine government encouraged investors by enacting laws that benefit technology-based companies. This has led to an increase in the number of companies employing AI, machine leadership and robots. In addition, federal regulations can help companies improve their product offerings and meet the growing demand for robots. These factors expand the growth of the industrial robot market in Argentina.

On the other hand, Chile is concentrating on robotics to improve productivity in multiple industries. Along with this, several international players are establishing business operations in the country, which will increase the adoption of robots. The adoption of robots accelerates the development of the industrial robot market in Chile.

According to the International Federation of Robotics-IFC (Müller et al., 2019) for 2019, within the first ten countries that implemented robotic technologies to the industrial sector, in ninth place of the first ten countries, is Mexico in the representation of the Latin American region.

According to the Inter-American Development Bank citing the IFC (Lee, 2019), Mexico had 27,010 units (64.25%), was the first country in the Latin American region to incorporate robots in its industry. Followed by Brazil had 12,373 units installed (29.43%).

Argentina had 2,238 installed units (5.32%). Also, Chile had 182 installed units (0.43%). Finally, Colombia had 149 installed units (0.35%). Robotization in Latin America is not an issue foreign to the region. The participation of Latin American countries is not insignificant and presents an opportunity for change.

CONCLUSION

The fourth industrial revolution affects the future development of Latin America. The impacts of this revolution in the region will probably be more disruptive than in other parts of the world. The challenge for the regional industry is enormous to adapt to this new revolution. Many jobs are at risk in Latin America. These jobs could replace by robots. If robots take over routine tasks in Latin America, regional companies can unleash innovation. The fourth industrial revolution is an opportunity to close the gap with advanced economies.

The Covid-19 pandemic can reconsider the relationship between architecture and medicine to promote healthy environments. Besides, the pandemic consequences pose a restructuring of the labour market and a change in the urban structure in the long term. On the other hand, the high demand for housing implies thinking about the better development of Latin American cities. The resilient design in buildings increases well-being and reduces vulnerability to new pandemics. Construction robotics is an opportunity to add resilient and sustainable strategies in the design of housing in Latin American cities.

Construction robotics adoption improves productivity in the construction industry. The technological development of construction robots is underway globally. This scenario is the axis for developing the construction robotics industry in Latin America. On other hand, automation technologies are implementing in the housing industry around the world. So, housing production is the main benefit to reduce costs in the Latin American construction industry.

Construction Robotics focuses on the implementation of automation to renew the construction industry and to stop its current low productivity. Because automation in construction could achieve what other manufacturing industries have already successfully shown. Then world construction industry would elevate the sector to the level of the rest of the manufacturing fields So, it is needed to develop smart housing and cities that integrate Automatic assistive technologies developed in a better sustainable environment for people. Robotic automatization could provide an innovative affordable housing solution to mitigate the impact of an ageing society.

The World Economic Forum proposes six main challenges for the construction sector at the international level. These challenges directly affect Latin America, because the region has not consolidated their industrial processes before the fourth industrial revolution. The Latin American construction industry requires capacities that will allow the construction sector in Latin America to develop solutions in virtual environments. Then, the construction of smart cities will allow the consolidation of sustainable buildings in the region.

Each construction firm's capabilities are different, that advisable to evaluate differentiated strategies for each construction firm. It seems a technology difference between Latin American construction firms. They generated a technological exclusion between these actors. That is one of the main barriers to overcome. It should carry out an adequate technology transfer to the construction sector in Latin America. Since we run the risk of technological exclusion for the Latin American city.

We should carry out an adequate technology transfer to the construction sector in Latin America. Latin America is full of opportunities for the technological development of the construction sector. For example, Mexico and Brazil have consolidated internationalized experiences with the robotization of their industries. It can transfer and adapted to other Latin American countries. On other hand, Latin American Universities plays an important role in this process. The academy must take an active position to develop the Construction robotics process for the production sector in each country of Latin America.

The exponential development of the adaptation of construction robotics will allow the implementation of physical-digital collaboration environments that will permit the region to take a great leap in its technological development processes. It will admit a significant reduction in environmental impacts in the region to a high degree. Some research works propose a concept of dynamic vertical urbanism, featuring constant vertical urban transformation by applying construction robotics technologies.

The concept of Robot-Oriented Design emphasizes the idea that all parameters shall have been considered at the earlier design and production stages. This concept could provide different actors with a new design tool for future Latin American cities. Other authors explain that construction faces problems after the pandemic construction firms need to provide a safe environment for working. It is an opportunity to give better work conditions to the construction workers in Latin America. Because construction firms can develop their worker's skills and give a high level of training.

Some studies explain the short and long-term trends of how the construction sector can emerge after the pandemic. These opportunities will allow the region to focus on the following changes 1) Short terms: Increased digitization. Supply chains resilient. 2) Long terms: Augmented consolidation. Vertical integration. Further investments in digital technology and building systems innovation. Increase in off-site construction. The acceleration toward sustainability.

Construction robotics can increase production rates in the construction industry. For example, these innovations have developed the concrete printing process in the industry. According to an international report digital transformation can help Latin America to recover more quickly from the COVID-19 crisis. The construction sector in Latin America got an additional challenge the climate change. So, It is crucial to increase the innovation processes that allow the execution of actions that respond to the effects of climate change. Therefore, the construction sector needs to develop collaborative management models from robotization. That would allow a quick technology innovation in the construction sector.

To Latin America, the concept of affordable housing goes beyond the cheap housing concept. If people cannot have cities services next to them, it cannot be said to be affordable. Latin American construction sector must consider different forms of tenure. They can offer more options to city residents' demands, for example, they include build-to-rent or co-housing.

In affordable housing production land and construction cost is often the biggest barriers in developing housing. Emerging construction digital technologies like construction robotics could bring down operational costs. Drive green technologies to support developing alternative construction materials. Finally, developing collaborative Public-private partnerships on training to address construction sector skills.

Latin American cities governments must define long-term plans for increasing affordable housing production to the increased population demand. Construction robotics provides quality and speed of production. It is a positive impact on the environment and city communities. This digital technology should develop large-scale affordable housing and reduce emissions in the city.

The COVID-19 pandemic has highlighted the shortcomings of current urbanization patterns in Latin America. The informal and formal developments require special attention on quality and affordability for low-income families. The industry must identify the low-income families' needs demand in Latin American cities to help for reducing social inequity and drive economic growth. Besides, the Construction industry, the government, and academia must work together to design public policies strategies to develop better cities in the region.

What is the Dynamic of Construction Robotics to Develop Affordable Housing in Latin American Cities?

In conclusion, the dynamic of construction robotics to develop affordable housing in Latin America follows the next steps (see figure 6):

- Robots will take over routine tasks, and regional companies will increase the innovation process.
- Construction robotics will add resilient and sustainable strategies to affordable housing production.
- Automation technologies will reduce costs and integrate digital technologies to build a better sustainable environment.
- The concept of Robot-Oriented Design (ROD) will push building systems like Production Kinematic Manipulation (PKM) and Single-task construction robots (STCR).
- The innovation processes will develop collaborative management models from robotization beyond the cheap housing concept.
- Green technologies will develop alternative construction materials. That will give the possibility to elaborate large-scale affordable housing and reduce the CO2 emissions in the city.

Finally, innovation in Latin America has been assumed as a strategy to improve the competitiveness of companies and products. The literature review identified a prevalence of studies in Brazil, Mexico, and Colombia. This study shows that innovation has been the right path for the region to promote a better standard of living and enhance the functioning of institutions in the region. The relationship between the variables that make up the innovation index and innovation capabilities is observed as a priority in the Latin American context.

Figure 6. Dynamic Construction Robotics in Latin America
Source: Own elaboration (2021).

Regarding the market outlook, the industrial robot market in Latin America will increase at an annual growth rate of 9.21% during the years evaluated from 2021 to 2028. Brazil, Mexico, and the rest of Latin America make up the overall market in the study region. Robotization in Latin America is not an issue foreign to the region. The participation of Latin American countries is not insignificant and presents an opportunity for change.

ACKNOWLEDGMENT

This research was supported by the Centro de Investigación para el Desarrollo y la Innovación - CIDI de la Universidad Pontificia Bolivariana [grant number 016C-01/18-71].

REFERENCES

Bannova, O. K., & Kristiansen, T. H. (2014). Technology transfer: Current trends in incorporating new technologies into the housing industry. *Proceedings of the International Astronautical Congress, IAC, 13*(November), 9751–9756.

Biörck, J., Blanco, J. L., Mischke, J., Ribeirinho, M. J., Rockhill, D., Sjödin, E., & Strube, G. (2020). *How construction can emerge stronger after coronavirus*. McKinsey & Company. https://www.mckinsey.com/business-functions/operations/our-insights/how-construction-can-emerge-stronger-after-coronavirus

Bock, T. (2007). Hybrid construction automation and robotics. *Automation and Robotics in Construction - Proceedings of the 24th International Symposium on Automation and Robotics in Construction, Isarc*. 10.22260/ISARC2007/0002

Bock, T. (2014). Ki-NEW-Matics. *The 31st International Symposium on Automation and Robotics in Construction and Mining (ISARC 2014) Ki-NEW-Matics*, 1–8.

Bock, T. (2015). Construction Robotics enabling innovative disruption and social supportability. *32nd International Symposium on Automation and Robotics in Construction and Mining: Connected to the Future, Proceedings*.

Bock, T., & Linner, T. (2015). *Robotics Industrialization - Automation and robotic technologies for customized componet, module, and building prefabrication*. Cambridge University Press.

Bock, T.-A. (2017). Robot-Oriented Design. *Proceedings of the 5th International Symposium on Automation and Robotics in Construction (ISARC)*. 10.22260/ISARC1988/0019

Castagnino, S., Rothballer, C., Abreu, J., & Zupancic, T. (2018). *6 ways the construction industry can build for the future*. World Economic Forum (WEF). https://www.weforum.org/agenda/2018/03/how-construction-industry-can-build-its-future/

Cities Alliance. (2020). *Affordable Housing and Land in Latin America*. Cities Alliance / Cities without Slums. https://www.citiesalliance.org/newsroom/events/affordable-housing-and-land-latin-america

Cubillos-González, R.-A. (2021). Barreras y oportunidades de la Construcción 4.0 frente al cambio climático. In M. E. Molar Orozco, M. G. Vázquez Jiménez, & M. J. Velázquez Lozano (Eds.), *Un mundo sostenible en el siglo xxi: Alternativas para el rescate del planeta* (pp. 125–153). Universidad Autónoma de CoaHuila. http://www.investigacionyposgrado.uadec.mx/publicacion/un-mundo-sostenible-en-el-siglo-xxi-alternativas-para-el-rescate-del-planeta/

Cubillos-González, R. A., & Cardoso, G. T. (2020). Clean technology transfer and innovation in social housing production in Brazil and Colombia. A framework from a systematic review. *Sustainability (Switzerland), 12*(4), 1–12. doi:10.3390u12041335

Darko, A., Chan, A. P. C., Ameyaw, E. E., He, B.-J., & Olanipekun, A. O. (2017). Examining issues influencing green building technologies adoption: The United States green building experts' perspectives. *Energy and Building, 144*, 320–332. doi:10.1016/j.enbuild.2017.03.060

de Soto, B. G., & Skibniewski, M. J. (2020). Future of robotics and automation in construction. In A. Sawhney, M. Riley, & J. Irizarry (Eds.), *Construction 4.0: An Innovation Platform for the Built Environment*. Routledge. doi:10.1201/9780429398100-15

Framil, L., & Ovanessoff, A. (2018). *Here's why Latin Americans will thrive in the age of the robots.* World Economic Forum (WEF). https://www.weforum.org/agenda/2018/03/here-s-why-latin-americans-will-thrive-in-the-age-of-the-robots/

García de Soto, B., Agustí-Juan, I., Joss, S., & Hunhevicz, J. (2019). Implications of Construction 4.0 to the workforce and organizational structures. *International Journal of Construction Management*, *0*(0), 1–13. doi:10.1080/15623599.2019.1616414

Hazem, N., Fatehi, A., & Bakr, A. (2021). reviewed paper Construction Automation: Post-Pandemic Integrated Robotized Construction Sites. *Journal of Construction Engineering and Management*, *134*(8), 201–211.

Hilkevics, S., & Hilkevics, A. (2017). The comparative analysis of technology transfer models. *Entrepreneurship and Sustainability Issues*, *4*(4), 540–558. doi:10.9770/jesi.2017.4.4(11)

Hu, R., Pan, W., & Bock, T. (2020). Towards Dynamic Vertical Urbanism. *International Journal of Industrialized Construction*, *1*(1), 34–47. doi:10.29173/ijic208

Jennifer Graeff, A. (2015). *Supporting Innovation : Case Studies from Latin America on Sustainable Housing and Community Development*. /www.planning.org/publications/document/9149230/

Kang, M., Choi, Y., Kim, J., Lee, K. O., Lee, S., Park, K., Park, J., & Seo, I. (2020). COVID-19 impact on city and region : What ' s next after lockdown? *International Journal of Urban Sciences*, *24*(3), 297–315. doi:10.1080/12265934.2020.1803107

Klöckner, M., Haage, M., Nilsson, K., Robertsson, A., & Andersson, R. (2020). Parallel kinematic construction robot for aec industry. *Proceedings of the 37th International Symposium on Automation and Robotics in Construction, ISARC 2020: From Demonstration to Practical Use - To New Stage of Construction Robot, Isarc*, 1488–1495. 10.22260/ISARC2020/0206

Lee, S. (2019). *Robots en América Latina: ¿cuántos son, dónde están y cuánto tributan?* BID Mejorando Vidas. https://blogs.iadb.org/gestion-fiscal/es/robots-en-america-latina-cuantos-son-donde-estan-y-cuanto-tributan/

Linner, T., Pan, W., Hu, R., Zhao, C., Iturralde, K., Taghavi, M., Trummer, J., Schlandt, M., & Bock, T. (2020). A technology management system for the development of single-task construction robots. *Construction Innovation*, *20*(1), 96–111. doi:10.1108/CI-06-2019-0053

López-Escamilla, Á., Herrera-limones, R., León-Rodríguez, Á., & Torres-García, M. (2020). Environmental Comfort as a Sustainable Strategy for Housing Integration: The AURA 1.0 Prototype for Social Housing. *Applied Sciences Article*, *10*(21), 7734. doi:10.3390/app10217734

McKinsey Global Institute. (2017). *Reinventing Construction: A Route To Higher Productivity.* McKinsey & Company. doi:10.1080/19320248.2010.527275

Melenbrink, N., Kassabian, P., Menges, A., & Werfel, J. (2017). Towards force-aware robot collectives for on-site construction. *Disciplines and Disruption - Proceedings Catalog of the 37th Annual Conference of the Association for Computer Aided Design in Architecture, ACADIA 2017*, 382–391.

Müller, C., Wyatt, S., Bieller, S., Qu, D., & Song, X. (2019). IFR Press Conference: Shanghai 18th September 2019. *International Federation of Robotics,* 1–32. https://ifr.org/downloads/press2018/IFR World Robotics Presentation - 18 Sept 2019.pdf

Ng, M. K. (2020). Sustainable Development Goals (SDGs) and Pandemic Planning. *Planning Theory & Practice*, *21*(4), 507–512. doi:10.1080/14649357.2020.1807130

OECD. (2020). *Latin American Economic Outlook 2020: Digital Transformation for Building Back Better*. OECD Publishing. doi:10.1787/e6e864fb-en

Pan, M., Linner, T., Pan, W., Cheng, H., & Bock, T. (2020). Structuring the context for construction robot development through integrated scenario approach. *Automation in Construction, 114*(June). https://doi.org/10.1016/j.autcon.2020.103174

Quintero Sepúlveda, I. C., Ospina Nieto, Y., Quiroga Parra, D., & Cubillo-González, R.-A. (2021). Relación entre Capacidad de Innovación e Índice de Innovación en América Latina. *Journal of Technology Management & Innovation*, *16*(3), 47–56. https://doi.org/https://doi.org/10.4067/S0718-27242021000300047

Rossato, A. (2018). *Is Latin America ready for the Fourth Industrial Revolution?* World Economic Forum (WEF). https://www.weforum.org/agenda/2018/03/industry-4-0-fourth-industrial-revolution-is-latin-america-ready/

Triton-Market-Research. (2021). *Latin America Industrial Robots Market 2021-2028*. Research and Markets. https://www.researchandmarkets.com/reports/5459459/latin-america-industrial-robots-market-2021-2028

Witthoeft, S., & Kosta, I. (2017). *Shaping the Future of Construction. Inspiring innovators redefine the industry*. World Economic Forum (WEF). https://www3.weforum.org/docs/WEF_Shaping_the_Future_of_Construction_Inspiring_Innovators_redefine_the_industry_2017.pdf

World Economic Forum. (2019). *Making Affordable Housing a Reality in Cities*. World Economic Forum. homeguides.sfgate.com/student-hous

Yahya, M. Y., Lee Hui, Y., Yassin, A. B. M., Omar, R., Robin, R. O., & Kasim, N. (2019). The Challenges of the Implementation of Construction Robotics Technologies in the Construction. *MATEC Web of Conferences, 266*. doi:10.1051/matecconf/201926605012

Chapter 5
Green Infrastructure as a Nature-Based Recovery Strategy for Natural Areas in Desert Developing Countries of the South Pacific Coastal Strip:
The Peruvian Case Study of Chimbote and Nuevo Chimbote

Aarón Alberto Abad Alva
Universidad Politecnica de Madrid, Spain

Estela Karem Samamé Zegarra
https://orcid.org/0000-0003-2317-5076
Universidad Nacional Santiago Antúnez de Mayolo, Peru

Roberto Alonso González Lezcano
https://orcid.org/0000-0002-6185-4929
Universidad CEU San Pablo, Spain

ABSTRACT

The present research is part of a wider qualitative study that aims to assist cities in the South Pacific developing countries with the recovery of natural areas through a green infrastructure-based approach, following a case study method. The overarching purpose of this study is to pinpoint relevant contributing elements for the successful implementation of the green infrastructure approach aiming at providing Peruvian coastal cities with novel sustainable environmental policies. To foster the conservation of the natural or semi-natural ecosystems converging with the cities, the following specific objectives were

DOI: 10.4018/978-1-6684-4030-8.ch005

set: 1) to carry out a comprehensive physical and spatial analysis of the natural areas of Chimbote and Nuevo Chimbote; 2) to review the Peruvian regulatory framework, at the national and municipal levels, as well as international standards on the conservation of natural areas; and 3) to explore a scenario featuring the disappearance of these Peruvian coastal natural spaces and its associated consequences.

INTRODUCTION

It is well known that Green Infrastructure is a natural life support system that interconnects cross-scale natural and semi-natural areas (Allen, 2012) while providing a wide array of ecosystem services. Moreover, it is considered to be a strategic tool for addressing current global environmental challenges such as climate change, which have been attracting the attention of many researchers (Caparrós et al., 2021). The substantial interest generated by Green Infrastructure emphasises the value of nature's benefits to human society and the need to mobilise investments to sustain and reinforce them ("Green Infrastructure (GI)", 2013; "The Multifunctionality of Green Infrastructure", 2012).

It is noteworthy that some European countries such as, *inter alia*, Spain, Germany, England, and the Netherlands, have been taking this issue increasingly seriously since 2013. These heighten the need for implementing comprehensive development policies in line with those established in the European Commission's White Paper and investing in Green Infrastructure-based urban planning models ("Green Infrastructure (GI)", 2013) where it plays a key and central role towards smart habitat conservation for the 21st century (Benedict & McMahon, 2002) rather than merely assuming a beneficial scenic or recreational value for human physical and mental health. Thus, the singling out of Green Infrastructure initiatives as a European investment priority comes as no surprise ("Green Infrastructure (GI)", 2013).

Europe offers an encouraging outlook on this issue based on the implementation of strategic green space networks across several cities such as Manchester in England ("A Green and Blue Infrastructure Strategy for Manchester", 2015) and Vitoria-Gasteiz in Spain ("The Urban Green Infrastructure of Vitoria-Gasteiz", 2014). Specifically, Vitoria-Gasteiz's urban green belt is a set of periurban parks of high ecological and landscape value arising from an ambitious restoration and recovering project of the city's outlying areas, with an environmental and social focus. This sustainable urban policy allowed for the recovery of the *Salburua* wetland park, which was subsequently recognised as a Wetland of International Importance by the Ramsar Convention in 2002.

A vast array of wetlands can be found across the globe, as is the case with Peru's coastal desert regions. As illustrated in Fig.1, this area exhibits protected wetlands, e.g. the *Paracas* National Reserve, the *Manglares de Tumbes* National Sanctuary, the *Pantanos de Villa* Wildlife Refuge, the *Lagunas de Mejía* National Sanctuary and the *Pacaya Samiria* National Reserve off the coast ("The Ramsar Convention on Wetlands", 1971), in addition to other wetlands which are yet to be covered by governmental protection. Yet, a worrying statistic reveals Latin America as the leading world nation in loss of wetlands, with urban growth as one of the main culprits (Rojas et al., 2020). In fact, previous research has shown that there is a significant correlation between residential development and wetland loss (Halls & Magolan, 2019).

Figure 1. Map of Peru displaying the Ramsar sites, the natural, artificial, and coastal river-mouths wetlands
(Source: own elaboration based on the document Wetlands of the Peruvian coast, 2010).

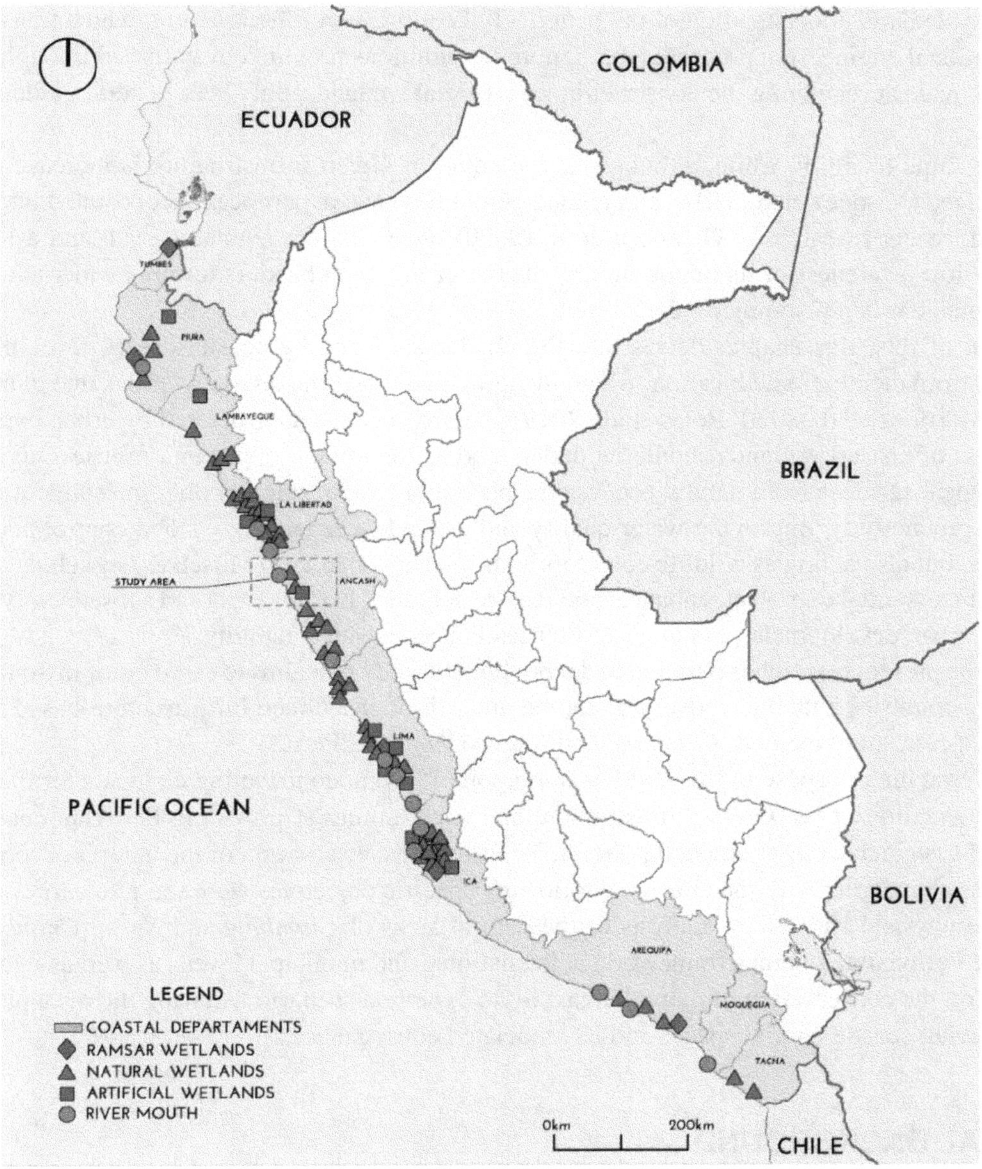

The current climatic, political, social, and economic conditions of the South Pacific coast vastly differ from European ones. The former is still home to developing countries such as Colombia, Ecuador, Peru, and Chile, all of which are particularly vulnerable to climate change. Yet, the shared geography and climate features of Peru and Chile make them unique: their location in a desert strip characterised by extreme temperatures combining lacking vegetation and overpopulation is conducive to urban grey infrastructure growth at exponential and uncontrollable rate.

Taking the case of Peru as an example, its horizontal urban growth leads to an ever-increasing need for developable land, resulting in the development of informal settlements and the change of land-use patterns in natural areas. Furthermore, the Peruvian expert Espinoza (2021) highlights the lack of common and systematic understanding of the benefits linked to Green Infrastructure, which translates into monofunctional engineering processes that ignore the solutions provided by nature. Although attempts have been made at regulating the construction of artificial wetlands, only very generic guidelines were produced.

By the same token, in Chile, although the potential of Green Infrastructure is increasingly being acknowledged (Vásquez et al., 2016) and related guidelines have in part been incorporated into national legislation, recent research by Giannotti et al. (2020) identified the regulatory gap and a low social priority or low awareness of its importance as the two preeminent barriers towards wider adherence of this sustainable urban planning model.

In light of this, this chapter delves into the challenges faced by coastal-desert developing cities stemming from their extreme location, where the few existing natural areas are at risk due to man-made hazards (Martínez et al., 2020; Rojas et al., 2019) and urban conflicts generated by urban expansion.

The loss of coastal wetlands should be understood as the missed opportunity for leveraging social and ecological services from natural ecosystems playing a crucial role not only in sediment retention, high-seas productivity, improving water quality and groundwater recharge, flood control and drought prevention, but also acting as wildlife corridors and waterfowl habitats (Mitsch & Gosselink, 2000a). It is, therefore, essential to protect, enhance, and restore degraded natural areas and subsequently integrate them into urban development planning to foster resilience and sustainability.

Thus, the present research is part of a wider qualitative study that aims to assist cities in South-Pacific developing countries with the recovery of natural areas through a Green Infrastructure-based approach, following a case study method: *Chimbote* and *Nuevo Chimbote*, Peru.

The overarching purpose of this study is to pinpoint relevant contributing elements for the successful implementation of the Green Infrastructure approach aiming at providing Peruvian coastal cities with novel sustainable environmental policies. To foster the conservation of the natural or semi-natural ecosystems converging with the cities, the following specific objectives were set: i. to carry out a comprehensive physical and spatial analysis of the natural areas of *Chimbote* and *Nuevo Chimbote*; ii. to review the Peruvian regulatory framework, at the national and municipal levels, as well as international standards on the conservation of natural areas; iii. to explore a scenario featuring the disappearance of these Peruvian coastal natural spaces and its associated consequences.

GENERAL BACKGROUND

Human evolution has been marked by a very strong Human-Nature bond. This has allowed him to obtain domestication know-how regarding animals and plants, as well as how to implement what could be designated as empirical urbanism by adapting the physical environment to meet his needs. This practice has transcended over the years and cultures worldwide, which then transmute through local assimilation.

Manuel Ruiz, in his research "Man and Nature", suggests that the actions of all living beings have a repercussion on the rest of the Earth system and, thus, the disappearance of a species may drastically impact the current balance of life and ecosystems. The elimination of a species or ecosystem will indeed affect its geographical area and generate a rupture of the Human-Nature bond. The turning point in the relationship between Man and his environment took place with industrialisation and its technological advances: natural areas were disposed of to make way for factories, worker housing, and ultimately, large cities. Under this premise, "the city is thus built in pieces, occupying areas of countryside, and leaving free spaces in between these pieces. But this continuous breaking up of the city into small parts does not generate solidarity spaces such as the old villages because each piece lacks to incorporate all vital functions, and on the contrary, the gap only grows wider and wider: between functions, social classes, even between spaces." (Fariña Tojo, 1998, 2002, p. 8).

Although undoubtedly tied with great achievements for mankind, the process of modernisation brought about important consequences for land, air, and water: with the expansion of big cities arose the increased demand for developable land. Nonetheless, the idea of harmoniously integrating nature into urban developments first emerged with Robert Owen and Charles Fourier, not seeing the light of day until 1898, when Ebenezer Howard introduced the concept of the "Garden City" in his landmark book "To-morrow: a peaceful path to real reform". The latter propelled the incorporation of green elements into urban planning proposals, such as those of Raymond Unwin and Barry Parker, who applied it for solving problems generated by industrialisation. By 1928, Clarence Stein came up with the Radburn Planning Concept, currently known as "superblock", which sought the harmonious blending of private space and open areas. For this purpose, each dwelling was equipped with large gardens. This proposal provided a stepping stone for furthering the development of small green spaces for the inhabitants of large cities.

From that point onwards, architects and urban planners began to include green and natural areas in their urban planning proposals to a greater extent, but it was not until the 1960s that the focus on natural areas peaked. Research emerged on the topic, outlining the "Environmental Corridor" concept (Lewis, 1964) and considering the balance between urban and rural spaces in "Design With Nature" (McHarg, 1969), which became a reference book for urban planners.

The term Green Infrastructure, whose identity metamorphosed throughout human evolution, was only formally coined in 2006 by Benedict and McMahon as "an interconnected network (natural life support system) of waterways, wetlands, forests, wildlife habitats, and other natural areas; greenways, parks, and other conservation lands; working farms and ranches; wilderness and other open spaces that support native species, maintain ecological processes, sustain air and water resources, and contribute to the health and quality of life of America's communities and people" (Cantó López, 2014, p. 12).

In the Pre-Inca era, the harmonious behaviour of human beings and their technology with the natural environment was noteworthy. Examples of said bond materialised in the sowing and harvesting of water conducted by the *Amunas* and the *Waru Waru*, the *qochas*, the aqueducts, the forested river systems, and the implementation of agricultural terraces (Ancajima, n.d.).

The extensive existing literature on Green Infrastructure has proven its recovery potential for natural areas and beneficial impact on people's quality of life, as happened with the emblematic case of Vitoria-Gasteiz's urban green belt, which could be extrapolated to extreme geographical conditions such as the South Pacific coastal areas.

RESEARCH PROBLEM

The Peruvian and Chilean desert-coastal areas are characterised by very little vegetation, combining zones with extremely high temperatures and lacking atmospheric rainfall with very-heavy rainfall ones, generating the El Niño phenomenon (from October to January). Moreover, these areas "exposed to natural forces such as tsunamis and floods and alarmingly threatened by man-made hazards are in a precarious state due to their climate change vulnerability arising from the delicate regional relationship with water resources and the morphological connection to the sea level" (Tabilo, Burmeister, Chávez, & Zöckler, 2016).

In the present Peruvian study area, the municipalities of *Chimbote* and *Nuevo Chimbote* are losing more and more natural areas each year that passes (Figure 2). The large natural extensions of their coastal wetlands have been affected by human activity in the form of land-use pattern alterations, water pollution, rubbish dumps, oxidation reactors, amongst others, which will be discussed in depth over the next sections. Both cities present serious urban planning challenges towards integrating natural ecosystems in urban spaces, with the excessive population growth and lack of developable land only exacerbating the situation further. This leads to land-use pattern alterations taking place to the detriment of natural areas, which despite being legally protected, are considered expendable and lose their protected status to benefit the expansion of urban space.

In view of these issues, a strategy aiming at implementing a Green Infrastructure system with the potential to help preserve these areas and then integrating them into the future urban planning of both cities should be contemplated.

PHYSICAL SPATIAL ANALYSIS: GEOMORPHOLOGICAL, WATERSHEDS, AND ECOSYSTEM COMPONENTS

As aforementioned, the study area is located on the northern Peruvian coast, specifically within the department of *Ancash*, in the province of *Santa*, in the districts of *Chimbote* and *Nuevo Chimbote* (Figure 1). Due to their geographical position and proximity to the coast, these districts have an extensive landscape diversity, encompassing beaches, deserts, dunes, wetlands, agricultural parcels, and mountains.

Spatial-Physical Analysis

With a location between the coast and the mountains, the municipalities of *Chimbote* and *Nuevo Chimbote* present a variety and diversity of elements that shape their physical territory, differing in regard to their latitudes, vegetation, and bodies of water, to name a few. In this sense, we could highlight two fundamental aspects of the morphology of the site.

The topography of the coastal strip of both municipalities exhibits the lowest elevation indices and geographical features, with the presence of deserts, dunes, flat territorial extensions, hills with rocky outcrops, and other slightly elevated landforms. In addition, "the shape of the bays points to an older and deeper erosion level than the current one, subsequently refilled with quaternary sediments" ("Gobierno Regional de Ancash", 2014). Indeed, the coastline experiences a modified surface as a result of seismic activity in the Pacific Ring of Fire, climate conditions-related soil erosion, and anthropogenic activities.

Likewise, the desert or coastal plains "have a greater presence near the coastal zone, gradually reducing and disappearing towards the East. These plains occasionally include other geomorphological units such as dunes or sandbanks and dissected surfaces" (2014, p. 32). In this way, the territorial spatial configuration undergoes substantial transformation, including geographical features, in addition to presenting a sandy soil type which is a limiting factor for the wild vegetation growth in the area.

Figure 2. Maps based on historical photographs from 1872 and GIS 2019 of natural areas in Chimbote and Nuevo Chimbote
(Source: own elaboration based on GIS).

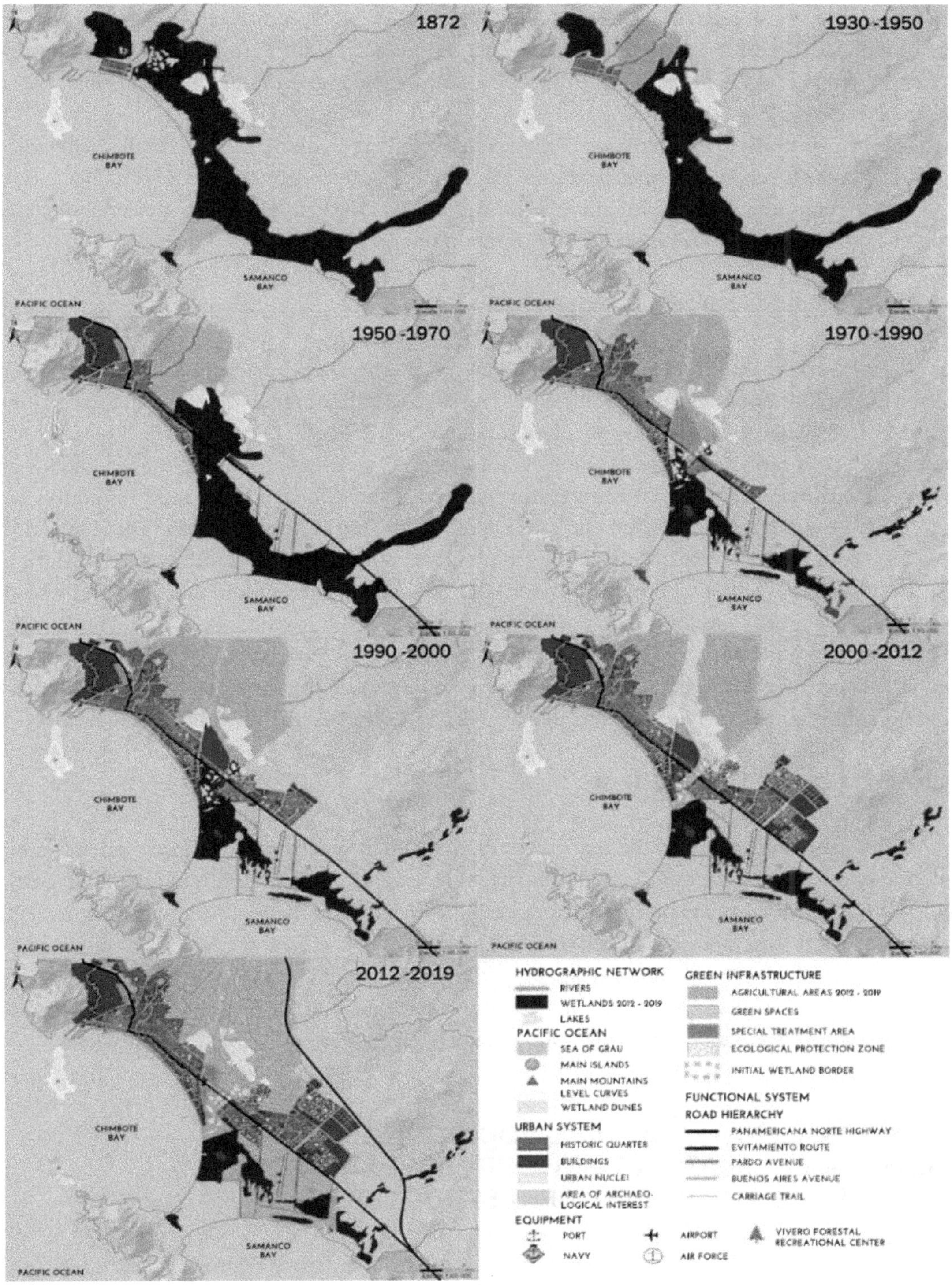

On the other hand, the area's hydrographical system is located within the so-called Pacific Slope (a set of watersheds that flow into the Pacific Ocean). The study area covers three hydrographic basins: *Santa* River Basin, *Lacramarca* River Basin, *Nepeña* River Basin, and two inter-basins. "The *Santa* River Basin is one of the largest out of the 53 existing Pacific slope basins, enjoying permanent water all year round. Moreover, it is characterised by vast biodiversity with 21 life zones and two protected natural areas (*Huascarán* National Park in the *Cordillera Blanca* and the *Calipuy* National Reserve in *La Libertad*)" (INAIGEM, 2016, p. 2).

The presence of these watersheds leads to vegetation outcrops along the riverbanks. However, due to low groundwater levels (freshwater from the Andes), these converge with the Pacific Ocean creating new and scarce ecosystems, including mangroves, wetlands, and lakes. The study area encompasses five wetlands, which make up the last green lungs of the municipalities.

Ecosystem Biotic Components

An ecosystem could be described as the set of all living and non-living elements within a physical territorial space and its components can be classified as Biotic (living organisms, e.g. fauna and flora) and Abiotic (non-living elements that shape its environment).

The case study's biotic fauna factors are "a total of 186 species, of which 90 are resident and 96 hail from other latitudes. Of the latter migratory species, which have a horizontal migration pattern, 48 originate from the Nearctic Region and eight come from the Southern Region. The Andean migratory species, accounting for ten breeds, come from the Peruvian Andes and follow a vertical migration scheme. Additionally, there are 30 non-migratory occasional species, seldom recorded and which have been established as Peruvian-coast natives" (2010, p. 34).

As far as the flora is concerned, "81 species of vascular plants have been identified, inclusive of native and introduced wetland species (...), which are also part of the interactions between the ecosystem components" (Pronaturaleza - Fundación Peruana para la Conservación de la Naturaleza, 2010, p. 34).

Ecosystem Abiotic Components

Abiotic components consist of all the non-living elements that have a physical presence or can be felt in the ecosystem, such as, *inter alia,* wind, climate, solar radiation. Within the study area, in particular, we find: sandy and rocky type soils, which under a 4% slope are named valleys or plains, over 8% are termed coastal plains or *pampas*, and between a 50 to 70% slope are known as low mountains or hills; water bodies, specifically the *Lacramarca* river and irrigation canals, coastal wetlands, among others.

The knowledge of the biotic and non-abiotic factors of an ecosystem is essential for Green Infrastructure implementation. Hereunder are some of elements of interest identified in the study area.

INTERNATIONAL, NATIONAL, AND LOCAL REGULATORY FRAMEWORK IN RELATION TO THE PRESERVATION OF NATURAL AREAS AND GREEN INFRASTRUCTURE

For the purposes of this research, a brief overview of the regulatory framework concerning Green Infrastructure and related natural elements is provided, focusing on coastal wetlands. Said regulatory framework gathers and includes legislation at different levels and scales, from international and national regulations to local or municipal ones.

International Regulations

The first one to mention is the RAMSAR Convention of 1971, which aims to protect and enhance the value of the most important wetlands, or those of interest to each member, for the preservation of species and ecosystems. In this sense, "for the purposes of this Convention, wetlands are defined as areas of marsh, fen, peatland or water, whether natural or artificial, permanent or temporary, with water that is static or flowing, fresh, brackish or salt, including areas of marine water the depth of which at low tide does not exceed six metres" ("The Ramsar Convention on Wetlands", 1971).

By virtue of the above, Peru has, at the present moment, thirteen RAMSAR-recognised wetlands, but none of which is included in the study area. In spite of not falling under the jurisdiction and protection of the RAMSAR Convention, if the relevant enhancement interventions were applied to the case study's wetlands, these would qualify for entering this category on the basis of their wide, native and migratory, flora and fauna biodiversity.

Peruvian Regulations

Four national regulations should be mentioned. The first one can be found in the **Political Constitution of Peru of 1993**, where in Title III of the economic regime of Chapter II of the environment and natural resources, the following articles are established: In Articles 66, 67, 68, and 69, it is established that: "Natural resources, renewable and non-renewable, are patrimony of the Nation. The state is sovereign in their use. The organic law establishes the conditions for their use and their concession to private individuals. The concession grants its holder a Right in Rem, subject to that legal norm" (1993, p. 16). It also establishes that the Peruvian government is obliged to promote and encourage the conservation of its natural spaces and biological diversity, as well as its protected areas.

Secondly, there is the **Legislation of Natural Protected Areas N° 26834 of 1997**, whose aim is to establish the management processes for Peruvian natural protected areas. With respect to Article 1, "the Natural Protected Areas constitute patrimony of the Nation. Their natural condition must be maintained in perpetuity, and the regulated use of the area and the exploitation of resources may be permitted, or the restriction of direct uses may be determined" (Congress of the Republic, 1997, p. 1).

In addition to the two previous ones, the **National Wetlands Strategy 2014** strives to prevent the degradation and loss of wetlands in Peru. This legislation establishes for the first time in a Peruvian legal instrument the definition of wetlands, as follows: "extensions or surfaces covered or saturated with water, under a natural or artificial, permanent or temporary, fresh, brackish or salty water regime, and which host typical biological communities that provide ecosystem services" (Ministry of the Environment of Peru (MINAM), 2014, p. 39). In this way, a policy framework is established to promote the sustainable use of wetlands, as well as to enhance the value of their natural resources and endemic biodiversity.

Finally, there is the **Forestry and Wildlife Law No. 29763 of 2015**, which sets out that all ecosystem services, whether forest or wild, must be included as part of the landscape. For that purpose, a legislative framework targeting the protection and conservation of all forest and wild ecosystems within the national scope is set up. Said ecosystems can be classified according to a forest zoning scheme:

- **Permanent production zones:** where the physical space allows for the implementation of elements such as forest use.
- **Ecological protection and conservation zones:** all fragile ecosystems, which are extremely complicated and highly unstable in the face of any natural or anthropogenic event.
- **Recovery zones:** all areas that require and need a recovery strategy or, failing that, renaturalisation or reforestation.
- **Special treatment areas:** all areas where actions are warranted in the cultural, geopolitical, social, and biodiversity fields, and therefore require further study and special treatment at the time of intervention.

Local Regulations

To the best of the authors' knowledge, there are three *Santa* Provincial Municipality (SPM) ordinances, whose description is provided below:

- **Municipal Ordinance N° 002 - 2000 - SPM:** The Metropolitan Park of *Chimbote* and *Nuevo Chimbote* is declared as an intangible, inalienable, and imprescriptible area. It is thus established that the Metropolitan Park becomes part of the recreational and environmental protection system for both *Chimbote* and *Nuevo Chimbote.*
- **Municipal Ordinance N° 007 - 2000 - SPM:** Creation of the *Chimbote* and *Nuevo Chimbote* Land Use Plan, classifying wetlands as recreational, ecological protection, and others.
- **Municipal Ordinance N° 010 - 2003 - SPM:** Creation of the special Metropolitan Park project of *Villa María.*

As mentioned earlier in this section, various local and international legislations protect coastal wetlands. For the purposes of this research, some of the main regulations have been used as a basis for identifying the degree of protection afforded to the wetlands in the study area. Table 1 shows the current regulations protecting these wetlands.

From Table 1 it can be asserted that despite the existence of at least five wetlands in the study area, most of them are unprotected, with the exception of the *Villa María* one, which has at least six laws protecting it and is currently one of Peru's protected areas.

Table 1. Summary table of international, national, and local regulations protecting the Villa María, P.P.A.O., San Juan, Southern and Peninsula wetlands in Chimbote and Nuevo Chimbote

CHIMBOTE AND *NUEVO CHIMBOTE* WETLANDS	INTERNATIONAL		NATIONAL				LOCAL		
	RAMSAR CONVENTION	CONVENTION ON BIOLOGICAL DIVERSITY	POLITICAL CONSTITUTION OF PERU	PROTECTED AREAS LEGISLATION N° 26834 / 1997	NATIONAL WETLAND STRATEGY 2014	FOREST AND WILDLIFE LAW N° 29763 / 2015	MUNICIPAL ORDINANCE N° 002 / 2000	MUNICIPAL ORDINANCE N° 007 / 2000	MUNICIPAL ORDINANCE N° 010 / 2003
VILLA MARÍA	-	-	X [1]	X	X	-	X	X	X
P.P.A.O. [2]	-	-	X	-	X	-	-	-	-
SAN JUAN	-	-	X	-	X	-	-	-	-
SOUTHERN	-	-	X	-	X	-	-	-	-
PENINSULA	-	-	X	-	X	-	-	-	-

([1]) The symbol (X) indicates that the regulation protects the wetland.

([2]) P.P.A.O.: Programa Piloto de Asentamientos Orientados.

EXISTING ANTHROPOGENIC CONFLICTS AND NATURAL CONFLICTS

Anthropogenic Conflicts

The district of *Chimbote*, in comparison with *Nuevo Chimbote*, has one of the most aggravating circumstances regarding man-made conflicts, whose severity depends on their degree of environmental impact, and which can be caused by a common pedestrian, or by bad public management and land-use planning. The current conditions or conflicts are the following:

- **Landfill sites:** the present research showed that the natural areas least impacted by contamination are those isolated from large urban centres, as well as difficult access areas, such as those with a high sedimentation rate and the presence of very tall vegetation, e.g. reeds or cattails.

In this sense, it was concluded that of the five wetlands in the study area, those of *San Juan*, the *P.P.A.O.* and *Villa María* are in a critical state due to significant soil contamination caused by rubbish dumps. The presence of micro-dumps, determined and evidenced in this research, has a damaging effect on ecosystems which can worsen over the years, favouring disease proliferation at the respiratory and dermatological levels.

- **Changing water regimes:** *Villa María* exhibited the highest level of hydrological alterations out of the five wetlands analysed, since it is crossed by the *Lacramarca* river, which flows down through the entire valley of the *Lacramarca* river basin. Water conflicts are generated by three types of pollutants:
    - **Wastewater:** coming from the upper part of the valleys, where the entire agricultural and livestock sector that supplies both municipalities is located. It bears a raft of waste discharges such as fertilisers, manure, pesticides, and mineral salts, on top of the influx of chemical products stemming from the irrigation canals that join the *Lacramarca* river. The findings also reveal that the *San Juan* wetland is heavily affected by this type of wastewater, as well as by domestic water.
    - **Urban centres:** when the *Lacramarca* River begins to cross the *Villa María* wetland, it encounters another challenge, i.e. the presence of small urban centres generators of domestic wastewater, which further aggravate the current river situation and the wetlands. Similarly, the P.P.A.O. wetland is also impacted by domestic wastewater.
    - **Industrial sector:** Despite the criteria for industrial establishment and related pollution control policies enforced by SPM, it is well known that this sector violates all local and provincial regulations, as factories dump heavy metals directly into the sea and into the *Villa María* wetland.
    - **Soil contamination:** the analysis evidenced that the surface of both districts became quite damaged through the large urban, agricultural, industrial, and commercial centres that have taken ownership of natural areas, including:
        - **The *Villa María* wetland:** directly affected by urban centres, rubbish dumps, and pollution from the industrial sector.
        - **The *San Juan* Wetland:** afflicted by rubbish dumps and domestic wastewater, generating alarming soil contamination levels.

    - **The P.P.A.O. Wetland:** mainly impacted by agricultural wastewater, rubbish dumps, and domestic wastewater. The soil has become quite dense due to its high levels of pollutants.
    - **The Southern Wetlands:** in a fairly acceptable state, due to their remoteness from urban centres. These have a fairly low level of soil contamination, which can be attributed to the presence of waste collection points.
    - **Peninsula Wetlands:** This wetland is out of danger from human-induced soil contamination.
- **Fragmentation of natural space:** the research found that the only wetland seriously affected by this condition is *Villa María*, as it is crossed by two major horizontal axes: the Pan-American Highway North (international traffic) and *José Pardo*/Pacific Avenue (inter-district traffic). These high-speed roads generate a rupture in the physical space of the wetland, which in turn damages the ecosystem and hampers the exchange of species.
- **Change of land use:** Despite establishing regulations for natural area protection, the SPM engages in contradictory behaviour by constantly changing the land use plan for the metropolitan area, with serious consequences for the *Villa María* wetland. This wetland is the most affected by land-use changes, combining multiple land uses, from protected areas to residential, commercial, recreational, and agricultural zones.

Table 2. Summary table of the identification of anthropogenic conflicts in the wetlands of Villa María, P.P.A.O., San Juan, Southern, and Peninsula in Chimbote and Nuevo Chimbote

WETLANDS	ANTHROPOGENIC CONFLICTS				
	LANDFILL	WATER REGIME	SOIL CONTAMINATION	FRAGMENTATION OF NATURAL AREAS	LAND USE CHANGE
VILLA MARÍA	X (1)	X	X	X	X
P.P.A.O.	X	X	X	-	-
SAN JUAN	X	X	X	-	-
SOUTHERN	-	-	-	-	-
PENINSULA	-	-	-	-	-

The symbol (X) indicates the existence of anthropogenic conflict.

In conclusion, based on the summary table of anthropogenic conflicts (Table 2), it can be inferred that the most damaged wetland is *Villa María*, as previously outlined. Wetlands closely located to urban centres are most likely to suffer major changes and detrimental impacts, as it has been demonstrated that the remote Southern and Peninsula wetlands are in a better conservation state. This highlights the imperativeness of improving legislative policies for the protection of these spaces. By the same token, the local governmental body should better manage its policies and ordinances so as not to be at variance with its own work.

The coastal wetlands together with other areas of interest such as: the forest nursery, agricultural plots, coastal edges, special treatment zone, metropolitan park and the vegetation components present in the urban centers (green areas and tree-lined streets) are absolutely compatible with the green infrastructure, although some of them present slight problems that must be corrected, others would need a rigorous intervention to renaturalize and protect them, such as the Villa María wetland.

With the general results presented, we can finally conclude that, according to the analysis, the implementation of a territorial green infrastructure among the municipalities of Villa María and Villa María, is a good idea. between the municipalities of Chimbote and Nuevo Chimbote, "IS VIABLE", since it has all the necessary elements for the that it has all the necessary elements to make this large network of systems a reality.

REFERENCES

Allen, W. L. III. (2012). Environmental Reviews and Case Studies: Advancing Green Infrastructure at All Scales: From Landscape to Site. *Environmental Practice, 14*(1), 17–25. doi:10.1017/S1466046611000469

Ancajima, R. (n.d.). *Tecnologías Ancestrales - Sistemas Hidráulicos Pre Incas e Incas* [conference session]. Obtained from: https://www.minam.gob.pe/diadiversidad/wp-content/uploads/sites/63/2015/01/resumen1.pdf

Benedict, M. A. (2002). ET Green Infrastructure: Smart Conservation for the 21st Century. *The Renewable Resources Journal, 20*(3), 12-17. Obtained from: https://www.merseyforest.org.uk/files/documents/1365/2002+Green+Infrastructure+Smart+Conservation+for+the+21st+Century.pdf

Birch, R. L. (1980). Radburn and the American Planning Movement. Departmental Papers (City and Regional Planning).

Cantó López, M. T. (2014). *La ordenación de la infraestructura verde en el sudeste Ibérico (Comunidad Valenciana, España).* Universidad de Alicante. doi:10.14198/cdbio.2014.45.03

Caparrós-Martínez, J. L., Milán-García, J., Martínez-Vázquez, R. M., & de Pablo Valenciano, J. (2021). Green Infrastructures and Grand Environmental Challenges: A Review of Research Trends by Keyword. *Agronomy, 11*(4), 782. . doi:10.3390/agronomy11040782

Centro de Estudios Ambientales (CEA). (2014). *The urban green Infrastructure of Vitoria- Gasteiz.* Obtained from: https://www.vitoria-gasteiz.org/wb021/http/contenidosEstaticos/adjuntos/eu/34/21/53421.pdf

de la República, C. (1997). *Ley de Areas Naturales Protegidas.* Gobierno del Perú.

Convención Relativa a los Humedales de Importancia Internacional Especialmente como Hábitats de Aves Acuáticas (RAMSAR). (1971). *¿Qué son los humedales?* Convención de Ramsar.

European Commission. (2019). *Green Infrastructure (GI)—Enhancing Europe's Natural Capital.* Communication from the Commission to the European Parliament, the Council, The European Economic and Social Committee and the Committee of the Regions.

Fariña Tojo, J. (2002). La ciudad y el medio natural. Madrid: Akal.

Fundación Peruana para la Conservación de la Naturaleza. (2010). *Documento base para la elaboración de una estrategia de conservación de los humedales de la costa peruana.* ProNaturaleza.

Giannotti, E., Vásquez, A., Galdámez, E., Velásquez, P., & Devoto, C. (2021). Green infrastructure planning for climate emergency: Learning from the "stgo+" project, Santiago de Chile. *Cuadernos de Geografia: Revista Colombiana de Geografia*, *30*(2), 359–375. doi:10.15446/rcdg.v30n2.88749

Gobierno del Perú. (1993). *Constitución Política del Perú*. Gobierno del Perú.

Halls, J. N., & Magolan, J. L. (2019). A methodology to assess land use development, flooding, and wetland change as indicators of coastal vulnerability. *Remote Sensing*, *11*(19), 2260. Advance online publication. doi:10.3390/rs11192260

Howard, E. (1898). *To-morrow: A peaceful path to real reform*. Swan Sonnenschein & CO.

INAIGEM. (2016). *Para la Formulación, seguimiento y evaluación del plan Operativo Institucional-Poi del Instituto Nacional de Investigación en Glaciares y Ecosistemas de Montaña- INAIGEM*. Obtained from: https://cdn.www.gob.pe/uploads/document/file/568900/23970810750073625520200318-16266-o1mrp9.pdf

Manchester City Counsil. (2015). *Manchester's great outdoors – a green and blue infrastructure strategy for Manchester*. Obtained from: https://www.manchester.gov.uk/downloads/download/6314/manchester_green_and_blue_strategy

Martínez, C., López, P., Rojas, C., Qüense, J., Hidalgo, R., & Arenas, F. (2020). A sustainability index for anthropized and urbanized coasts: The case of Concón Bay, central Chile. *Applied Geography (Sevenoaks, England)*, *116*, 102166. doi:10.1016/j.apgeog.2020.102166

McHarg, I. L. (1995). *Desing With Nature*. Wiley Blackwell.

Rojas, C., Aldana-Domingue, J., Munizaga, J., Moschella, P., Martínez, C., & Stamm, C. (2020). Urban Wetland trends in three Latin American Cities during the latest decades (2002): Concón (Chile), Barranquilla (Colombia) and Lima (Perú). *Wetlands Science & Practice*, *37*(4), 283–293.

Ministerio del Ambiente del Perú (MINAM). (2014). *Estrategia Nacional de Humedales - Document for public consultation*. Obtained from: http://www.minam.gob.pe/consultaspublicas/wp-content/uploads/sites/52/2014/02/RM-N%C2%B0-051-2014-MINAM.pdf

Municipalidad Provincial del Santa. (2012). Obtained from: https://eudora.vivienda.gob.pe/OBSERVATORIO/INSTRUM_GESTION/PAT/ESTUDIOS/SANTA/01-PAT%20SANTA-RESUMEN%20EJECUTIVO.pdf

Rojas, C., Munizaga, J., Rojas, O., Martínez, C., & Pino, J. (2019). Urban development versus wetland loss in a coastal Latin American city: Lessons for sustainable land use planning. *Land Use Policy*, *80*, 47–56. doi:10.1016/j.landusepol.2018.09.036

Science for Environmental Policy, In Depth Report, The Multifunctionality of Green Infrastructure. (2012). Obtained from: https://ec.europa.eu/environment/nature/ecosystems/docs/Green_Infrastructure.pdf

Vásquez, A., Devoto, C., Giannotti, E., & Velásquez, P. (2016). Green Infrastructure Systems Facing Fragmented Cities in Latin America - Case of Santiago, Chile. Procedia Engineering, 161, 1410–1416. doi:10.1016/j.proeng.2016.08.602

Ventura, S., Niemann, B., Miskowiak, D., & Heinzel, M. (1998). *Environmental Corridors: Land Use Planning Guide*. Land Information & Computer Graphics Facility - University of Wisconsin – Madison.

Tabilo, E., Burmeister, J., Chávez, C., & Zöckler, C. (2016). *Humedales y aves migratorias en la costa árida del Pacífico Sudamericano. Etapa I: Evaluación Ecológica Rápida*. Centro Neotropical de Entrenamiento en Humedales y Manfred Hermsen Stiftung de Alemania.

William, L., & Allen, I. I. I. (2012). Environmental Reviews and Case Studies: Advancing Green Infrastructure at All Scales: From Landscape to Site. *Environmental Practice*, *14*(1), 17–25. doi:10.1017/S1466046611000469

Chapter 6
Green Knowledge and Technology and Their Implications in Organizational Green Innovation, Urban Green Innovation Spaces, and Green Roofs

José G. Vargas-Hernández
https://orcid.org/0000-0003-0938-4197
Posgraduate and Research Department, Tecnológico Mario Molina Unidad Zapopan, Mexico

ABSTRACT

This study aims to analyze the implications of green knowledge and technology in organizational green innovation, urban green innovation, and green roofs. The analysis is supported by the assumption that green sharing knowledge and technology is basic to organizational green innovation and urban green innovation areas practices, operations, and activities. The methods employed are based on the analytical-reflective and descriptive supported with the review of theoretical and empirical literature. The analysis concludes that green knowledge sharing is relevant to create and develop the green technology with positive implications for organizational green innovation, urban green innovation areas, and green roofs.

INTRODUCTION

The COVID-19 pandemic is a sanitary crisis that questions many activities to become greener and more sustainable. Sustainable and green are two concepts increasingly used to mean the same, however, sustainability refers to the persistence and indefinite future of necessary and desired characteristics of the human subsystem within and the ecosystem (Hodge, 1997).

DOI: 10.4018/978-1-6684-4030-8.ch006

Organizational environmental and green knowledge learning is linked to green technology for the environmental protection to stimulate organizational green innovation, urban green innovation areas and green roofs. Sustainable development is related to decision making in the economic and social effects (Wiering, Liefferink, Boezeman, Kaufmann, Crabbé, & Kurstjens, 2020). Sustainable development meets the needs of the present without compromising the ability of future generations to meet their own needs (Brundtland, 1987).

Urban sustainability including social and economic has been highlighted as one of the leading features of cities (VisitBerlin (Ed.) 2017). Some green trends are proposed as smart city, sustainable city, and so forth. Use of green and ecofriendly technology offers more sustainability environment with zero gas emissions to the environment and some other opportunities and challenges.

People living in dense communities tend to use public urban green innovation spaces and public parks more frequently to have more relaxation time and may travel more to the countryside for leisure. Experimenting and testing new greener ways of conducting organizations concentrated on environmental sustainability, enable to introduce green innovations (Anderson et al., 2010; Stubbs and Cocklin, 2008). Organizations embracing the concept of saving money, by creating recycling programs and monitoring thermostats, focusing on environmental sustainability. Organizations may contribute to support nongovernmental agencies involved in environmental troublesome areas.

The frequency of use of a green space living environment support individuals to be satisfied with public spaces and improves the social space and the mental health (Hadavi, 2017). People using public and private urban green innovation spaces attach meanings, identity, and psychological experiences to different types of green spaces, as described in the Place identity integrated model and environmental representation (Bernardini and Irvine 2007). Research on green organization identity has focused on individual level (Chen, Chang, 2013; Chen, 2011). Green organizational identity supports individual tasks of organizational members related to the organizational environmental activities and strengthens the ability to cope with organizational green oriented conditions. Landscape connectivity has some differences between urban green innovation roofs and urban open space management.

Organizations are facing challenges regarding the compliance with green sustainability strategy. The deployment of a green strategy to face the negative effects that industrialization has on the environment. Organizational business sustainability considers the green growth, green branding, and green sustainable reporting. Branding sustainability enhance the ability of the organization to appeal customers concerned about the environment.

Green sustainable development combined with economic growth, social justice and progress and environmental security concepts are relevant issues in research such as green entrepreneurship in organic farming (Gupta, and Vegelin, 2016; Mohd, and Norhidayah, 2016; Savickiene, and Miceikiene, 2018; Ihnatenko, and Novak, 2018; Kucher, 2019; Shevchenko, and Petrenko, 2020, Skydan, Nykolyuk, Pyvovar, and Martynchuk, 2020). Not much research has addressed the integration of green roofs and urban green innovation space.

Organizations engage in green initiatives to develop sustainable competitive advantages and competitiveness (Wysocki, 2021; Galdeano-Gómez, Céspedes-Lorente, & Martínez-del-Río, 2008; López-Gamero, Claver-Cortés, & Molina-Azorín, 2008). Sustainable components of organizations and business achieve competitive advantage (Namkung and Jang, 2012; Chang and Fong 2010; Bansal, 2005 and Barnet, 2007). Organizational business systems are the foundation for any kind of sustainable organizational business. Organizations are formulating and implementing green initiatives to attain sustainability and competitiveness (Chuang & Huang, 2015). The ecological partnership between organizations and society provides sustainable competitiveness.

The organizational ecological and sustainable procedures follow a strategic orientation (Fernandes & Solimun, 2017). Sustainability has been considered and an added cost, but nowadays is viewed as a sustainability strategy and tool to derive value and leads to innovation (Van Holt et al., 2020). An organizational shared vision enables development of green behaviors, making meaningful contributios (Larwood, Falbe, Kriger, Miesing, 1995; Tsai, Ghoshal, 1998; Oswald, Mossholder, Harris,1994).

GREEN KNOWLEDGE AND TECHNOLOGY

New knowledge in green organizational and business models are uncertain and volatility in business ecosystem conditions, challenging risk management and decision making. Technological knowledge of economic, social, and environmental benefits of constructing green roofs enhances the good construction of green roofs without any problems of maintaining the roof vegetation. Organizational green knowledge shares issues adopting a multilevel framework to make contributions to green technology. Organizational management must formulate green knowledge and green technology.

Knowledge management of environmental issues is a source for learning, solving problems and creating competitiveness (Liao, Chang, Cheng, Kuo, 2004). Green knowledge management is an effective tool for organizational improvement in transformation environment (Raudeli ¯ unien˙ e, Davidaviˇcien ˙ e, Jakubaviˇcius, 2018). The dynamic complexity of organizational green innovation requires environmental knowledge learning (Li et al., 2019) and green technology. Organizational environmental green knowledge learning drives green innovation. Organizational environmental knowledge learning is linked to green technology for the environmental protection to stimulate green innovation (Redman, 2014).

Knowledge sharing is the basis for knowledge management (Zhang, Sundaresan, 2010). Green knowledge sharing in terms of sustainable development goals such as eliminating pollution, environmental protection, and SDGs, etc.) with other organizational members is good (Bock, Zmud, Kim, Lee, 2005). Green knowledge sharing is the behavior of organizational members who pass information, data and knowledge about green concerns and issues to other members and create new knowledge and learning opportunities to encourage others (Norton, Parker, Zacher, Ashkanasy, 2015; Chen, Chang, 2013).

Green Knowledge Sharing. GKS is the organizational members behavior which is keen to pass on personal information and new knowledge learning opportunities about green issues to other members (Bock, Zmud, Kim, Lee, 2005). Green property psychological ownership and green knowledge sharing are influenced by green organizational identity. Knowledge sharing of people under pressure of time is reduced (Chang, Chen, Yeh, Li, 2020).

Organizational knowledge sharing focuses on creating an atmosphere of environmental protection to create green knowledge sharing related to green issues. Green technology is a socio-environmental healing technology capable to reduce environmental damages and degradation while conserving natural resources. Sustainable green technologies used in processes and applications do not create footprint Aithal & Shubhrajyotsna, (2016).

Urban green innovation technology can be transferred by public government and private business pursuing inclusive green growth and ensuring social integration by eradicating poverty of vulnerable groups and maintaining the footprint of humanity within the ecological boundaries. Building an inclusive and green organization of available and retained talent is an arduous task (Goulden, Mason & Frasch

2011; Pell, 1996; Brands & Fernandez-Mateo, 2017). Green roof technology is used to tackle social and environmental concerns to enhance climate resilience. Green plant based green chemistry principles are used for green synthesis, nanomaterials, and nanoparticles (Maas, and Hox, 2005). Green and eco-friendly nanotechnology based on green chemistry techniques tends to decrease the risks associated to industrial applications

Green eco-friendly nanotechnology solutions used in sustainable development goals reduces the technification threat. Environmental degradation and climate pose a threat to global sustainable development and the Sustainable Development Goals. The concept of achieving sustainable development goals by means of using and developing proven green ecofriendly nanotechnology processes further accelerates the spread and growth of many other systems and devices. Green nanotechnology solutions are critical in realizing sustainable development goals by eliminating the threat technification development processes.

Green nanotechnology treats with environmentally friendly processes of manufacturing and industrial use of nanomaterials able to minimize potential risks of environmental degradation and reduce health hazards. Green nanotechnology and nanomaterials solve many socio economic and environmental problems and improve the quality of life supporting the United Nations sustainable development goals which ca be realized by using green nanotechnology.

Green nanotechnology is based in miniature sized communication and computation devices and nano sensor devices, high density, and memory chips. Ensuring ecological communication among the stakeholders disclose the relevance of environmental concerns (Abimbola, Lim, Hillestad, Xie, & Haugland, 2010). Sustainable green nanotechnology principles and processes in the primary sector of the economy is related to the extraction of raw materials from nature but also in manufacturing and services. Green nanotechnology is used for renewable energy management systems in generation, transmission, and storage.

Long-term effects of human food are required by green manufactured nanomaterials designed according to principles of green nanotechnology complemented with current regulations aimed to address sustainable development of green technology. Manufactured and processed inorganic food has increased the cases of illnesses leading to healthy food safety. Green design and development of health and environment using nanoscale materials is a concern to improve sustainable solutions. Green synthesis methods can be used to take care of adverse effects on nanomaterials on the environment and user health. Complementarity of mixed methods to detect contradictions, improve the results of one method using the results of the other method and enhance the scope and breath for different components of research (Greene et al. 1989).

Manufacturing green nanoelectronics devices involves complexity in the shifting from silico-based to molecular nanomaterials-based devices. Green nanomaterials enable the network for smart city communications. Green nanotechnology develops miniaturized drones and bots equipped with artificial intelligence and bees, nanosatellites, and nano sensors to provide more visibility and nano-nuclear biological and chemical weapons. Adding nanoparticles to sustainable green nanotechnology improves the properties of construction materials. Green nanotechnology provides some special effect paints changing their color at different spaces and times. Green nanotechnology supports the design in multiple properties of the fashion industry.

Green nanotechnology in the service industry is intangible in nature and affected directly and indirectly. Green nanotechnology innovates in educational technology through higher quality and low-cost and ubiquitous online education. ICCT technologies working with green nanotechnology provide intelligent services in artificial intelligence, internet of things, cloud and quantum computing, 3D printing, etc. Green nanotechnology supports ICCT technologies to develop super-intelligent machines and human beings. Green nanotechnology improves the durability, speed, and reachability of digital entertainment instruments (Aithal & Shubhrajyotsna Aithal, 2018).

ORGANIZATIONAL GREEN INNOVATION

Organizational green business innovations follow the principles, values and norms of economic concern, societal parity, and ecological accountability (Alawattage & Fernando, 2017). New green organizational innovations are created either in-house or external. Organizational green innovation activities are stimulated by green contexts-oriented factors. Organizations have attached more relevance to environmental issues based on green organizational identity that enhances organizational green innovation and competitiveness and ensures organizational competitive advantages (Geraie, Rad, 2015). Organizational identity theory and psychological ownership theory are combined to integrate a conceptual model.

Organizational green process innovation is friendly to the environment through developing processes of manufacturing goods according to their cycle (Chiou, Chan, Lettice, Chung, 2011; Kam-Sing Wong, 2012). Organizational green innovation and actions are designed to reduce adverse environmental consequences. There is a relationship between organizational green innovation strategy and actions because the green innovation behaviors carried out under a strategy.

Motivational mechanisms stimulate innovation to facilitate organizational green innovation. Motivated organizational stakeholders to pursue collective efforts towards green innovative outcomes with their teammates lead with values of organizational green innovation Norton, Parker, Zacher, Ashkanasy, 2015; Jabbour, Santos, Fonseca, Nagano, 2013). Meta-analytical research contends that prosocial motivation is a mechanism to link contextual factors and innovative and creative outcomes (Liu, Jiang, Shalley, Keem, Zhou, 2016).

Organizational green innovation addresses the potential risks and challenges of implementing innovative resources and inputs supported by green efficacy (Liu, Chen, Tao, 2015). Green organizational innovation is related to individual tasks of organizational environmental activities to strengthen the organization. Some demographic variables are related to green organizational innovation behaviors such as age. Older people are less likely to perform employee green behaviors using more innovative technologies and involving changing habits and interacting with others.

Sustainable green organizations encourage employees to participate in social and environmental. This often means encouraging employees to participate and get involved in the social and environmental initiatives such as promoting to refuse, reduce, repair, reuse, and recycle. Society gives relevance to environmental issues and concerns where green organizational identity enhancing organizational creativity and innovation and ensuring sustainable competitive advantages (Geraie, Rad, 2015). Social acceptance enables the organization to attain and develop sustainable organizational green competitiveness capability.

The green competitive advantage is the concept describing the conditions to hold a position in sustainable and ecological management and innovation of the firm which cannot be imitated by others (Lin & Chen, 2017). Resources and capabilities that are not imitable provide sustainable competitive advantage (Barney, 1991). The resource-based view of organizational research focuses on the characterization of organizations (Pereira & Vence, 2012) and organizational resources and capabilities (Fong & Chang, 2012; Marchi, 2012) such as organizational redundancy, innovation capabilities, network, embedding, strategy orientation and green innovation strategy as the drivers of green innovation practices.

Organizations depend on intangible resources to enhance environmental sustainability (Singh, Del Giudice, Chierici, & Graziano, 2020). In a resource intensive sector, a green pioneer organization must have an organizational model (Elkington,1994) with a social and environmental impact.

Human resources green management deals with valuable organizational resources and assets considered based on environmental sustainability. There is evidence that the workforce engages strongly about the environmental sustainability development and become more organizational committed and satisfied. Human resource green management is engaged in organizational environmental sustainability management (Wirtenberg, Harmon, Russell, & Fairfield, 2007; Shoeb and Nisar 2015).

Green dynamic capability is in the nature within the organization processes. The green dynamic capabilities are permanently reliable in the organizations (Chen, 2008).

URBAN GREEN INNOVATION SPACES AND GREEN ROOFS

The rapid growth of the urban population and the low rate of economic development due to the pandemic, are leading to serious civilization problems, among others the lack of urban greener innovations spaces, pollution, and lack of health. Smart growth is supported by sustainable growth based on adoption of greener, efficient technologies, inclusive growth, fostering employment and social cohesion (Gunn, & Mintrom, 2016). Organizations must take an approach to stay in inclusive green organizations embracing elements such as equity and social justice needed to create goals, formulate, and implement strategies and develop public commitment (Johnson, 2017).

Organizational green innovation refers to the innovation that emphasizes the implementation of organizational green culture, environmental prevention of pollution and waste reduction to enhance organizational sustainability. Urban green innovation areas are increasing in green metropolis driven by the preservation and expansion of sustainable environmental development (Kalandides, Grésillon, 2021).

Urban green innovation spaces must be defined by the limits of construction and protection. The interactions of people and urban green innovation spaces are divided in public and private (Vijayaraghavan, 2016). Public and private urban green innovation spaces co-exist and can be incorporated green roofs as open green spaces aimed to use multi-functionality framed in an effective landscape management. Public and private green spaces have differences in perception (Coolen and Meesters 2012) although they should co-exist (Mesimaki, et al. 2017).

People have different perceptions on private and green urban spaces with social identity and environmental sustainability. Green behaviors are defined as the actions and behaviors on which employees engage linked with and contribute to environmental sustainability (Ones & Dilchert, 2012a, p. 87). This concept includes voluntary green behaviors beyond the job tasks referred as organizational citizenship, also considered as task-related environmental behavior (Boiral & Paillé, 2012; Bissing-Olson, Iyer, Fielding, & Zacher, 2013). Sustainable development and green jobs guidelines are adopted by the international labor organization (ILO) for a transition towards environmentally sustainable societies and economies (ILO, 2015).

Public parks contribute to develop more community social interactions and livability while private gardens are paces for freedom and enjoying life (Coolen and Meesters 2012). Residents attach to open and green spaces removed from urban plans of the sustainable, green, and intelligent city. The minimum green urban areas should be 50 m2 per urban resident (Russo & Cirella, 2018).

The disappearance of green urban areas due to residential and commercial purposes, enhanced by the urban sprawl, leads to loss of habitats, biodiversity, and recreation areas, and decrease of natural resources and elements (Burszta-Adamiak, Fiałkiewicza, 2019). Urban green spaces with high social and environmental vulnerability are the most urbanized areas with high land temperature might be prioritized for the implementation of urban green innovation areas and green roofs.

Traditional streets can be transformed into green urban connecting areas considered as an open space system and park lands. An urban green innovation belt prevents urban expansion and urban sprawl by merging with nearby settlements. Greenbelt conserves accessibility of open spaces along roadways and connect to a regional trail. Nevertheless, greenbelt was not successful to prevent urban expansion and sprawl but facilitates the access to open spaces (Maruani and Amit-Cohen 2007).

Intensification of investments to develop accessible green spaces conceived to cover economic, social and leisure needs, and activities contribute to create the sense of community and respond to the pressures of construction. Urban green innovation areas are open to enhance urban green innovation space networks and extend the green infrastructure with the functions of green roofs (Fung, 2018). Green infrastructure is the urban physical environment in a network of green open spaces that brings economic, social, and environmental benefits to residents of communities in cities, towns, and villages (TEP 2005, 1).

Green infrastructure is a way to create an attractive environment, increase the urban resilience benefits by better managing stormwater, mitigating the urban heat island effect, purifying air. Green roofs are considered passive open green spaces that have environmental benefits and provide open spaces for more urban livability (Mesimaki, et al. 2017). Eco roofs are required for new city infrastructures and facilities. Multi-functional urban green innovation infrastructure should allocate funding for green initiatives of development.

The benefits of urban green innovation infrastructure include multiple socio-ecological, and economic benefits. Green roofs have environmental and social benefits used as multidimensional green infrastructure. Green roofs have some comprehensive benefits. Green roofs environmental benefits are the measure to enhance climate resilience and energy efficiency and provide development opportunities. Development of urban green innovation roofs is driven on aesthetic reasons, climate resilience and to increase property values of buildings. Green building plays a critical role in dealing with socioenvironmental issues serving as an organizational platform for financial savings. Suitable buildings for green roofs could be improved by including roof decks and courtyards. Urban green innovation roofs developments mismatch in some places and areas where are needed to be prioritized.

Scholars have applied quantitative and qualitative analysis to address the needs and perceptions of people on green roofs. The environmental effectiveness of green roofs should improve public health and connect the social needs of vulnerable people (Vijayaraghavan 2016). Green roofs and solar and living roofs considered a multi-benefit asset bringing significant economic benefits beyond stormwater management, energy efficiency for cooling, improving air quality, reducing urban heat island. Loss of green spaces intensify heat island effect and storm water runoffs. The development of urban green innovation spaces and roofs are required to comply with regulations on stormwater control. A motivation program for the installation of eco roofs provides incentives to better management of the stormwater.

Green roofs have historically evolved from the hanging gardens in Babylon embracing the aesthetic values and facilitating the human interactions and increase the insulation. Le Corbusier applied green roofs to modern architecture. (Berardi, GhaffarianHoseini and GhaffarianHoseini 2014). In Germany followed by France and Switzerland occurred the research and implementation of green roofs. In Portland, USA in 2005 requires the city-owned buildings to build green roof while Toronto passed a bylaw requiring all the new developed buildings. In Tokyo, the new constructed buildings must have green roofs (Vijayaraghavan 2016).

Urban green innovation areas should provide spaces for green infrastructure solutions such as the building green roofs, which have beneficial influence on the urban environment. The stage of operation of green infrastructure is the maintenance solutions. The green infrastructure is single function derived from the ability to effectively manage stormwater and focusing on ecological landscape conservation than social and economic benefits (Mell 2010). Urban green innovation roofs integrated into open space management have multifunctional opportunities to develop green infrastructure and benefits in urban communities and neighborhoods. Private and public green roofs as open spaces a multi-functional green infrastructure and should accommodate the urban open space needs of residents (Meerow and Newell 2017)

Well-developed urban green innovation infrastructure needs to be supported by a multi-faceted approach. The increasing awareness of green roofs to alleviate global and local environmental impacts. Vegetated green roofs reduce CO2 in the atmosphere naturally sequestered through the photosynthesis processes (Rowe 2011). It has been confirmed the capacity of green roofs to retain rainwater, reduce air pollutants and improve microclimate that have a positive influence in urban heat island effect (Shafi que, Kim & Rafi q, 2018; Burszta-Adamiak, Stańczyk & Łomotowski, 2019).

There are different types of intensive and extensive green roofs which include vegetation with different soil depth and weight, irrigation, plant species and maintenance. Urban green innovation projects include urban gardening, permaculture, guerrilla gardening, smart city and building architecture. Productive urban green innovation combines urban agriculture, allotment gardens and self-sufficiency culture ecologically motivated (Plattform produktives Stadtgrün 2020; Chen, Lin, Lin, Hung, Chang, Huang, 2020).

Mitigation of greenhouse gas emission technique to improve the environmental impact of agriculture needs to consider animal welfare the largest anthropogenic contributor (Llonch, Haskell; Dewhurst; Turner, 2017; Reisinger, Clark, 2018; Kucher, Heldak, and Orlenko, 2018). Wasteland spaces can be converted into a place of productive greenery and developed into natural and green innovation areas that preserve and increase the ecological value and be used also as place for joint creation and learning, (Nachhaltiges Berlin, 2020; Green Berlin, 2020).

Rehabilitation of buildings using energy optimization should consider the creation of new alternative green spaces transformed into green innovation areas for free recreational purposes. Green roofs provide leisure spaces and enhance aesthetical values of buildings (Sutton, 2014) reduce carbon footprint in urban areas (Ugai, 2016) and direct water footprint (Fialkiewicz et al., 2018). Matching urban places needing urban green innovation areas and green roofs in suitable buildings to determine opportunities to implement in communities and neighborhood

Administrative and technical support are necessary in designing, constructing, developing, and maintaining urban green innovation areas and green roofs. This task recommendation enhances the allocation of urban green innovation areas based on the structural design of buildings and the surrounding environmental characteristics in the community. The design and construction of green roofs need to consider the plant species and the development of small fauna.

The construction and application of green roofs can be achieved in large urban areas are located near another urban constructed areas and should not be single investments. Stovin, Vesuviano & Kasmin (2012) calculate that the surface of roofs account for almost 50% of the sealed urban areas with the potential for urban green innovation roofs. Green roofs decrease the costs for discharging snowmelt and stormwater. The infrastructure of green storm water initiative reduces greenhouse gas emissions and improves the air quality. The development initiatives of green roofs are linked with storm water management to meet the flood management control required

Some studies evaluate the functionality and usefulness of constructing green roofs, such as the Forschungsanstalt Landschaftsentwicklung Landschaftsbau (FLL) guidelines (FLL, 2002). The FLL guidelines are used as the basis for the construction of green roofs. Vegetation on green roof infiltration reduces the stormwater runoff. Vegetated surfaces may utilize public facilities to demonstrate a mandate of a green buffer surrounded by parking facilities that may also function as a stormwater infiltration. Green stormwater infrastructure reduces the demand of grey stormwater infrastructure, enhancing the climate resilience.

Investments in green roofs increases the biologically active areas in urban areas and towns to mitigate heat island and reduce carbon emissions and improve the flood control. Cities have created incentives and regulatory programs to encourage implementation of urban green innovation areas and green roofs. The incentive programs for green roofs focus on the local level should be tied to storm water management and combined sewers overflow control, improve water quality and quantity. Local government uses green roofs as infrastructure to control stormwater and reduce water pollution.

Local building codes must allow green roofs exempted from the building floor area and related to open spaces. The incentives for zoning to encourage the implementation of the urban green innovation roof should include the development of parking landscapes as one measure adopted in a stormwater management plan that may provide solutions (Murray 2017).

The development of green roofs value investments as an additional incentive for potential customers. Meeting the ecological needs of customers (Jain and Kaur 2004). There is a trend of organizations to shift operations towards the direction of environment-friendly processes despite the challenges of customer cynicism regarding organizational green actions (Kumar & Christodoulopoulou, 2014). Co-financing and local legal regulations for the construction of green roofs are most often used incentives. Green development finance has been assigned to Multilateral Development Banks, alongside public-private partnerships (PPPs) and new forms of finance, blended finance, bond instruments, green bonds, social impact bonds and development impact bonds. Urban green innovation roofs are the result of the implementation of motivational tools and other incentives for the different stakeholders.

The implementation of tools to motivate public and private investors to construct green roofs requires a program to gather information to identify future incentives and to formulate recommendations for the construction of green roofs in green urban areas. The results of reasonable investments in urban green innovation innovation areas and green roof should be to receive the return on investment within certain period in relation with regulations. Efficiency of operations in cost-savings are the concern of green investments.

The growing interest in green roofs in urban areas comes with the introduction of incentives. However, the limited incentives for green roof investments rise the construction costs. Introducing incentives for green roofs may increase biologically active areas by introducing nature-based solutions. The implementation of incentives is leading to the realization of green roofs in local spatial development plans.

Documenting the existing locations of urban green innovation areas and green roofs and compiling to determine spatial development patterns including details of purposes, types, size, locations, etc., focusing more on the comparison of social and environmental vulnerability among the different places. Local governments are relevant actors in promoting environmentally friendly solutions by granting financial and non-financial subsidies, regulate eligibility and increasing the surface area of urban green innovation roof spaces.

The existence of types of barriers encourages local authorities to improve the use of incentives for the creation and functioning of urban green innovation areas and urban green innovation roofs. The most commonly use of incentives are the direct ones and legal regulations for the construction of regulations (Mentens, Raes & Hermy, 2006). Incentives must be translated from sufficient interest into motivation for the excessively strict requirements for the construction of green roofs such as the minimum surface area, to become eligible for co-financing and tax allowance.

Market demand-related incentives such as promotional instruments, are tools to determine demand for buildings with greenery in comparison to traditional. The implementation of financial and non-financial incentives is an important factor for the green roofs as a relevant element of urban environment (Brudermann & Sangkakool, 2017). Financial incentives include subsidies or donations granted to reimburse costs of investments that are aimed to support potential investors.

The non-financial incentives include instruments of gratification. Other financial incentives are real property tax allowances to green roofs constructed depending on the surface area of buildings. The financial and non-financial incentive programs must be complementary and flexibility in terms of the different options of greenery design as the determining the subsidy amount of creating surfaces that are water-permeable along with incentives in tax allowances in snowmelt and stormwater.

The development of urban green innovation areas and green roofs are limited to regulations in local spatial development plans on incentives not only limited to co-financing. Some indicators to measure the adaptability of urban green innovation areas and green roofs are more effective to tackle environmental concerns such as the land temperature and stormwater. Environmental vulnerability indicators in stormwater flow, air quality and pollution are considered in the analysis to acknowledge the priorities of all actors and stakeholders in any urban community.

The development of rural and urban green innovation tourism close to nature and based on settlements is becoming more relevant for the impact on green economy and ecology, with the self-employment of residents to provide environmentally accommodation and food to the visitors (Tomashuk, Baldynyuk, 2021, Kolomiets, Tomashuk, 2021, Mazur, Tomashuk, 2019). The green economy development and the pursuit of environmental protection is bringing profits to the organizations (Chen, Lai, Wen, 2006).

CONCLUSION

This study analyzes the organizational environmental and green knowledge learning is linked to green technology for the environmental protection to stimulate organizational green innovation, urban green innovation areas and green roofs.

Organizations transform to mitigate and neutralize the environmental impact and to adapt environmental sustainability. Green transformation is a multifaceted process incorporating interconnected and overlapping processes posing several managerial challenges. Environmental community involvement promotes and improve the natural environment aiming to sustainable growth of society, inspiring for voluntary participation in social activities. The green government should make and secure access to affordable sustainable housing in urban green innovation spaces and open buildings to create an urban social cohesion process consolidated in urban social movements.

Organizational networks are vital to connecting each other in the space of a green movement. The environmental movements along the nongovernment organizations and environmental foundations lack of racial diversity despite the efforts to increase diversity and inclusion (Johnson, 2019) Residents also express their intentions to be involved in discussions and debates to develop a green urban framework.

Adding green components to technology can become sustainable green technologies able to avoid degradation and provide clean environment. Green roof technology is used to tackle social and environmental concerns to enhance climate resilience. Green roof is related to be linked to resilience, restorative, and accessible. Green nanotechnology represents opportunities and challenges in industry to encourage growth by supporting nanotechnology usage.

Gain competitive advantage through the implementation of a sustainable organizational strategy is relevant for the survival. The organizational green shared vision should involve sustainable business goals. Long-term organizational green innovation strategy must stimulate followers to implement green operations. Organizational socio ecological strategies are formulated an implemented to achieve green competitiveness (Wang, Hu, Dai, & Burns, 2021).

REFERENCES

Abimbola, T., Lim, M., Hillestad, T., Xie, C., & Haugland, S. A. (2010). Innovative corporate social responsibility: The founder's role in creating a trustworthy corporate brand through "green innovation". *Journal of Product and Brand Management, 19*(6), 440–451. doi:10.1108/10610421011085758

Aithal & Aithal. (2016). Nanotechnology Innovations and Commercialization – Opportunities, Challenges & Reasons for Delay. *Proceedings of National Conference on Changing Perspectives of Management, IT, and Social Sciences in Contemporary Environment*, 14, 1-12. 10.5815/ijem.2016.06.02

Aithal & Aithal. (2018). Nanotechnology based Innovations and Human Life Comfortability –Are we Marching towards Immortality? *International Journal of Applied*.

Alawattage, C., & Fernando, S. (2017). Postcoloniality in corporate social and environmental accountability. *Accounting, Organizations and Society, 60*, 1–20. doi:10.1016/j.aos.2017.07.002

Anderson, R., Amodeo, M., & Harzfeld, J. (2010). Changing business cultures from within. The World watch Institute.

Bansal, P. (2005). Evolving sustainably: A longitudinal study of corporate sustainable development. *Strategic Management Journal, 26*(3), 197–218. doi:10.1002mj.441

Barnet, M. L. (2007). Stakeholder influence capacity and the variability of financial returns to corporate social responsibility. *Academy of Management Review, 33*(3), 794–816. doi:10.5465/amr.2007.25275520

Barney, J. (1991). Firm Resources and Sustained Competitive Advantage. *Journal of Management, 17*(1), 99–120. doi:10.1177/014920639101700108

Berardi, U., GhaffarianHoseini, A. H., & GhaffarianHoseini, A. (2014). State-of-the-art analysis of the environmental benefits of green roofs. *Applied Energy, 115*, 411–428. doi:10.1016/j.apenergy.2013.10.047

Berlin, G. (2020). *Capital of Green Trends: How Berlin leads the Way in Urban Sustainability*. Available online: https://about. visitberlin.de/en/greenberlin

Berlin, N. (2020). *Stadt der grünen Trends-Wie Berlin den Weg der urbanen Nachhaltigkeit geht.* Available online: https://about.visitberlin.de/nachhaltiges-berlin

Bernardini & Irvine. (2007). The 'nature' of urban sustainability: private or public greenspaces? *Transactions on Ecology and the Environment,* 661-673.

Bissing-Olson, M. J., Iyer, A., Fielding, K. S., & Zacher, H. (2013). Relationships between daily affect and pro-environmental behavior at work: The moderating role of pro-environmental attitude. *Journal of Organizational Behavior, 175*(2), 156–175. doi:10.1002/job.1788

Bock, G. W., Zmud, R. W., Kim, Y. G., & Lee, J. N. (2005). Behavioral intention formation in knowledge sharing: Examining the roles of extrinsic motivators, social-psychological forces, and organizational climate. *Management Information Systems Quarterly, 2005*(29), 87–112. doi:10.2307/25148669

Boiral, O., & Paillé, P. (2012). Organizational citizenship behaviour for the environment: Measurement and validation. *Journal of Business Ethics, 109*(4), 431–445. doi:10.100710551-011-1138-9

Brands, R. A., & Fernandez-Mateo, I. (2017). Leaning out: How negative recruitment experiences shape women's decisions to compete for executive roles. *Administrative Science Quarterly, 62*(3), 405–442. doi:10.1177/0001839216682728

Brudermann, T., & Sangkakool, T. (2017). Green roofs in temperate climate cities in Europe - An analysis of key decision factors. *Urban Forestry & Urban Greening, 21,* 224–234. doi:10.1016/j.ufug.2016.12.008

Brundtland, G. (1987). *Our Common Future: The World Commission on Environment and Development.* Oxford University Press.

Burszta-Adamiak, E., & Fiałkiewicza, W. (2019). Review of green roof incentives as motivators for the expansion of green infrastructure in European cities. *Scientific Review – Engineering and Environmental Sciences, 28*(4), 641–652. doi:10.22630/PNIKS.2019.28.4.5

Burszta-Adamiak, E., Stańczyk, J., & Łomotowski, J. (2019). Hydrological performance of green roofs in the context of the meteorological factors during the 5-year monitoring period. *Water and Environment Journal: the Journal / the Chartered Institution of Water and Environmental Management, 33*(1), 144–154. doi:10.1111/wej.12385

Chang, N. J., & Fong, C. M. (2010). Green product quality, green corporate image, green customer satisfaction, and green customer loyalty. *African Journal of Business Management, 4*(13), 2336–2344.

Chang, T. W., Chen, Y. S., Yeh, Y. L., & Li, H. X. (2020). Sustainable consumption models for customers: Investigating the significant antecedents of green purchase behavior from the perspective of information asymmetry. *Journal of Environmental Planning and Management, 2020,* 1–21.

Chen, Y. S. (2008). The driver of green innovation and green image–green core competence. *Journal of Business Ethics, 81*(3), 531–543. doi:10.100710551-007-9522-1

Chen, Y. S., & Chang, C. H. (2013). The determinants of green product development performance: Green dynamic capabilities, green transformational leadership, and green creativity. *Journal of Business Ethics, 2013*(116), 107–119. doi:10.100710551-012-1452-x

Chen, Y. S., Lai, S. B., & Wen, C. T. (2006). The influence of green innovation performance on corporate advantage in Taiwan. *Journal of Business Ethics, 2006*(67), 331–339. doi:10.100710551-006-9025-5

Chen, Y. S., Lin, S. H., Lin, C. Y., Hung, S. T., Chang, C. W., & Huang, C. W. (2020). Improving green product development performance from green vision and organizational culture perspectives. *Corporate Social Responsibility and Environmental Management, 2020*(27), 222–231. doi:10.1002/csr.1794

Chiou, T. Y., Chan, H. K., Lettice, F., & Chung, S. H. (2011). The influence of greening the suppliers and green innovation on environmental performance and competitive advantage in Taiwan. *Transp. Res. Logist. Transp. Rev, 2011*(47), 822–836. doi:10.1016/j.tre.2011.05.016

Coolen, H., & Meesters, J. (2012). Private and public green spaces: Meaningful but different settings. *Journal of Housing and the Built Environment, 27*(1), 49–67. doi:10.100710901-011-9246-5

Elkington, J., 1994. Towards the sustainable corporation: Win-win-win business strategies for sustainable development. *Cal. Manag. Rev., 36*(3), 90-100.

Fernandes, A. A. R., & Solimun. (2017). The mediating effect of strategic orientation and innovations on the effect of environmental uncertainties on the performance of business in the Indonesian aviation industry. *International Journal of Law and Management, 59*(6), 1269–1278. doi:10.1108/IJLMA-10-2016-0087

Fialkiewicz, W., Burszta-Adamiak, E., Kolonko-Wiercik, A., Manzardo, A., Loss, A., Mikovits, C., & Scipioni, A. (2018). Simplified direct water footprint model to support urban water management. *Water (Basel), 10*(5), 630. doi:10.3390/w10050630

Fong, C. M., & Chang, N. J. (2012). The impact of green learning orientation on proactive environmental innovation capability and firm performance. *African Journal of Business Management, 6*(32), 727–735.

Forschungsanstalt Landschaftsentwicklung Landschaftsbau (FLL). (2002). *Dachbegrünungsrichtlinie. Richtlinien für die Planung, Ausführung und Pfl ege von Dachbegrünungen* [Green roof policy. Guidelines for the planning, execution, and maintenance of green roofs]. Bonn: Forschungsanstalt Landschaftsentwicklung Landschaftsbau.

Fung, K. L. (2018). *Expanding the green network on rooftops: A study of integrating green roofs as a part of urban green innovation space planning* (Order No. 10932095). Available from ProQuest One Academic. (2124999351). Retrieved from http://wdg.biblio.udg.mx:2048/login?url=https://www.proquest.com/dissertations-theses/expanding-green-network-on-rooftops-study/docview/2124999351/se-2?accountid=28915 https://www.proquest.com/docview/2124999351?pqorigsite=gscholar&fromopenview=true

Galdeano-Gómez, E., Céspedes-Lorente, J., & Martínez-del-Río, J. (2008). Environmental performance and spillover effects on productivity: Evidence from horticultural firms. *Journal of Environmental Management, 88*(4), 1552–1561. doi:10.1016/j.jenvman.2007.07.028 PMID:17825476

Geraie, M. S., & Rad, F. M. (2015). Mediator role of the organizational identity green in relationship between total quality management and perceived innovation with sustainable competitive advantage. *International Journal of Biology, Pharmacy and Allied Sciences, 2015*(4), 266–276.

Goulden, M., Mason, M. A., & Frasch, K. (2011). Keeping women in the science pipeline. *The Annals of the American Academy of Political and Social Science*, *638*(1), 141–162. doi:10.1177/0002716211416925

Greene, J. C., Caracelli, V. J., & Graham, W. F. (1989). Toward a conceptual framework for mixed-method evaluation designs. *Educational Evaluation and Policy Analysis*, *11*(3), 255–274. doi:10.3102/01623737011003255

Gunn, A., & Mintrom, M. (2016). Higher Education Policy Change in Europe: Academic Research Funding and the Impact Agenda. *European Education*, *48*(4), 241–257. doi:10.1080/10564934.2016.1237703

Gupta, J., & Vegelin, C. (2016). Sustainable development goals and inclusive development. *International Environmental Agreement: Politics, Law and Economics*, *16*(3), 433–448. doi:10.100710784-016-9323-z

Hadavi, S. (2017). Direct and indirect effects of the physical aspects of the environment on mental well-being. *Environment and Behavior*, *2017*(49), 1071–1104. doi:10.1177/0013916516679876

Hodge, T. (1997). Toward a conceptual framework for assessing progress toward sustainability. *Social Indicators Research*, *1997*(40), 5–98. doi:10.1023/A:1006847209030

Ihnatenko, M., & Novak, N. (2018). Development of regional programs for the development of agrarian enterprises with organic production based on the European and international experience. *Baltic Journal of Economic Studies*, *4*(4), 126–133. doi:10.30525/2256-0742/2018-4-4-126-133

ILO. (2015). *News*. https://www.ilo.org/global/topics/green-jobs/news/WCMS_422575/lang--en/index.htm

Jabbour, C. J. C., Santos, F. C. A., Fonseca, S. A., & Nagano, M. S. (2013). Green teams: Understanding their roles in the environmental management of companies located in Brazil. *Journal of Cleaner Production*, *2013*(46), 58–66. doi:10.1016/j.jclepro.2012.09.018

Jain, S. K., & Kaur, G. (2004). Green marketing: an Indian perspective. *Decision*, *31*(2), 161–209.

Johnson, S. K. (2017). What 11 CEOs have learned about championing diversity. *Harvard Business Review.*

Johnson, S. K. (2019). *Leaking Talent How People of Color are Pushed Out of Environmental Organizations How People of Color are Pushed Out of Environmental Organizations.* www.diversegreen.org https://diversegreen.org/research/leaking-talent/

Kalandides, A., & Grésillon, B. (2021). The Ambiguities of "Sustainable" Berlin. *Sustainability*, *2021*(13), 1666. doi:10.3390u13041666

Kam-Sing Wong, S. (2012). The influence of green product competitiveness on the success of green product innovation: Empirical evidence from the Chinese electrical and electronics industry. *European Journal of Innovation Management*, *2012*(15), 468–490. doi:10.1108/14601061211272385

Kolomiets, T.V., & Tomashuk, I.V. (2021). Entrepreneurship and development of rural areas in Ukraine. *Colloquium-Journal*, *9*(96), 29-42.

Kucher, A. (2019). *Sustainable soil management in the formation of competitiveness of agricultural enterprises.* Academic Publishing House «Talent». . doi:10.13140/RG.2.2.19554.07366

Kucher, L., Heldak, M., & Orlenko, A. (2018). Project management in organic agricultural production. *Agricultural and Resource Economics*, *4*(3), 104–128. doi:10.22004/ag.econ.281753

Kumar, V., & Christodoulopoulou, A. (2014). Sustainability and branding: An integrated perspective. *Industrial Marketing Management*, *43*(1), 6–15. doi:10.1016/j.indmarman.2013.06.008

Larwood, L., Falbe, C. M., Kriger, M. P., & Miesing, P. (1995). Structure and meaning of organizational vision. *Academy of Management Journal*, *1995*(38), 740–769.

Liao, S. H., Chang, J. C., Cheng, S. C., & Kuo, C. M. (2004). Employee relationship and knowledge sharing: A case study of a Taiwanese finance and securities firm. *Knowledge Management Research and Practice*, *2004*(2), 24–34. doi:10.1057/palgrave.kmrp.8500016

Lin, Y. H., & Chen, Y. S. (2017). Determinants of green competitive advantage: The roles of green knowledge sharing, green dynamic capabilities, and green service innovation. *Quality & Quantity*, *51*(4), 1663–1685. doi:10.100711135-016-0358-6

Liu, D., Jiang, K., Shalley, C. E., Keem, S., & Zhou, J. (2016). Motivational mechanisms of employee creativity: A meta-analytic examination and theoretical extension of the creativity literature. *Organizational Behavior and Human Decision Processes*, *2016*(137), 236–263. doi:10.1016/j.obhdp.2016.08.001

Liu, J., Chen, J., & Tao, Y. (2015). Innovation Performance in New Product Development Teams in China's Technology Ventures: The Role of Behavioral Integration Dimensions and Collective Efficacy. *Journal of Product Innovation Management*, *2015*(32), 29–44. doi:10.1111/jpim.12177

Llonch, P., Haskell, M. J., Dewhurst, R. J., & Turner, S. P. (2017). Current available strategies to mitigate greenhouse gas emissions in livestock systems: An animal welfare perspective. *Animal*, *2017*(11), 274–284. doi:10.1017/S1751731116001440 PMID:27406001

López-Gamero, M. D., Claver-Cortés, E., & Molina-Azorín, J. F. (2008). Complementary resources and capabilities for an ethical and environmental management: A qual/quan study. *Journal of Business Ethics*, *82*(3), 701–732. doi:10.100710551-007-9587-x

Maas, C. J., & Hox, J. J. (2005). Sufficient sample sizes for multilevel modeling. *Methodology*, *2005*(1), 86–92. doi:10.1027/1614-2241.1.3.86

Marchi, V. D. (2012). Environmental innovation and R&D cooperation: Empirical evidence from Spanish manufacturing firms. *Research Policy*, *41*(3), 614–623. doi:10.1016/j.respol.2011.10.002

Maruani, T., & Amit-Cohen, I. (2007). Open Space planning models: A review of approaches and methods. *Landscape and Urban Planning*, *81*(1-2), 1–13. doi:10.1016/j.landurbplan.2007.01.003

Mazur, K. V., & Tomashuk, I. V. (2019). Governance, and regulation as an indispensable condition for developing the potential of rural areas. *Baltic Journal of Economic Studies*, *5*(5), 67–78. doi:10.30525/2256-0742/2019-5-5-67-78

Meerow, S., & Newell, J. P. (2017). Detroit, Spatial planning for multifunctional green infrastructure: Growing resilience. *Landscape and Urban Planning*, 62-75.

Mell, I. C. (2010). *Green infrastructure: concepts, perceptions, and its use in spatial planning* [Doctoral Thesis]. School of Architecture, Planning and Landscape Newcastle University.

Mentens, J., Raes, D., & Hermy, M. (2006). Green roofs as a tool for solving the rainwater runoff problem in the urbanized 21st century? *Landscape and Urban Planning*, *77*(3), 217–226. doi:10.1016/j.landurbplan.2005.02.010

Mesimaki, M., Hauru, K., Kotze, D. J., & Lehvavirta, S. (2017). Neo-spaces for urban livibility? Urbanities' versatile mental image of green roofs in the Helsinki metropolitan area, Finland. *Land Use Policy*, *61*, 587–600. doi:10.1016/j.landusepol.2016.11.021

Mohd, H. A., & Norhidayah, S. (2016), Sustainable food production: insights of Malaysian halal small and medium sized enterprises. *International Journal of Production Economics, 181*(B), 303–314. . doi:10.1016/j.ijpe.2016.06.003

Murray, N. (2017). Urban disaster risk governance. A systemic review. PPI-Centre. UCL Institute of Education.

Namkung, Y., & Jang, S. (2013). Effects of restaurant green practices on brand equity formation: Do green practices really matter? *International Journal of Hospitality Management*, *33*(2), 85–95. doi:10.1016/j.ijhm.2012.06.006

Norton, T. A., Parker, S. L., Zacher, H., & Ashkanasy, N. M. (2015). Employee green behavior: A theoretical framework, multilevel review, and future research agenda. *Organization & Environment*, *2015*(28), 103–125. doi:10.1177/1086026615575773

Ones, D. S., & Dilchert, S. (2012a). Employee green behaviors. In D. S. S. E. Jackson (Ed.), *Managing human resource for environmental sustainability* (pp. 85–116). Jossey-Bass.

Oswald, S. L., Mossholder, K. W., & Harris, S. G. (1994). Vision salience and strategic involvement: Implications for psychological attachment to organization and job. *Strategic Management Journal*, *1994*(15), 477–489. doi:10.1002mj.4250150605

Pell, A. N. (1996). Fixing the leaky pipeline: Women scientists in academia. *Journal of Animal Science*, *74*(11), 2843–2848. doi:10.2527/1996.74112843x PMID:8923199

Pereira, Á., & Vence, X. (2012). Key business factors for eco-innovation: An overview of recent firm-level empirical studies. *Cuadernos de Gestión*, *12*, 73–103. doi:10.5295/cdg.110308ap

Plattform Produktives Stadtgrün. (2020). Available online: https://www.berlin.de/gemeinschaftsgaertnern/

Raudeliuniene, J., Davidaviˇciene, V., & Jakubaviˇcius, A. (2018). Knowledge management process model. *Entrep. Sustain. Issues*, *5*, 542–554.

Redman, C. (2014). Should sustainability and resilience be combined or remain distinct pursuits? *Ecology and Society*, *19*(2), 190–202. doi:10.5751/ES-06390-190237

Reisinger, A., & Clark, H. (2018). How much do direct livestock emissions contribute to global warming? *Global Change Biology*, *2018*(24), 1749–1761. doi:10.1111/gcb.13975 PMID:29105912

Rowe, D. B. (2011). Green Roofs as a means of pollution abatement. *Environmental Pollution, 159*(8-9), 2100–2110. doi:10.1016/j.envpol.2010.10.029 PMID:21074914

Russo, A., & Cirella, G. (2018). Modern compact cities: How much greenery do we need? *International Journal of Environmental Research and Public Health, 15*(10), 2180. doi:10.3390/ijerph15102180 PMID:30301177

Savickiene, J., & Miceikiene, A. (2018). *Sustainable economic development assessment model for family farms* (Vol. 64). Agricultural Economics – Czech. doi:10.17221/310/2017-AGRICECON

Shafique, M., Kim, R., & Rafiq, M. (2018). Green roof benefits, opportunities, and challenges– A review. *Renewable and Sustainable Energy Reviews, 90*, 757-773.

Shevchenko, A., & Petrenko, O. (2020). Current state of micro and small agribusiness in Ukraine. *Agricultural and Resource Economics, 6*(1), 146–160. doi:10.51599/are.2020.06.01.10

Shoeb, A., & Nisar, T. (2015). Green Human Resource Management: Policies and practices. *Cogent Business & Management, 2*(1). https://www.tandfonline.com/doi/full/10.1080/23311975.2015.1030817 doi:10.1080/23311975.2015.1030817

Singh, S. K., Del Giudice, M., Chierici, R., & Graziano, D. (2020). Green innovation and environmental performance: The role of green transformational leadership and green human resource management. *Technological Forecasting and Social Change, 150*, 119762. doi:10.1016/j.techfore.2019.119762

Skydan, O., Nykolyuk, O., Pyvovar, P., & Martynchuk, I. (2020). Methodological approach to the evaluation of agricultural business system flexibility. *Management Theory and Studies for Rural Business and Infrastructure Development, 41*(4), 444–462. doi:10.15544/mts.2019.36

Stovin, V., Vesuviano, G., & Kasmin, H. (2012). The hydrological performance of a green roof test bed under UK climatic conditions. *Journal of Hydrology (Amsterdam), 414*, 148–161. doi:10.1016/j.jhydrol.2011.10.022

Stubbs, W., & Cocklin, C. (2008). Conceptualizing a sustainability business model. *Org.Env., 21*(2), 103-127.

Sutton, R. (2014). Aesthetics for green roofs and green walls. *The Journal of Living Architecture, 1*(2), 1–20. doi:10.46534/jliv.2014.01.02.001

TEP. (2005). *Advancing the delivery of green infrastructure Targeting Issues in England's Northwest.* Issue Paper, England Northwest.

Tomashuk, I.V., & Baldynyuk, V.M. (2021). Identification of problems and prospects of rural infrastructure development of Ukraine. *Economic Sciences*, 13. doi:10.24412/2520-6990-2021-13100-58-70

Tsai, W., & Ghoshal, S. (1998). Social capital and value creation: The role of intrafirm networks. *Academy of Management Journal, 1998*(41), 464–476.

Ugai, T. (2016). Evaluation of sustainable roof from various aspects and benefits of agriculture roofing in urban core. *Procedia: Social and Behavioral Sciences, 216*, 850–860. doi:10.1016/j.sbspro.2015.12.082

UNDP. (2020). *Environmentally Sustainable Operations.* https://www.undp.org/accountability/social-and-environmental-responsibility/sustainable-operations

Van Holt, T., Statler, M., Atz, U., Whelan, T., van Loggerenberg, M., & Cebulla, J. (2020). The cultural consensus of sustainability-driven innovation: Strategies for success. *Business Strategy and the Environment, 29*(8), 3399–3409. doi:10.1002/bse.2584

Vijayaraghavan, K. (2016). Green roofs: A critical review on the role of components, benefits, limitations, and trends. *Renewable & Sustainable Energy Reviews, 57*, 740–752. doi:10.1016/j.rser.2015.12.119

VisitBerlin. (Ed.). (2017). *12 mal Berliner Leben, 12 mal Berlin Erleben. Konzept für einen stadtverträglichen Berlin-Tourismus 2018+.* VisitBerlin. Available online: https://about.visitberlin.de/tourismuskonzept-2018

Wang, Y., Hu, H., Dai, W., & Burns, K. (2021). Evaluation of industrial green development and industrial green competitiveness: Evidence from Chinese urban agglomerations. *Ecological Indicators, 124*, 107371. doi:10.1016/j.ecolind.2021.107371

Wiering, M., Liefferink, D., Boezeman, D., Kaufmann, M., Crabbé, A., & Kurstjens, N. (2020). The Wicked Problem the Water Framework Directive Cannot Solve. The Governance Approach in Dealing with Pollution of Nutrients in Surface Water in the Netherlands, Flanders, Lower Saxony, Denmark, and Ireland. *Water (Basel), 12*(5), 1240. doi:10.3390/w12051240

Wirtenberg, J., Harmon, K. D., Russell, W. G., & Fairfield, K. D. (2007). HR's role in building a sustainable enterprise. *Human Resource Planning, 30*, 10–20.

Wysocki, J. (2021). Innovative green initiatives in the manufacturing SME sector in Poland. *Sustainability, 13*(4), 2386. doi:10.3390u13042386

Zhang, Z., & Sundaresan, S. (2010). Knowledge markets in firms: Knowledge sharing with trust and signalling. *Knowledge Management Research and Practice, 2010*(8), 322–339. doi:10.1057/kmrp.2010.22

Chapter 7
Landscape and Water Networks:
Impact on Health for the Smart City – Case Study: El Pardo and the Manzanares River Basin

Guadalupe Cantarero-García
https://orcid.org/0000-0003-3169-8547
Universidad San Pablo CEU, Spain

ABSTRACT

Historically, water has played an essential role in choosing the location and settlement of a habitat and, consequently, in the configuration of the landscape, building, and urban context. The quality of the inhabitants' hygiene and health depends on water and sewage treatment. Water supply and sanitation are crucial to achieving end-user quality and enjoyment of the home and the city. This case study focuses on the city of Madrid, intrinsically related to water from its origin and name. Therefore, the "mayra" abounds in Madrid and is the "mother of water." The latter refers to Madrid's location surrounding a large fountain that produced a stream that flowed into the Manzanares River. It recovers photographs of the landscape found in unpublished historical and military archives with the intention of showing certain excavations that affected the El Pardo Woodlands, such as Janini's artesian wells and other missing projects in the Manzanares River Basin.

INTRODUCTION

The water challenge that the GWP (Global Water Partnership) proposes is that managing the world's water resources is foundational to development. If we intend to contribute to poverty reduction, human health, and economic prosperity, we need to pay attention to water. The key principal requirements to achieve hydric health and safety in countries are: keeping watersheds and aquifers in balance, keeping rivers clean, achieving total coverage of drinking water, drainage and sanitation, and making population settlements safe from floods caused by extreme precipitation and hurricanes. Fueled by possible effects of climate change as well as future alterations in parameters such as precipitation, temperature, and hu-

DOI: 10.4018/978-1-6684-4030-8.ch007

midity will modify the balance of our ecosystems and perhaps expand the range of pest vectors. It could also impact water dependent species that make up the biodiversity and cities may be affected as well.

This case study focuses on the city of Madrid, intrinsically related to water from its origin and name. Madrid was originally a Visigoth settlement called "Matrice," a Latin word meaning "womb." Later, the Muslim Madrid received the name "Mayra," which means "mother of water" or "water vent."

Finally, the word acquired the Mozarabic suffix "it," which resembles the Latin "etum" indicating abundance. Therefore, the "mayra" abounds in Madrid and is the "mother of water." The latter refers to Madrid's location surrounding a large fountain that produced a stream that flowed into the Manzanares River. The Madrid so-called *viajes de agua* or "water trips" are water channels that were used to provide water to the city in the 17th century. Now, Madrid's water supply is managed by the Canal de Isabel II, a public entity created in 1851. This canal is responsible for the supply, depurating wastewater, and the conservation of all the natural water resources of the Madrid region.

Despite the critical role of the Canal de Isabel II, the issue lay in how to improve the sanitation of the Manzanares River. Adequate waste channeling was imminent due to the stench of stagnant water in the river. The first action of the Green Railway Corridor introduced urban renewal processes, resulting in architectural renovation at formal and social levels, which had a positive impact on the area's image. The subsequent large-scale intervention in Madrid was spatially parallel to the previous Green Railway Corridor action and provided the renovation that the south of Madrid required.

BACKGROUND

Magerit means "land rich in water." This is how the Arabs called this area on the central plain of the Iberian Peninsula, close to Sierra de Guadarrama, where King Phillip II of Spain later established the royal court. Later on, it grew into the large city we know today.

The first historical record of Madrid dates back to the year 865 when Emir Muhammad I commissioned the construction of a fortress in the village of Mayrit on the banks of the Manzanares River. *Mayrit* means "plenty of waterways." For eleven centuries, the original Arabic "qanat" underwent successive extensions to increase its capacity as the population grew (Figure 1).

Figure 1. Hamman Al Ándalus. Atocha Street in Madrid and Hamman Al Ándalus location at Texeira Plan location. Source: Archivo Histórico de la Villa. Madrid

Various studies show the subsoil of Madrid as "a great basin of 3,700 square kilometers of groundwater." In turn, hydrogeologists explain that the alternation of absorbent hills and permeable troughs allow for upwelling in wetlands and ponds. The *viajes del agua* or "water trips" in historic Madrid follow the "qanat" and were built between the 8th and 9th centuries during the Arab domination of the citadel and the primitive *Mayrit* (or *Majerit*) (Figure 2).

Figure 2. Manzanares River basin. Source: urban-e.aq.upm.es
"Viajes del agua" or "water trips." Source: madrid.es

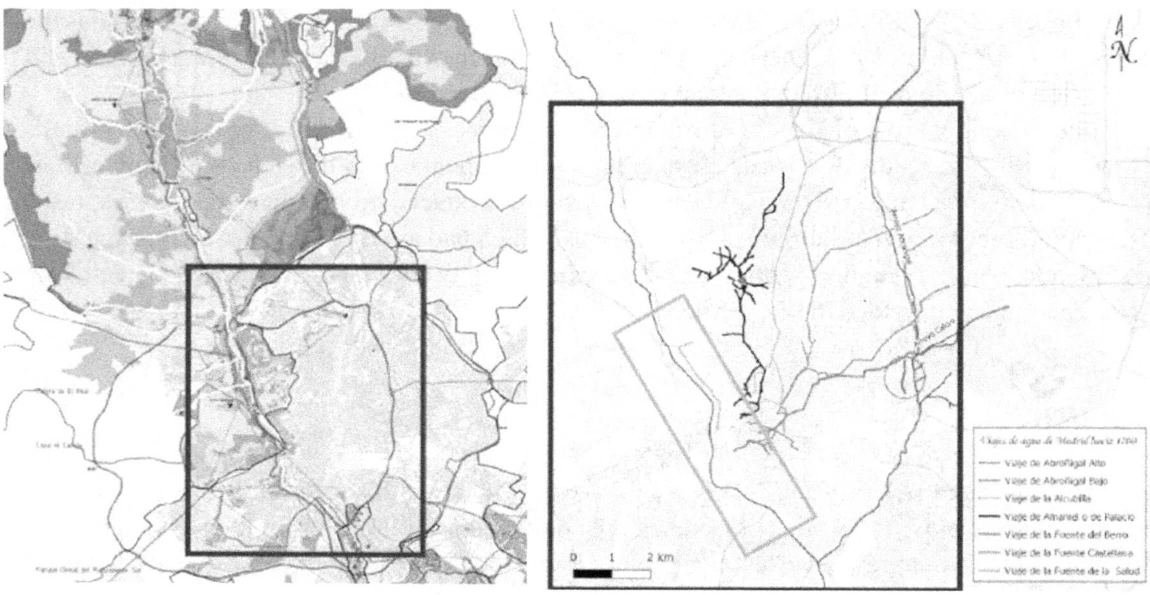

The Manzanares River route comes from one of the purest natural places in the capital city and it is important to analyze its course. The El Pardo Woodlands is known to be the largest natural setting in the Community of Madrid and it is the oldest Royal Site in the Spanish capital. These hunting grounds were declared a Special Protection Area for Birds in 1987. The National Heritage has granted land usage for various sporting and socio-recreational activities as well as research centers that are considered of special social interest. One important intervention developed at El Pardo Woodlands is the artesian wells constructed by engineer Janini between 1910 and 1920. The other one is the construction of swimming pools on the Manzanares River and other things that affected the river during these centuries. The meticulous studies and field work in agricultural experiments by the engineer, Janini Janini, led to the first hydrodynamic testing canal called CEHIPAR in Spain, which is in the El Pardo Royal Site. Other engineers also carried out agricultural investigations such as Maluquer and Mesa (Maluquer and Salvador and Mesa and Ramos 1909), who worked on new techniques for artesian wells in the United States. Janini's work inspired the implementation of a modern water supply network and urban sanitation in El Pardo that later favored a new installation by Méndez in the 1940s.

The history of interventions on the Manzanares Canal is a collection of incomplete plans shaped around political premises since the beginning of the 18th century. Initially, the idea was to connect the river with Toledo but Carlos Lemaur planned to connect it with Seville. However, this proposal was not viable due to its high cost. Furthermore, there was not much knowledge of the project that Jacques Hardoin-Mansart de Sagone presented in 1768 because he lost his patents two years later. Charles III entrusted the project to the businessman Pedro Martinengo and a less ambitious project followed.

The goal of connecting Madrid with the sea led to the construction of dams, wharfs, locks, bridges, and mills, but it was not until 1826 that Miguel de Inza succeeded in implementing navigation through the river. For a historical comparison of the Madrid skyline from the southwest, we have the view that artist Antonio Joli painted in 1753, showing the Manzanares River from the Bridge of Segovia[1].

The Royal Palace stands out on the left, as well as the Ermita de la Virgen del Puerto (built in 1716) at the foot of the river and what would eventually be called La Isla on the Manzanares River. Several domes and towers of the many churches and convents of the capital can be seen in the distance (Figure 3). Francisco de Goya's view is taken from another angle further south than Joli's, from the Toledo Bridge[2] below the San Isidro Hermitage (Figure 4).

Figure 3. Picture of the Bridge of Segovia (called "puente segoviana" here) and Manzanares River oil on canvas by Antonio Joli. Source: Royal Academy of Fine Arts of San Fernando in Madrid (1753) and picture of washing at the Bridge of Segovia on the Manzanares River. Source: Otto Wunderlich's Archive (1917)

Figure 4. San Isidro Prairie oil on canvas by Francisco de Goya. Source: Museo del Prado de Madrid (1788) and Pradera de San Isidro - San Isidro Prairie. Source: Otto Wunderlich's Archive (1917)

There are several researchers who have lent their studies to the Madrid landscape, highlighting images and plans of the current context of Madrid based on the historical context (Rodríguez Romero, 2018) and so, there are new ways of researching using graphic language that favors the understanding of the context of Madrid for its improvement. "It becomes necessary to examine the concept of peri-urban landscape in relation to current challenges of sustainable development, as well as the benefits of urban, green infrastructure in the contour of the city...After identifying the green infrastructure that act as thresholds in the city of Madrid, we focus on the south-east diagonal of the capital in order to reaffirm its importance in the construction of the image and identity of the city. We defend the importance of urban, green infrastructure to and from the city, suggesting the necessity of a supra-municipal planning tool to change the peri-urban landscape, usually perceived as subsidiary, to deem the proximity vision of the city as relevant for its design" (Santo Tomás, 2020).

In turn, Muñoz de Pablo, researcher and expert in drawing and documentation of architecture and cities, specifically in El Ensanche de Madrid called, The Madrid´s Expansion, wrote, "The Castellana stream is the only one that, in addition to collecting rainwater from the slopes that flow into it, has channeled water from the Castellana natural spring located in the northern lands about 1,000 m from the fence. It ran through the land in a north-south direction as far as the Abroñigal stream, to finally join the Manzanares.

The Leganitos stream ran from the street named after it in a southwesterly direction towards the San Vicente slope and entered into the Palace Park. The Arenal and Minillas streams flowed into it, running through the north-south watercourse located to the west of Príncipe Pío Mountain. Segovia Street, located in the hollow of the valley of San Pedro, runs along the bed of the old stream that flowed into the Manzanares River at the Bridge of Segovia" (Muñoz de Pablo 2006).

At the end of the 19th century, and due to the fact that the Manzanares River was used insalubriously, some regulations preventing the digging of wells at less 30 meters from the river and making the cleaning of the facilities in the early hours of the morning became necessary. The regulations also established that lifeguards be present in all bathing areas[3] and for children under ten years of age to be accompanied when bathing[4]. "Dyers, brass workers, and leather workers" were prevented from washing their work tools in the bathing areas as well (Municipal Ordinances Project 1886).

Therefore, and as described below, the purpose of this study is to put in order all these historical interventions that precede what happened in the 20th century and the hygienic improvements that were implemented in the 21st century and have favored the environmental quality of the banks of the Manzanares River in the capital of Spain.

OBJECTIVE AND WORK METHOD

The purpose of this case study is to compile the most relevant historical hydraulic actions that have been carried out in Madrid from the 19th to the 21st century. It shows the water purification and collection systems of the Spanish capital that were designed to overcome water shortages during the reign of Isabel II. These systems provided engineering references for other cities and also showed how subsequent engineering actions modified the terrain, the landscape, and the course of the Manzanares River with the sole purpose of making Madrid a safe, healthy, and cutting-edge city in terms of water supply and purification.

The topographic actions of the artesian wells of engineer Janini that began in the Royal Site of El Pardo at the beginning of the 20th century and together with the CEHIPAR have also been a reference. Other references are the pools of the Manzanares River of the mid-20th century, the modifications to the railway line, and the new building implementation that began in the late 20th century with the Green Railway Corridor project. The most important topographical actions took place at the beginning of the 21st century with the creation of storm tanks and the Madrid-River to bury the traffic lanes. Hence, the case study in this research analyzed the similarities between these interventions that are separated by a century using graphic and photographic restitution. The aim is to understand how the banks of the Manzanares River have changed and how they have adapted to technologies and new requirements in the transformation from a historical city to a Smart City.

The analysis is focused on the interventions that have made the banks of the Manzanares River what it is today. Without all these actions, which seemed to be the result of specific needs within specific spaces and time periods, the Manzanares River would not have provided Madrid with the current urban quality. The mere fact of establishing the chronology of all these interventions implies a study of the question that requires researching different archive resources since different fields of study are involved (topographical, geological, aeronautical, hydraulic, landscape, and building).

Also, this study recovers photographs of the landscape found in unpublished historical and military archives with the intention of showing certain excavations that affected the El Pardo Woodlands, such as Janini's artesian wells.

In the development of this study, the proposed objective is to examine what occurs in the subsoil of the protected area of the El Pardo Woodlands and how that has transformed the gardening, landscape, and urban development into what it is today. This case study also brings to light interventions that facilitated the river's channeling, use, and enjoyment in the 20th and 21st centuries. These interventions have modified the natural terrain and, for different reasons, have impacted the area from the subsoil to the surface. These engineering works conditioned subsequent interventions in the design of water networks, sanitation, and other infrastructure, which had an impact on the urban layout of El Pardo and the capital city of Madrid in the 20th century.

The method proposes to start by first understanding the materials used and the unpublished information found in archives and, secondly, to give an overview of what exactly was done to obtain the data as well as requesting information outside the chronological framework of the goals of study in order to find starting or conclusive points within documents that were previously unpublished, non-digitized, or unsolicited. In this study, photographs have been recovered and oral sources who participated in the analysis of certain photographs were used. Part of the published documentation reviewed that was used as bibliographic references has been relocated due to changes in the management of some archives. These documents include the Map Library of the General Military Archive of Madrid and the Central Military Library. Also, there have been updates to other documents in places such as the National Heritage Palace General Archive. Therefore, updating some data has been part of the objectives of the study, thus speeding up the search for future research. Diagrams are used to explain the research methodology of sources detailed in this chapter.

To establish the chronology, it was essential to contrast photographic documentation with historic plans and analyze the layouts. Also, superimposing the urban layouts helps to understand building and residential development.

CASE STUDY: EL PARDO AND THE MANZANARES RIVER BASIN IN MADRID

With the purpose of carrying out a study on the impact of water networks on the landscape, it is necessary to first review the chronology of the actions involved. It is important to take into account the current concept of a Smart City, which is included in the new Spanish regulations drafted by AENOR[5].

The origins of these interventions must be analyzed before a more in-depth analysis of the impact of the water networks on the landscape. The origins date back to the mid-18th century with the first designs of the Manzanares River[6], followed by the construction of the Isabel II Canal in the mid-19th century up to the Madrid-Río Project, which was inaugurated in 2011. The second step involves overlaying the routes of actions that refer to the banks of the Manzanares River from Monte de El Pardo to the Madrid-Río Park on top of each other as well as identify the storm tanks and the water paths that affect the riverbank called the Journey of the Health Fountain.

Projects Carried Out on the Manzanares River in the 19th Century

1. Canal de Isabel II

Canal de Isabel II in English,[7] was created when Madrid was planning a future Ensanche de Madrid. Since its inception, the canal has faced the challenge of providing water to a growing population. The origins date back to the construction of the Ponton de la Oliva Reservoir. The dam's stone foundation was laid on August 11, 1851. The waters destined to supply Madrid were to come from this dam[8]. Following years of work and many difficulties, on June 24, 1858 the inauguration of the arrival of water from the Lozoya River to Madrid took place.

Figure 5. Inauguration of the Canal de Isabel II on June 24, (1858) and the Tanque de Luis Moya
Source: canaldeisabelsegunda.es. First water tank. Source: alumnoshistoriasdevida.blogspot.com

The water flowed from the Pontón de la Oliva to what is today known as the First Reservoir of the Campo de Guardias[9]. The first water tank was built between 1907 and 1911 by engineers Luis Moya and Ydígoras González (Capitel A., García-Gutiérrez Mosteiro J. 2000) and Ramón Aguinaga. Today, it is no longer in use and is now home to the Canal de Isabel II Art Room (Figure 5). The tower was made of stone, brick, and iron and was thought to be a polyhedral shape with twelve sides. It was a complex feat of engineering that could hold more than 58,000 cubic meters of water[10]. It was constructed to commemorate the arrival of water from the Lozoya River to Madrid.

By November 1850, Charles Clifford[11] was running a photographic portrait gallery in Madrid and constructed aqueducts that were almost eighty kilometers long. It was crowned one summer afternoon in 1858 with the spectacular arrival of the waters from the Lozoya valley to the source that then presided over the Glorieta de San Bernardo. From that moment on, the modern intakes of the canal were replacing the trips. The Amaniel water trip (Figure 6) was a water trip from the city of Madrid that was built between 1614-1616. It was on royal property and was created during the reign of Felipe III to supply the *Alcázar*[12], as well as three other public fountains: the Matalobos, the Priest, and the fountain in the square of the armory.

The Royal Heritage was passed on to the Madrid Water Board in 1954. In 2018, restoration work began to allow educational visits. Since 2019, it has been possible to visit the interior of the galleries on Calle Juan XXIII.

"Despite everything, Amaniel's water trip still remained operational for almost a century. A multitude of administrative files date back to the late 19th and early 20th centuries that reveal that the growth of Madrid to the north, now known as the neighborhoods of Fuencarral, Tetuán, and Chamartín, was gradually contributing to its dismantling. Among them we observed the withdrawal of many neighbourhoods that hindered the passage in the new streets, the excavation of wells near the trip to supply recently built properties, or the progressive abandonment of small branches" (Muñoz de Pablo 2006).

Figure 6. Construction of the Amaniel Canal or Viaduct. Source: Charles Clifford's Archive (1856)
Amaniel Canal or Viaduct today along the Pablo Iglesias de Madrid Street. Source: author's picture (2013)

2. Artesian Well Projects by Engineer Rafael Janini Janini and the Impact on the Landscape as a Result of Topographic Interventions

In 1904, Alfonso XIII encouraged agriculture and livestock usage within the Royal Site of El Pardo. As a result, the Royal Palace cultivated large plots in the woodlands with around five hundred hectares of non-irrigated land. A wire fence was built around it to protect the crops from both large and small game. Numerous wells were dug on several plots that reached about one hundred and ten meters in depth. These wells, near Queen Victoria's Bridge, yielded about eight hundred liters of water per minute and could shoot out of the ground more than twenty meters high. The El Pardo Woodlands (Figure 7), a protected natural area, was first a hunting ground used by the Spanish monarchy. Later, the palace was located within the reserve and the area was deemed a Royal Site. These woodlands were declared a Special Protection Area for Birds in 1987 (ZEPA)[13].

Figure 7. "Royal Road Project"
Source: Sebastián de Rodolphe, in Archive: (A.G.P., 1741)

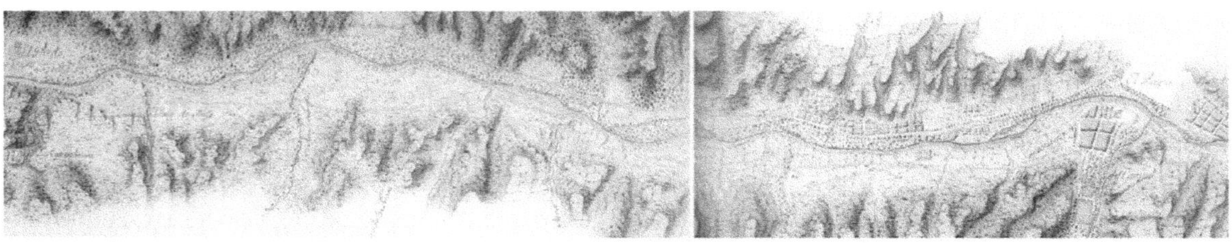

Historical underground interventions in the area, including the artesian wells (Figure 8) at the El Pardo Woodlands, are now undetectable. Although these elements have not yet been considered in previous scientific investigations, artesian wells are important factors in landscaping that has been conditioned by the soil and its treatment throughout history. These interventions have modified the natural terrain and, for different reasons, have impacted the area from the subsoil to the surface. These engineering works conditioned subsequent interventions in the design of water networks, sanitation, and other infrastructures, which had an impact on the urban layout of El Pardo and the capital city of Madrid in the 20th century.

Figure 8. Manzanares River Source: Otto Wunderlich's Archive (1917)
Janini at artesian well number 4 in the Telegraph Military Regiment plot in El Pardo. Source: in Archive (BCM, 1913)

"In terms of recreation, landscaping changes have been made with the creation of hiking and biking trails. Hence, surveyors and engineers write articles, essays, and books, which are sometimes accompanied by cartography based on planimetry as well as cuts or profiles. Intrinsic values of the El Pardo Woodlands are recorded in a descriptive, practical, and visual style, almost to lure in visitors and outdoor recreationists, many of whom are foreigners…in order to demonstrate a specific case of interior well-being in relation to the landscape, two different exterior contexts in Madrid have been studied in accordance with the study factors of this research, which are tellurism, sacred geometry, and landscape. The method is described in the section titled 'Work Method.' These two placements are the El Pardo Royal Site and the Carabanchel neighborhood" (Cantarero 2021).

There were also four electrohydraulic pump installations. Two of the most powerful of these pumps could draw out 3,300 and 6,000 L. of water per minute, respectively. With these installations, it was expected that the total extension of irrigated land would reach 187 hectares. Grains, legumes, and vegetables such as wheat, barley, oats, rye, grass peas, beans, chickpeas, carob, potatoes, alfalfa, corn, turnips, beets, and other foods that are currently consumed were harvested in these farmlands. Crops from non-irrigated gardens were worked using modern, agricultural technology of the time such as moldboard plows, rollers, harrows, seeders, cultivators, mowers as well as powerful steam powered threshers. Manure and chemicals were used as fertilizer.

The cultivation plots (in green) are named and numbered from 1 to 26. The total area is estimated to be 411.49 hectares. All of the plots are outside the Royal Site except numbers 25 and 26, which correspond to "Mielga below the Convent next to the Manzanares" (0.16 hectares) and "Pataca below the Convent next to the Manzanares" (0.17 hectares). In this plan we can see crop area number 12, that was called Mingo Rubio, which in the 1950s was affected by the urbanization of the area known as "Colonia de Mingorrubio" (Figure 9).

3. CEHIPAR Waterway Project in El Pardo

The meticulous studies and field work in agricultural experiments made by engineer, Janini Janini, led to the first hydrodynamic testing canal, called CEHIPAR, in Spain, which is in the El Pardo Royal Site. Several hydraulic projects have arisen from the development of hydrodynamic experimentation, as seen in the previous section on artesian wells.

Significant hydraulic works were carried out in the Zarzuela barracks, as well as new pipelines, the installation of a very powerful modern waterwheel, and the creation of a military stud farm, all supported by irrigation from artesian wells. The land between Puerta de Hierro, the road, and the Manzanares was given to the Madrid City Council for nurseries and gardens. The land was thus greatly embellished, bringing water to the mountain, taking care of the trees, and forming meadows for the practice of Drid-Polo, golf, cricket, soccer, tennis, etc. In the National Heritage General Archive [A.G.P], authorization signed on May 16, 1930 by Luis Perla, General Administrator of the Royal Heritage Administration, authorization of the execution on March 20, 1930 by the Ministry of the Navy, and the authorization to the General Intendant of the Royal Household and Heritage can be seen.

This file contains, in turn, an unpublished plan presented here (Figure10) detailing the location of the tank on the land where the canal is located. Initially estimated to be an area of about 3.7 hectares (i.e., 370m x 100m. parallel to the river).

Figure 9. El Pardo by Rafael Janini Janini
Source: in Archive (BCM, 1913)

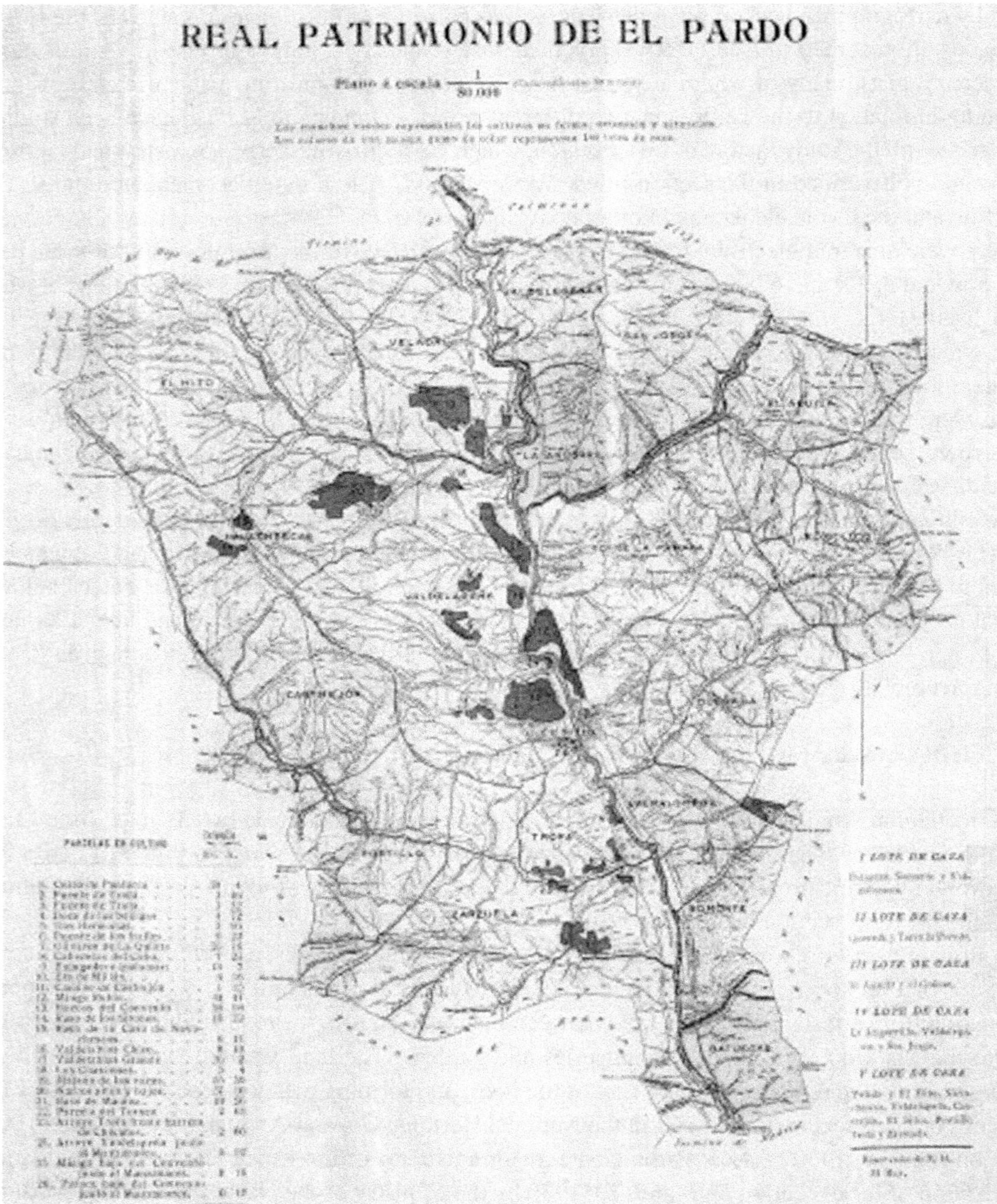

Figure 10. Land of the El Pardo Royal Heritage Site required for the construction of the tank
Source: Previously unpublished plan provided by the author. (A.G.P., 1930)

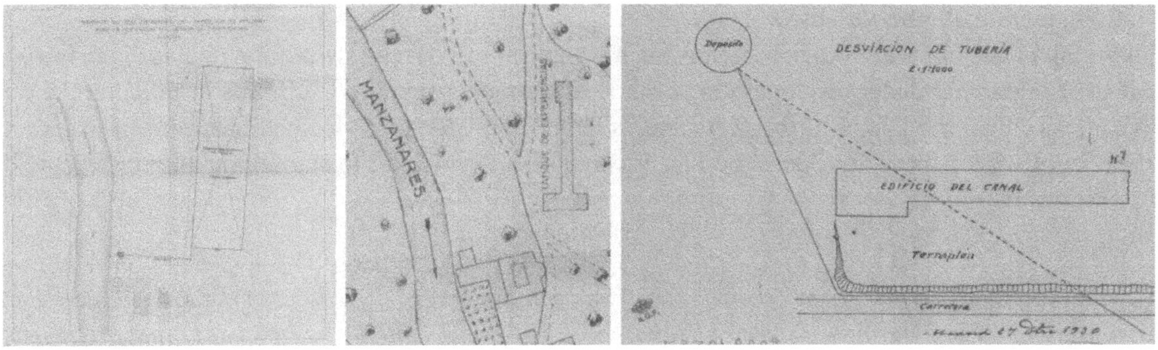

The buried pipeline projected perpendicular to the Manzanares River and is marked in red. It is diverted to a 10 x 10m water reservoir then used as irrigation for the orchards. At the same time, a pump is installed next to the river at the point where the development of the *Jardín del Príncipe* ends. CEHIPAR[14] is currently used for nautical studies such as the Calm Water Canal and other naval experiments (Figure 11).

Figure 11. "Canal de Aguas Tranquilas" inside the Hydrodynamic Experiment Canal, CEHIPAR, in El Pardo
Source: author's picture (2013)

Projects Carried Out on the Manzanares River in the 19th Century

1. Manzanares River Pools

During the year 1835, several ideas for the creation of baths and a swimming school on the Manzanares River were put forward[15]. As mentioned in the background section of this study, at the end of the 19th century, and due to the insalubrious usage that was being made of the Manzanares River, it was necessary to adopt some regulations. These were drafted in the Municipal Ordinances Project of 1886 and, from then on, numerous interventions were carried out on the riverbank.

The first third of the 20th century marked the re-encounter of the city with the Manzanares. In the words of architect Enrique Domínguez Unceta, "Madrid had never before devoted so much attention to its river." It was cleaned up and channeled (the first channeling was carried out between 1914 and 1925) and was the subject of a general plan that was drafted in 1930, which contemplated the creation of bathing areas and other facilities.

A young Luis Gutierrez Soto had already been commissioned for the swimming pool project. In 1944, the ISLA Sports Club, later headquarters of the Real Canoe Swimming Club, was installed at the pool of what was called the Island of the Manzanares River (the Island was located inside the river exactly at the foot of what is now the Cuesta de San Vicente at the Príncipe Pio station, next to the Puente del Rey) (Figure 12).

Figure 12. La Isla en el Plano de Texeira (1656) and photograph of the Island of the Manzanares River where architect, Luis Gutiérrez Soto, built the pools that were inaugurated in 1931, reconstructed in 1947, and demolished in 1954. Aerial military photograph (A.H.M., 1944-1954)

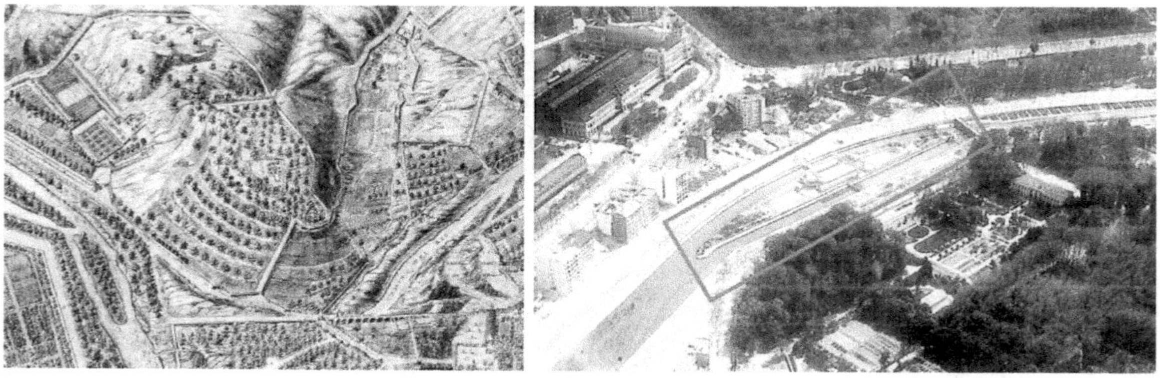

The island's water came from the river but leaked out before reaching the pool facilities. The Manzanares Island area was about six thousand square meters and it was decided to respect the first channeling of the river (between 1914 and 1925). Gutiérrez Soto designed the Madrid swimming pools in a building that emulated a ship, with curved facades and large glass surfaces, portholes, and an almost total absence of decoration in accordance with the precepts of rationalism. Simple and functional shapes prevailed in the materials used such as concrete, steel, glass, and concrete. During the years of war, a shell affected the structure and, following the end of the war, the facilities were rebuilt. However, in 1947 the overflowing of the Manzanares River affected the pools again. Moreover, as it also constituted an obstacle to the new locks system plan, it was decided to close them in 1954[16].

2. Railway Infrastructure and Building Growth Project in the Green Railway Corridor

Despite the critical role of the Canal de Isabel II, the issue lay in how to improve the sanitation of the Manzanares River. Adequate waste channeling was imminent due to the stench of stagnant water in the river. The first action of the Green Railway Corridor project introduced urban renewal processes, resulting in architectural renovation at formal and social levels, which had a positive impact on the area's image. The subsequent large-scale intervention in Madrid was spatially parallel to the previous Green Railway Corridor action and it provided the renovation that the south of Madrid required (Figure 13).

Figure 13. Green Railway Corridor
Source: publicspace.org

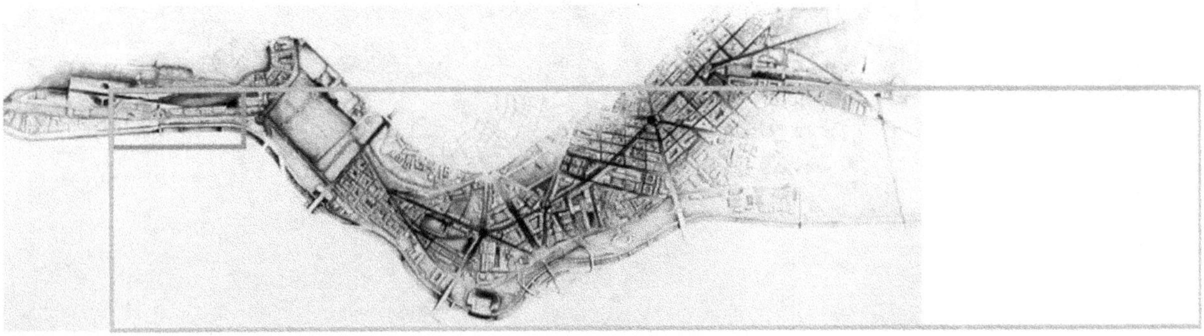

It is apparent that it gained wide social acceptance and its social benefit was analyzed by the Institute of European Studies of the Autonomous University of Barcelona (Riera, 1995). As an integral urban operation, it achieved a unitary design of all the elements that affect the formalization of the space, understanding that they are a single entity made up of urban roads, railway tunnels, subway parking lots, definition of building plots, the design of paving slabs, and the ceramic cladding of the plinths of the vertical walls of the station platforms. In the latter, the material is reworked to reflect the idealized profile of the Madrid cornice, whose image served as the basis for the consortium's corporate image.

Projects Carried Out on the Manzanares River in the 19th Century

1. Storm Tanks Project

The Real Estate boom or bubble at the beginning of the 21st century lasted only the first seven years of the century and with the arrival of the crisis in 2007, the change was appreciated in architecture studies and in engineering as well. However, it is worth noting that in 2007, Madrid promoted major engineering projects. In 2007, the Madrid Environmental Council also launched the Municipal Water Demand Management Plan and the Waste Water Reuse Plan, building the world's largest storm tank to be completed in 2009[17].

This underground tank was necessary to store the first rainwater, which is the most polluting water (even more than sewage) because it carries all the dirt accumulated on the streets and asphalt. The tanks prevent the treatment plants from exceeding their maximum flow and having to discharge the untreated surplus into the waterways.

Due to sediment, vegetation, and fauna, large rivers can naturally take on the self-purification of part of the wastewater that reaches them. However, this is not the case of the Manzanares, therefore, a complete network of storm tanks was necessary. The region of Madrid has 63 installations that are capable of retaining up to 1.46 cubic hectometers of water. It can be said that the two largest storm tanks in the world are located in Madrid[18].

Along with the Arroyofresno (Figure 14), the capital's most important tanks are that of Butarque with the capacity to retain 400,000 cubic meters, Abronigales (206,000 cubic meters), and China (136,429 cubic meters).

Figure 14. Arroyofresno storm tanks
Source: canaldeisabelsegunda.es

In addition to these, there are up to 21 other installations, which are strategically distributed on both sides of the Manzanares River. The M-30 road[19] has smaller tanks to prevent flooding on the beltway. These are the Cantarranas, Cuesta de San Vicente, Ermita del Santo, Delicias Viñuelas, and San Ambrosio tanks. Municipalities such as Parla, Fuenlabrada, Getafe, San Fernando de Henares, Pozuelo de Alarcon, Loeches, and Navalcarnero also have such infrastructures.

Researchers say, "In addition to protecting channels and aquifers from the discharge of water with a pollutant load, storm tanks serve as an instrument for the recovery of drainage flows and for the optimization of the operation of wastewater treatment plants (WWTP)." (García Calvo, Carlos Joaquín and VVAA, 2016). Other researchers reached conclusions that advanced the essential use of hydrographs and polutograms to analyze the contamination level. They said, "The availability of hydrographs and polutograms has made it possible to carry out pollution balances and to estimate the mobilization of pollution generated by flows of runoff water. The instruments used and the sampling techniques developed have proven to be robust and suitable for the characterization of flows from rain in complex, large installations and in aggressive conditions. The information obtained through this methodology will allow optimizing the operation and maintenance strategies of the tanks and their complementary infrastructures to minimize the pressure on the Manzanares River during rainy weather" (2017, Lastra, A and VVAA, 2017).

Before reaching the tanks, the water passes through a series of filters to retain solid contaminants such as plastic bottles or other objects. Many of the solid objects that arrive with the rainwater accumulate at the bottom of the waterway. They are subsequently removed by different cleaning systems. The biological oxygen demand (BOD) parameter, which evaluates the amount of organic matter in the water, indicates that the contamination from the first rainwater is twice as high as that of the wastewater in dry weather.

Other cities adopted this model as a reference and have taken advantage of subway entrances (as in Seville in Alameda, which was created in 1982).

The Infrastructure Plan for the Improvement of Water Quality on the Manzanares River, launched in 2005, also included the construction of 34 new kilometers of collectors and 27 storm tanks to eliminate all types of pollutants before they reach the riverbed. The storm tank in China was opened on February 6, 2007 and is also part of this plan (Figure 15).

Figure 15. Location of the tanks and storm tanks
Source: canaldeisabelsegunda.es

2. Madrid River Basin - Madrid-Río Project

Alongside the Green Railway Corridor intervention described above, the second large-scale intervention in Madrid provided the sanitation that the south of Madrid was requesting. On the right bank of the Manzanares, the neighborhoods of Alto de Extremadura, Carabanchel, and Usera remained disadvantaged from the center of the capital and mistreated by the impact of a river, which suffered the harshness of the M-30 highway.

The plan was very ambitious and entailed a large economic investment for Madrid, which today proves to be very beneficial for the capital in several ways. Thanks to this costly intervention, the connection of the southern beltway was improved by burying a conflictive traffic point that intercepted with the river that had terrible conditions of maintenance, diversion, and hygiene. The winning project was led by architect Ginés Garrido[20] and solved a major pollution issue not only in terms of gas emissions but also in terms of noise and visual pollution. This area of Madrid has been given a natural space that, together with Casa de Campo and Retiro Park, has a recreational and social function (Figure 16).

Figure 16. Madrid -Río Project
Source 1: Architectural platform. cl Source: burgos-garrido.com

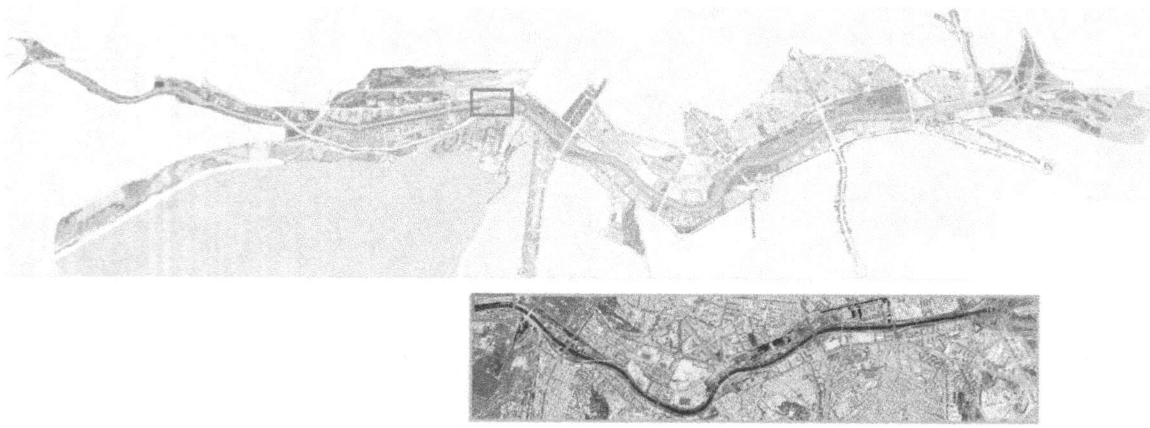

During the archaeological excavation process (between 2005 and 2009), vestiges of the Manzanares Royal Canal were found. They were carefully studied[21] and evidence the model of public works undertaken and developed during the reigns of Charles III, Charles IV, and Ferdinand VII[22]. The history of interventions in the Manzanares Canal has been an accumulation of incomplete plans forged on the basis of political premises since the beginning of the 18th century.

SOLUTIONS AND RECOMMENDATIONS

We continue to project Smart Cities as an ideal that meets the desires of a changing and sometimes chaotic society. Thus, attempting to subject ourselves to an artificial planimetric order that plays against the unforeseen and generates anxiety about what is to come, will always be better, especially when through the eyes of human positivism.

The latest actions to improve the salubrity and hygiene of the Manzanares River are projects for the remote control of hydraulic infrastructures and new ecological trails.

The works created a new centralized system allowing the efficient management of all the dams, as well as video surveillance of the dams and water discharge points into the river, such as spillways from the storm tanks and wastewater treatment plants. The entire system is informatively connected to the new control center[23]. This involves video surveillance of dams and spillway and treatment plant discharge points. Also, automation of dam gates, remote control of dam gates, remote monitoring of gauging of the Manzanares River, remote monitoring of the water table in the M-30 area, remote monitoring of the pluviometric network, and the installation of an operations center[24].

Three new ecological trails were created along the Manzanares River[25] for pedestrians and cyclists to enjoy leisure activities in a natural setting without having to leave the municipality of Madrid._This action provided connection and continuity with the paths used by pedestrians and cyclists beyond the urban sections of Madrid-Río and the Manzanares linear park[26]. The Villaverde path is only 360 meters long but allows the Novosur roundabout to be joined, which is next to a municipal walkway on the river, with the road that already existed on the left bank and ended in a loop next to the South Highway (A-4). The project recovered a significantly altered and deteriorated area by going under several railway platforms and bridges on the A-4 highway.

The path that runs between the Manzanares and the Villaverde electrical substation is located on Avenida de los Rosales 143. The trail extends 540 meters and gives continuity to the existing tracks on the right bank without forcing walkers and cyclists to cross to the other side of the riverbed. This route crosses the plots that the city council has reforested to make them a carbon sink and to comply with the Kyoto agreements, which will be replaced by the Paris agreements in 2020. In addition, the trails have several paved areas equipped with garbage cans, bicycle racks, and benches for walkers and cyclists to rest.

FUTURE RESEARCH DIRECTIONS

This study opens up different research lines, which are presented below:

- -The first line requires the study of historical interventions as a settlement of future bases and the possible adverse impact on the city's landscape.
- -The second future line of research is the comparative study of rivers passing through the capitals. The positive and negative factors of channeling networks, water sanitation, and social functions carried out in the basin of these rivers would be analyzed.
- -The third line of research questions and opens up criticism concerning the treatment of Smart City regulations that are currently being drafted. There is a need to unify the rules since the current projects and regulations that are proposed deal separately with the different infrastructures such as pedestrian and vehicular traffic, railway lines, electricity, sewerage, and water supply. As a result, there are inconsistencies in the overall solutions for the city to function optimally.

CONCLUSION

The first conclusion summarizes that if a city is smart because it responds to the current needs, it may have an expiration date based on the temporary advance of technologies and networks. However, it is evident that old technology continues to meet our needs today and is even being used again as optimal engineering systems. What is called intelligent is thus inherent from the beginning, from a project and historical basis (political and social, but also geological and logical) of the land and the city that uses it. Not everything that is useful today will be obsolete tomorrow, rather it can be put to good use. Therefore, dismissing the history of old interventions is a mistake when they show ways for improvement. Moreover, using them again is a way to understand the key to evaluating sustainable operations.

Another conclusion presented here is that the new regulations drafted to contextualize water and sanitation network infrastructures in a Smart City are based exclusively on their control without connecting with electricity, mobility, accessibility, and telecommunications regulations. Therefore, we may be missing an opportunity to include all of them in order to channel real sustainability.

Madrid is a complex city in terms of topography, orography, landscape, and urban aspects. As in other cities with great historical and heritage content, Madrid's architecture is rich in language and culture. Its diversity makes the city a complicated place when it comes to intervention.

This research, and in relation to the case study, presents several new conclusions. For the first time the historical interventions on the Manzanares Canal, the artesian wells, and the El Pardo road are connected with the major projects of the 20th and 21st centuries (the Green Railway Corridor, the storm tanks, and the Madrid-Río Project).

The ideal of the Manzanares Canal could not be implemented due to the creation of the railroad between Madrid and Aranjuez in 1851 also because it is difficult for railroad and hydraulic communication to coexist due to crossing points that are complex to solve and cause urban and rural interstices that are difficult to deal with. Furthermore, public and private territorial boundaries and competencies are not easy to manage when it comes to implementing a single idea of growth.

On the other hand, the infrastructure plan designed to improve the waters of the Manzanares River (with tanks and storm tanks for water purification) did not have the media coverage it deserved and is completely unknown to the citizens of Madrid, as are the water trips and how they are linked to the Spanish capital.

In conclusion, it is necessary to preserve or maintain all the plans that are initiated in cities given that the capitals are major tourist centers and international forums for various political, economic, and cultural events. Maintenance contracts must provide for the following services:

- The clearing of channels and easements defined in the technical requirements.
- Interventions on existing vegetation.
- Water quality control.
- Monitoring and surveillance of fish and wildlife.
- Care for the city's image. The possible landscaping impact that new facilities may have on the city's skyline and its urban and natural surroundings.
- The correct operation and maintenance of the electromechanical flow regulation equipment in the channel to deal with possible malfunctions.
- Maintenance of remote control and remote monitoring networks, as well as surveillance of the river's infrastructures.

Regarding the actions carried out with reclaimed water, for the purification and reuse of water from the autonomous community of Madrid, infrastructures have been continued for watering the main green spaces of Rivas Vacíamadrid and of a tertiary filtration system to improve the functioning and the quality of the reclaimed water supplied in the Soto Gutiérrez WWTP.

Among the new strategic plan actions to modernize and improve the sewage networks are the Network Plan to standardize the materials in the distribution networks and thus improve the quality of water, the Smart-Region Plan whose objective is to achieve 100% smart meters by 2030, and the Solar Plan, which focuses primarily on the creation of new photovoltaic plants in infrastructures of the company to boost the generation and self-consumption of clean energy. In terms of plans for environmental commitments, the most significant investments are related to the Improvement Plan for the sewerage treatment system and the extension of the supply of reusable water. Other investments mainly include technical compliances.

The Canal de Isabel II Company continues with its actions to automate its operations, increase the security of the computer systems, and the assembly of equipment and infrastructure in the telecommunications network. Also, we should mention the installation of new remote-control points in different locations for the acquisition and transmission of data through GPRS and digital trunking, and the 3rd phase of process automation and integration of the WWPS of the remote-control system. Finally, the Ministry of the Environment, Spatial Planning and Sustainability of the Autonomous Community of Madrid, through Canal de Isabel II, announced in 2020 the launch of an investment program of upwards of €1,700 million for the 2020-2030 period. This constantly growing investment program not only represents the clear management advances made through the modernization of the facilities, but also encourages economic reactivation and the generation of wealth, wellbeing, and stable employment.

These investments in new infrastructure (and the refurbishment of old ones) improve the citizens' health and well-being, as well as adapting to the new demands of use and the application of new regulations of infrastructures in the city of the future or the Smart City.

ACKNOWLEDGMENT

This research is part of the financed project, *El paisaje periurbano de Madrid: visiones desde la memoria hacia la nueva ciudad*, translated into English as The Peri-Urban Landscape in Madrid: Visions from Memory to the New City (financed by *El Ministerio de Ciencia e Innovación en la convocatoria* 2019 Proyectos de I+D+i - RTI Tipo B. Translated into English as the Ministry of Science and Innovation of Spain, code PID2019-110693RB-100).

This research has been translated by Diana Claveria (Architect from the University of Westminster, London, RIBA member and translator: www.dyfconsultants.com).

REFERENCES

Coca, J., & Fernández, F. (2011). La Renovación del Manzanares: Transformaciones y Reciclajes Urbanos. In Proyecto, Progreso, Arquitectura. Universidad de Sevilla.

Franchini, T. (Dir.) & Raventós, T. (Coord.). (2018). Temas de planeamiento urbano. Madrid, Spain: CEU Ediciones.

González Capitel A., & García-Gutiérrez Mosteiro, J. (2000). *Luis Moya Blanco. Arquitecto 1904-1990.* Ministerio de Fomento.

MAGRAMA. (2014). Manual nacional de recomendaciones para el diseño de tanques de tormenta. Equipo de redacción: Aqualogy.

Maluquer & Salvador. (2014). Manual Nacional de Recomendaciones para el diseño de Tanques de Tormenta. Ministerio de Agricultura, Alimentación y Medioambiente.

Marín Perellón, F. J., Ortega Vidal, J., & Aragoneses López, D. (2009). *El Canal Real de Manzanares.* Ayuntamiento de Madrid.

Martínez & Muñoz de Pablo. (2018). El palimpsesto del Ensanche de Madrid = The Madrid´s expansion palimpses. *EGA. Revista de expresión gráfica arquitectónica, 23*(33), 118-129.

Rieira, P. (2018). *Beneficio social del Pasillo Verde Ferroviario de Madrid.* Instituto Universitario de Estudios Europeos. Universidad Autónoma de Barcelona. Editorial Noesis. (Original work published 1995)

Rocío Santo-Tomás & Rodríguez Romero. (2020). *City thresholds. The role of urban green infrastructures in Madrid.* ZARCH 14. doi:10.26754/ojs_zarch/zarch.2020144289

Rodríguez Romero, E. J. (Dir.). (2018). Paisajes de aproximación a la ciudad de Madrid. Conarquitectura.

Suárez, J., & Puertas, J. (2005). Determination of COD, BOD, and Suspended Solid Loads During CSO Events in Some Combined Catchments in Spain. Ecological Engineering, 24, 199-217.

ADDITIONAL READING

Carl Troll, C. B. (1974) *Das Quanatsystem von Madrid um 1750* or *Water Trips in 1750.* Retrieved from: https://www.madrid.es/UnidadesDescentralizadas/Agua/DeInformacionsobreAgua/RecursosHidricosAlternativos/viajesAgua/Carl%20Troll%20y%20Cornel%20Braun%201974.jpg

KEY TERMS AND DEFINITIONS

Architecture: Knowledge of art, science, technology, and humanity. A general term used to describe buildings and other physical structures. The design activity performed by an architect.

Smart City: An urban area that uses different types of electronic Internet of things (IoT) such as sensors to collect data and use of insights gained from that data to efficiently manage assets, resources, and services efficiently.

Storm Tanks: Huge underground tanks designed to store the first rush of rainwater, which is also the most polluting, even more than sewage, because it carries away all the dirt accumulated on streets and pavements. Once purified, the water can be discharged back into the rivers without threatening the ecology of the water flow.

Tellurism: Refers to the earth and how it behaves above and below the surface.

Urban Landscape: Everything you can see when you look across an area of land, including buildings, bridges, and structures.

Urbanism: The study of how inhabitants use urban areas, such as towns and cities, and how they interact with the built environment. It is a direct component of disciplines, such as urban planning, which is a profession that focuses on the physical design and management of urban structures and urban sociology or the academic field of the study of urban life and culture.

Water Trips: The water trips or water travels in Madrid are part of the whole historical underground network of channels or "qanat" created to ensure water supply in Madrid from the Muslim period.

ENDNOTES

1 Juan de Herrera Project (1582-1584) ordered by King Felipe I. Formerly known as *Puente Segoviana*, the first references of this construction date from the 14th century, when King Alfonso XI of Castile authorized its construction in 1345.

2 Pedro de Ribera Project (1718-1732). Philip IV commissioned Juan Gómez de Morato design the first project, and it was built by José de Villarreal between 1649 and 1660. It was known as Puente Toledana.

3 Lifeguards will be female in the women bathing areas.

4 It was also forbidden for drunks and, as indicated in the regulation, for people "deprived of reason" to enter the baths.

5 Smart Cities. Infrastructures. Public Service Networks. Part 1: Water Networks. Rule or Norma-UNE-178101-1-2015-n0055581.

6 The Royal Manzanares Canal was a very ambitious infrastructure project and due to its great technical and natural difficulties and exorbitant costs, many considered it unfeasible. It was intended to link Madrid with Aranjuez through a waterway with the canal splitting to go to Lisbon through El Tajo or through another channel to Seville, according to the project of Isidro Gonzalez Velazquez.

7 The Y in the abbreviated form of the company's name is from the old spelling Ysabel for Queen Isabel II, during whose reign a modern water supply was provided for Madrid. A canal was constructed to provide water from the catchment of the river Lozoya.

8 Andrés Martí, was a military engineer (mid-century XVIII) that presented a project in which to build a two-kilometer-long canal, which ran from a dam in the Jarama to the Manzanares. The project was very controversial and controversial. One of the opponents Vicente Alonso Torralba presents a similar plan with certain modifications. One of the precursor projects in the mid-18th century is the one directed by Jorge de Sicre y Béjar, which mentions the use of the Lozoya, Jarama and Guadalix rivers.

9 Today it is an urban space delimited by the streets of Bravo Murillo, Cea Bermúdez, Boix and Morer, and Avenida Filipinas. At that moment (mid-century XIX) it was as a place of execution for prisoners sentenced to death.

10 On the exterior, the building has an exposed brick facade, adorned on the Bravo Murillo side with a late Romantic classicist style fountain with a representation of the River Lozoya sculpted by Sabino de Medina, flanked by allegories about Agriculture and Industry, by sculptors Andrés Rodríguez and José Pagnucci, respectively.

11 Clifford was, together with the French photographer, Jean Laurent, one of the leading photographers of his day in Spain.

12 Spanish palace or fortress of Moorish origin.

13 With the reference ZEPA ES0000011.

14 Center of Hydrodynamic Experiences of El Pardo.

15 This was called Project Portici.

16 The pool suffered some damage during the Civil War, which was repaired during the early post-war years. It was in service until February 1954, when it was demolished due to the works of the second canalization of the Manzanares. The island that acted as a base was also eliminated, since it constituted an obstacle for the planned new system of locks. In the following link, https://www.pasionpormadrid.com/, José Antonio Arroyo Minguez says the following, "Despite my young age, since I was born in November 1949, I have photos and memories of the swimming pool on the island, where my father used to take me by bicycle. At that time everyone belonged to Canoe, the dean of swimming in Madrid, and in 1944 a group of members, including my father, founded a new swimming club, the Grupo Deportivo ISLA (ISLA Sports Group). He was Vice President of the Club and its true promoter until his premature passing in 1970."When La Isla was closed, the Club that was founded there continued its sporting career in the Municipal pool of the Casa de Campo, where I have been a swimmer, coach, manager, and president at different stages.

17 Storm tanks are a very particular system that allows rainwater to be retained before it reaches the purification stations.

18 These are the facilities of Arroyofresno (under the Club de la Villa de Madrid) and Butarque. Each tank can store up to 400,000 cubic meters of water. Out of the 65 built in the entire Community of Madrid, 28 run through the capital city. Spain has a total of 470 tanks of very different sizes, but the three largest are located in Madrid and have the same capacity as all the other tanks combined.

19 The M-30 is a benchmark for mobility in the Spanish capital. It is beltway around the city with the characteristics of a motorway.

20 The winning proposal was that of the team led by Ginés Garrido and formed by an association of architecture studies of Madrid: Mrio arquitectos (Burgos & Garrido, Porras & La Casta and Rubio & Alvarez-Sala) in collaboration with the Dutch studio West 8.

21 By Maria Rosa Dominguez, who directed the excavation work of the area at the head of the Canal.

22 It also shows the qualified participation of architects and engineers such as Isidro González Velázquez, Miguel de Inza, Pedro Nolasco de Ventura or Martín López Aguado. In: Marin Perellón, José and Javier Ortega Vidal, *El Canal Real de Manzanares. Historia de una obra pública olvidada,* Madrid, Ayuntamiento de Madrid, p.11, 2009.

23 Located today in the premises of the building of Bustamante Street, 16. General Directorate of Water.

24 Total Budget: 2,841,593 euros. Start date: February 2011. Completion date: June 2013.

25 These routes had a budget of 474,268.46 euros.

26 It is now possible to walk or cycle along the stretches between the Franceses Bridge and the Green Cycle Ring to the north and the M-45 to the 630-meter-long Moncloa-Aravaca path, which begins on the existing path on the right bank of the Manzanares River next to the Franceses Bridge and is accessed via a ramp located at the intersection of Ribera del Manzanares and Santa Pola streets.

Chapter 8
Methodology for the Evaluation of Fire Safety in Existing Urban Residential Areas in Spain:
The Case of Social Housing in Pamplona (1940–80)

María Fernández-Vigil
Universidad de Navarra, Spain

Carmen Corrales
Universidad de Navarra, Spain

Raquel Garin
Universidad de Navarra, Spain

Juan Bautista Echeverría
https://orcid.org/0000-0002-4438-6449
Universidad de Navarra, Spain

ABSTRACT

Residential buildings are the most common scenario of fatal fires in Spain, and over the past decade, there has been no significant decrease of fatalities in this kind of accident, even a slight increase in recent years. In this context, a systematic methodology to evaluate the risk level of residential buildings, previously grouped into building typologies, is presented. Its application to different housing typologies allows covering large urban ensembles, providing an important information to regulators and fire services. The factors that most increase the level of fire risk are identified, analyzing them both by their construction stage and by their morphology.

DOI: 10.4018/978-1-6684-4030-8.ch008

INTRODUCTION

Between 2010 and 2017, there were 1359 fatal victims in fires in Spain (Fundación Mapfre & APTB, 2010, 2011, 2014, 2015, 2016, 2017, 2018), that is, an average of about 170 people per year. During this period, deaths in residential buildings represented 77% of casualties, meaning that houses are the most common scenario for this kind of fatal fires. 60.1% of these victims were more than 65 years old, this value contrasting with the percentage of people over this age in Spain, 18.3% (Instituto Nacional de Estadística, 2011). In other words, the risk of dying in a domestic fire is multiplied by 3.3 for the elderly people.

This is particularly relevant attending to demographic forecast for the coming decades: in 2064 people over 65 will be 38.7% of the entire Spanish population, which means that the elderly will account for more than the third part of the society (Instituto Nacional de Estadística, 2014). By the same date, life expectancy will be 95 years for women and 91 years for men; and the largest population group will be those between 85 and 89 years old (Instituto Nacional de Estadística, 2014).

The situation is therefore critical since, as the number of elderly people increases, it is expected that the number of deaths in residential fires will also increase. In fact, the downward trend that could be appreciated in the reduction of fatalities in domestic fires had a turning point in 2013, and it is still growing (Fundación Mapfre & APTB, 2018).

The main objective of this research is the creation of a systematic and regular methodology to evaluate and quantify the fire safety of a large number of residential buildings, and the identification of the most dangerous elements. Safety in these blocks built in the second half of the 20th century is often critical. For this reason, the conclusions of this assessment should be taken into account when considering the unavoidable urban regeneration that many Spanish cities will undergo. This renovation will be a great opportunity to solve the safety problem.

BACKGROUND

The first challenge when addressing fire safety in residential buildings in Spain is the lack of official data, what makes impossible to know what the risk factors behind fire events are and, therefore, design the more efficient measures to fight against them. At the present time, there is no nation-wide, systematic approach to collecting, analyzing and presenting fire loss data in Spain, in spite of being an essential tool for fighting against those accidents (Fernández-Vigil & Echeverría, 2019). The statistic treatment of the fire departments interventions in Spain is poorly regulated, and it does not have homogeneity. In 1985, the Spanish Royal Decree 1053/1985 about the statistical treatment in the Fire and Rescue Services was published ("Real Decreto 1053/1985, de 25 de Mayo, Sobre Ordenación de La Estadísticas de Las Actuaciones de Los Servicios Contra Incendios y de Salvamento," 1985). However, since 1994, official statistics have not been published. Every Fire Service has its own system for fire data collection, according to its resources and the regulation established in each community, region or municipality, as appropriate (Fernández-Vigil & Echeverría, 2019).

There are only few documents elaborated by MAPFRE Foundation in collaboration with APTB (Professional Association of Firefighters, *Asociación Profesional de Técnicos de Bomberos* in Spanish) collecting data on fatal victims since 2010 (Fundación Mapfre & APTB, 2010, 2011, 2014, 2015, 2016, 2017, 2018). Despite the utility of these documents, the information they present is limited and only allows us to draw a very general panorama of the problem. Recently, the University of Navarra built a database based on news, collecting all the incidents and victims, injured and casualties, of residential fires in 2016 (Fernández-Vigil & Echeverría, 2019). The research concluded that the advanced age, living in old housing units and having a low-income profile are important factors increasing the risk of dying in a domestic fire.

In this context, and for the analysis, social housing blocks in Pamplona´s suburban neighborhoods built between 1940 and 1980 were selected. These case studies were chosen due to several reasons: On the one hand, nearly 50% of the total housing stock was built in these decades (Table 1). In the case of Pamplona, 42% of these buildings are social housing, a consequence of the processes of generation of the city: large residential neighborhoods were created in order to give immediate response to the needs of the large masses of immigrant population displaced by the incipient industrial development. This process started on the 1950s, was consolidated in the 1960s and it lasted until the end of the 1970s (SAVIArquitectura, 2016). In addition, the considered blocks were built before current building regulations and, especially, any fire protection requirement, with criteria in which quantity prevailed over quality (SAVIArquitectura, 2016). Therefore, their safety conditions are usually far below what is required.

Table 1. Dwellings in Spain, Navarre and Pamplona according to year of construction. (Instituto Nacional de Estadística, 2011)

Year of construction	Number of dwellings in Spain	Number of dwellings in Navarre	Number of dwellings in Pamplona
Before 1940	2387395 (9.5%)	43445 (14.1%)	8375 (9.3%)
1940-1980	11249500 (44.6%)	123465 (40.0%)	48045 (53.3%)
After 1981	10882810 (43.2%)	136265 (44.2%)	32335 (35.9%)
Unknown	688915 (2.7%)	5435 (1.8%)	1415 (1.6%)
Total	25208620 (100%)	308610 (100%)	90170 (100%)

On the other hand, nearly 60% of fatalities in residential buildings between 2010 and 2016 lived in blocks, against a 40% of fatal victims who lived in single-family houses (Fundación Mapfre & APTB, 2010, 2011, 2014, 2015, 2016, 2017). This is reasonable considering that only 21% of the Spanish population lives in semi-detached houses and only 12.7% in detached houses (Eurostat, n.d.).

In addition, as it was mentioned above, the age of the dwellings, the age of the occupants and the socioeconomic profile are risk factors in case of fire. For that reason, buildings located in suburban neighborhoods have been selected as case studies and blocks of similar characteristics in central areas of the city have been excluded, since they are highly revalued due to their location. The selected dwellings have suffered a gradual physical decay, together with their occupation by low-income sectors of the population due to their low prices (SAVIArquitectura, 2016).

The present analysis allows us to identify which typologies and elements pose a higher fire risk. The conclusions drawn should be taken into account when addressing the urban regeneration of social neighborhoods, not only from an energy perspective, but also focused on safety.

METHODOLOGY

For this research, first of all a series of housing blocks grouped by typologies has been selected. Then, their fire risk level has been evaluated, comparing their current performance with the one required by the Fire Document, "DB-SI", in the Spanish Building Code, "CTE" (Ministerio de Fomento, 2019b). Finally, the case studies were assessed with the Fire Risk Assessment Method MEREDICTE © (Pérez-Martín, 2014), in order to verify the consistency of the results. The different stages of the research are explained below and are shown in Figure 1.

Throughout this chapter, several references to different regulations, abbreviations and codes for the typologies are used. For that reason, a collection of all these abbreviations is presented in Table 8, Appendix 1.

Figure 1. Diagram of the methodology used in the research

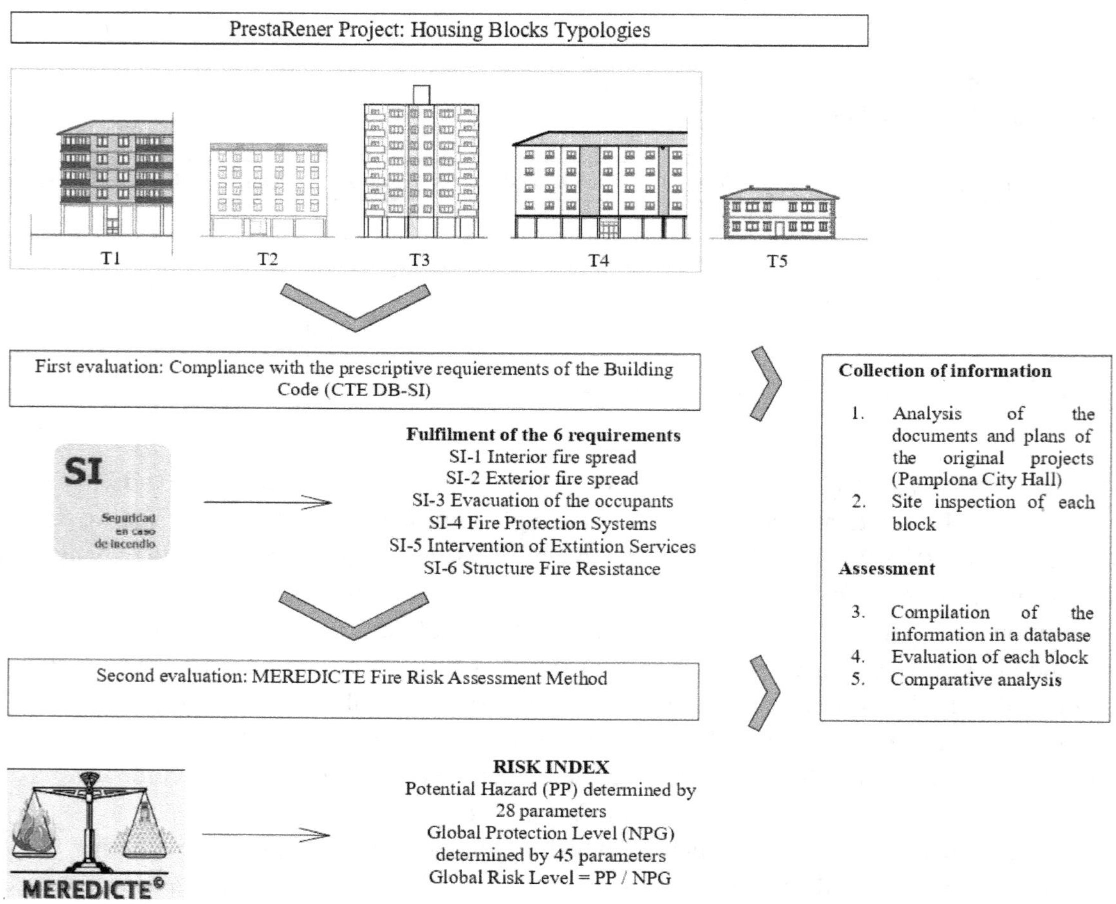

Identification of the Typologies: PrestaRener Project

To carry out the analysis in an efficient way, grouping the blocks in typologies becomes a key factor. This allows covering a bigger number of buildings, identifying those with the same problems and improvement needs, and generating comparably results. The systematic analysis based on a classification by typologies has proved successful in building research (Domínguez Arribas et al., 2015; Ortega Madrigal et al., 2013; Rubio del Val, 2011; SAVIArquitectura, 2016).

In this case, the typologies identified in the prestaRener project, which was funded by the Ministry of Economy and Competitivity in Spain, have been used (SAVIArquitectura & Worcester Polythecnic Institute, 2016).

In that research, the blocks in Pamplona have been grouped by typologies according to the construction date, their morphology and height, and in sub-typologies according to the construction characterization of their components (SAVIArquitectura, 2016), all of them key factors for fire risk analysis. The repetitiveness of the identified typologies allows to simplify the risk characterization.

Table 2 shows the division in typologies and sub-typologies. The sample size has been calculated assuming a confidence level of 90%, a sample error of 5% and a variability of 0.5. The final number of blocks to be studied is shown in the last column of the table.

The initial research also included the typology number T5, single-family house, which was divided in two sub-typologies: single-family house with ground floor (T5.1), and single-family house with ground floor plus one (T5.2). However, it was discarded in this analysis since, as it was already mentioned, most of residential fires happen in blocks, which have the additional risk of the evacuation through the stairs when there are several stories.

Table 2. Total number of dwellings and sample size according to the PrestaRener typologies (SAVIArquitectura, 2016)

Typology	Sub-typology	Number of dwellings	Percentage	Sample Size
T1 Linear Block	T1.1 Linear block with nearby building environment	432	16.8%	41
	T1.2 Non-clustered linear block	576	22.4%	55
	T1.3 Linear block with one orientation	33	1.3%	3
	T1.4 Linear block with minimum courtyard	60	2.3%	6
T2 H Block	T2.1 H-block	202	7.9%	19
	T2.2 H-block, height above Ground + 4	59	2.3%	6
	T2.3 H-block with minimum courtyard	60	2.3%	6
T3 Tower	T3.1 H-tower	99	3.9%	9
	T3.3 Cross Tower	6	0.2%	1
T4 T Block	T4.1 T-block, 3 dwellings per floor	28	1.1%	3
	T4.2 T-block, 2 dwellings per floor	69	2.7%	7
T5 Single-family houses (excluded in the analysis)	T5.1 Single-family house, ground floor	78	3.0%	7
	T5.2 Single-family house, ground floor + 1	868	33.8%	82
Total		2570	100%	244

Figure 2 shows a typical floor of each typology, and Table 3 shows the main characteristics of each sub-typology.

Figure 2. Typical floor of the typologies analyzed in the study

Table 3. Main characteristics of each sub-typology of the PrestaRener Project (SAVIArquitectura, 2016)

Typology	Sub-typology	Year of construction	Minimum Height	Maximum Height	Dwellings per floor	Residential use on the ground floor	Average surface area of the dwellings (m^2)
T1	T1.1	1940-1979	Ground+1	Ground+5	2	Some cases	80
	T1.2	1940-1978	Ground+1	Ground+11	2	Some cases	80
	T1.3	1950-1971	Ground+3	Ground+10	4	Some cases	85
	T1.4	1962-1979	Ground+4	Ground+6	2	Some cases	95
T2	T2.1	1959-1978	Ground+4	Ground+4	4	No	70
	T2.2	1959-1978	Ground+5	Ground+11	4	No	85
	T2.3	1954-1979	Ground+4	Ground+8	4	No	75
T3	T3.1	1961-1975	Ground+4	Ground+10	4	Some cases	80
	T3.2	1966	Ground+10	Ground+10	4	No	80
T4	T4.1	1964-1978	Ground+3	Ground+7	3	No	80
	T4.2	1963-1973	Ground+4	Ground+8	2	No	100

At the end of this section, it will be presented which typologies and which characteristics imply a higher risk of fire.

Fire Risk Assessment

The Building Code (CTE) establishes basic requirements to satisfy the objectives of habitability and safety of the buildings. According to Article 5.1, Part I, of the CTE (Ministerio de Fomento, 2019c), the designer has two options for justifying that a building complies with these requirements: either to adopt the solutions set out in the Basic Documents (prescriptive design), or to propose alternative measures whose performance is at least equivalent to those indicated in the Basic Documents (performance-based design).

In the case of Fire Safety, the objective consists in reducing to acceptable limits the risk that the building users suffer damage from an accidental fire. The Basic Document about Fire Safety, henceforth DB-SI, specifies objective parameters and procedures whose fulfillment ensures the satisfaction of the basic requirements (Ministerio de Fomento, 2019c).

For all these reasons, in order to evaluate the fire risk level in the selected buildings, the degree of fulfilment of the six requirements has been studied. If they do, it can be assumed that their risk level is acceptable. Each of the six sections refers to a basic requirement, namely: SI-1, limitation of the interior fire spread; SI-2, limitation of the exterior fire spread; SI-3, provision of adequate means of escape to ensure that the occupants evacuate the building safely; SI-4, provision of adequate fire protection systems: detection, control, extinction and alarm transmission; SI-5, means to allow the intervention of rescue and extinction services; SI-6, design of the load-bearing structure with the necessary fire resistance.

Working Method

Once the number of case studies was calculated, the working method consisted in selecting a representative sample of buildings that have been inspected, finding a coherent distribution of the affected neighborhoods for each sub-typology.

To proceed to the comparative analysis between the building characteristics and the prescriptions in the code an inspection document was elaborated. The document is divided into seven sections: On the one hand, general data, which includes basic information such as the address, typology, year of construction, number of stories, number of dwellings and areas, among others. On the other hand, the construction characteristics, where the materials of the structure, facade, roof, floor slabs, and interior partitions are described. The remaining five sections refer to the basic requirements SI-1 to SI-5. The analysis of the requirement SI-6, structural fire resistance, is a really difficult task, so it was approached with a different method, as it will be explained in the "Results" Section.

The collection of all the required information is a complex process, especially for dwellings built more than 40 years ago. To guarantee the reliability of the introduced data, the inspection had two stages.

The first one consisted in the analysis of the plans and written documentation of the original projects, in the Pamplona City Hall Archives. The review of the technical reports, as well as the economic budget and the graphic information, made it possible to extract data about the type of structure, materials, surface areas, uses and distribution of the dwellings.

However, the age of the documents made it necessary to carry out a second stage of data collection, which consisted of a site inspection of each selected block. This second phase has three objectives: to collect the information which was not provided by the original plans (especially necessary in older projects); to verify the information shown in the plans; and finally, to check whether there have been subsequent interventions. In that case, it is possible that the building has been adapted to the current fire regulations (for example, with the installation of fire protection systems) or that there have been changes which affect the behavior of the building in case of fire (such as a retrofitting of the envelope, with its consequent change of materials).

Once all the buildings were inspected, the information was compiled in an Excel data base. Then, the evaluation of the risk level of every block was completed, through the analysis of the compliance with the basic requirements. Finally, a comparative analysis by typologies and construction date completed the work.

MEREDICTE © Method

The case studies were assessed in a second evaluation, in order to verify the obtained results. For this purpose, they were analyzed with the Fire Risk Assessment Method MEREDICTE©. This method was part of the CTE between 2014 and 2020. The Royal Decree 732/2019, of December 20, incorporated modifications to the DB-SI. For that reason, the MEREDICTE© is no longer considered applicable to the current version of the DB-SI (Ministerio de transportes movilidad y agenda urbana, 2020). However, it is still an interesting risk analysis to study the adequacy of the constructed buildings to the requirements of the Spanish regulations.

MEREDICTE© is a Risk Index that aims to objectively quantify the level of danger and protection of buildings in case of fire. It assumes that a building will always have an acceptable level of risk if the fire hazard is compensated by the means of protection available [1].

$$Global\,Risk\,Level\left(NRG\right) = Potential\,Hazard \;/\; Level\,of\,Global\,Protection \quad [1]$$

It is considered that the risk is acceptable when the result is under 1. The level of Potential Hazard (PP) [2] is determined from 28 parameters, including the sector capacity (AS), parameters related to the fire tetrahedron (T), occupant characteristics (CO) and architectural characteristics (CA).

$$Potential\,Hazard\left(PP\right) = AS\times\left(0,4\times T + 0,3\times CO + 0,3\times CA\right) \quad [2]$$

The level of Global Protection (NPG) [3] is determined from 45 parameters, including the presence of a self-protection plan (NPA), the fire resistance of the structure (RFE), the interior and exterior fire spread (P), the evacuation of the occupants (EO), the fire protection systems (IPCI) and the intervention of the fire services (IB).

$$NPG = NPA\times\left[RFE\times\left(0,24\times P + 0,4\times EO + 0,24\times IPCI + 0,12\times IB\right)\right] \quad [3]$$

The components of these equations are also weighted with several coefficients that allow to quantify the level of danger and protection, as well as to know the relative importance of each of the hazards and safety measures (Pérez-Martín et al., 2010).

The risk level calculated by MEREDICTE© always refers to the safety of the building users, and never to the integrity of the building.

Once the first analysis was conducted, the second assessment using the MEREDICTE© Method was performed. For this purpose, the coefficients were introduced in a new Excel spreadsheet and the coherence between both results was verified.

RESULTS AND DISCUSSION

Before showing the obtained results, some clarifications are needed.

In some cases, the original projects covered several blocks at the same time. As it is quite difficult to give a definition of what a single building is, for the purpose, every housing part with a single entrance hall has been considered as a unit.

On the other hand, as the project is focused on housing the other establishments (commercial or others) integrated in the buildings have not been considered. Only the exterior spread requirements have been checked for compliance.

Interior Spread

The first requirement of the Fire Safety Document, DB-SI-1, establishes that buildings must be divided in fire compartments, according to variable maximum surface areas depending on the use. In the case of residential buildings, the fire compartments cannot exceed 2500 m^2.

88% of the analyzed blocks does not exceed 2500 m^2 of floor area, therefore, they do not need to be divided into compartments. However, the remaining 12%, which exceeds the surface limit, does not comply with the requirements: all of them have an open staircase that connects the stories of the block, creating a single fire compartment. The blocks that are not correctly compartmentalized are high-rise buildings, between 7 and 12 stories, mostly belonging to typology T2.1 (H-block, height above Ground + 4) or T3 (Tower), and they were built along the decade of 1970s.

In order to limit the fire spread between dwellings, the DB-SI document indicates that the separating elements must have at least 60 minutes of fire resistance (EI-60). This requirement is met in every analyzed block, since in all cases the interior partitions are made of variable thickness brick, with plaster in both sides. The construction solutions used to separate dwellings can be divided into five types: double-air brick wall (7 cm), double-air brick wall (12 cm), double wall of single-air brick (5+5 cm), with or without air chamber, and double wall of double-air brick (7+7 cm). According to the DB-SI document, the fire resistance of all these brick elements varies between 90 and 240 minutes.

However, in most cases the project documentation does not contain detailed information on the floor slabs. For this reason, it is not possible to guarantee that their fire resistance meets the requirements.

Regarding the special risk locals of the case studies, it can be appreciated that the blocks built in the 1940s only have electricity meter rooms. Later, spaces for the elevator machinery, storage rooms, cleaning and waste storage rooms, and boiler rooms also appear. In 28% of the cases, this special risk areas are not correctly compartmentalized. Sometimes, a low special risk area does not have a fire-resistant door, and sometimes a medium special risk area does not have an independent vestibule.

According to the section 4 of the DB-SI-1 document, the construction elements used in walls and ceilings in the occupied areas must have a fire reaction equal or greater than C-s2, d0, and floors must be E_{FL}. The requirement excludes the interior of the dwellings; therefore, the materials of the vertical and horizontal elements of the hallways and common areas were analyzed, only if they exceed 5% of the total surface of the enclosure.

All the studied blocks have a plaster finish on walls and ceilings of the staircases area, whose fire reaction class is A1, without the need for testing ("Real Decreto 842/2013, de 31 de Octubre, Por El Que Se Aprueba La Clasificación de Los Productos de Construcción y de Los Elementos Constructivos En Función de Sus Propiedades de Reacción y de Resistencia Frente Al Fuego," 2013). The cladding used in floors are hydraulic tiles, granite, or terrazzo, which are also class $A1_{FL}$ without the need for testing. Therefore, in all the cases, the materials used in the communication areas comply with the regulations.

There are a greater variety of finishes in the doorways, for instance, ceramic or marble coatings, used in both vertical and horizontal surfaces. These materials are also class A1 or $A1_{FL}$. In 16 buildings, the doorway walls were covered with wood. Seven of these cases are renovations made after 2006, when the current regulation was approved, so it can be assumed that the wood has been adequately treated in order to comply with the requirement. The remaining nine cases have not been modified after the construction, so they do not comply with the regulation, since the reaction to fire of the wood is class D-s2, d0.

The DB-SI document also establishes that the coverings used on walls of non-watertight concealed spaces, except those located inside dwellings, must be at least class B-s3, d0; B_{FL}-s2 for floors. In this regard, a conflictive area may be the space between the top floor slab and the roof. Table 4 shows the construction solutions for roofs according to the decade of construction.

In construction solutions C5, C7 and C8, the roof is directly supported by the top floor slab. However, in the remaining solutions a non-watertight concealed space is created, where a fire could start. Therefore, these areas must have the required fire reaction class, B-s3,d0.

Table 4. Percentage of buildings using different roof construction solutions, by decade of construction

Roof Construction Solution	Percentage of blocks			
	1940	1950	1960	1970
C1: Wooden substructure + shingle board + curved ceramic roof tile	100%			
C2: Wooden substructure + ceramic board + curved ceramic roof tile		39%		
C3: Supporting brick partitions + ceramic board + Concrete compression layer + curved ceramic roof tile		61%	92%	27%
C4: Supporting brick partitions + ceramic and concrete board + Concrete compression layer + curved ceramic roof tile			5%	21%
C5: Reinforced concrete floor slab + sloped concrete + fiber cement panel			3%	
C6: Supporting brick partitions + ceramic board + fiber cement panel				30%
C7: Reinforced concrete floor slab + sloped concrete + asphalt cloth + EPS insulation (3 cm) + Gravel				12%
C8: Reinforced concrete floor slab + sloped concrete + asphalt cloth + EPS insulation (3 cm) + XPS (7 cm) + Gravel				9%

Solution C3, C4 and C6 are made of double air brick walls, where a shingle board is sustained, with or without concrete, and a ceramic tile or a fiber cement panel as exterior cover. The fired clay pieces belong to a fire reaction class A1, without the need for testing, so they comply with the requirement.

By contrast, in solutions C1 and C2, which are present in 48 buildings of the sample, the roof is supported by a wooden substructure and a ceramic or shingle board. They do not comply with the requirement, since wood without fireproofing treatment belongs, at best, to class D-s2,d0. As it can be seen in Table 4 this solution is used in every block built in the 1940s, it loses prominence along the 1950s in favor of the roof supported by brick partitions, and it completely disappeared in later buildings.

75% of the blocks with solutions C1 and C2 belongs to the typology T1.2 (non-clustered linear block). These solutions also are present in the typologies T1.1 (Linear block with nearby building environment), T2.1 (H-Block) and T4 (T-Block).

Exterior Spread

The second section of the DB-SI document establishes that the vertical walls between buildings must have a fire resistance equal to or greater than 120 minutes. As it was explained before, every block with a single entrance hall has been considered as a unit. In order to evaluate compliance with this requirement, the construction solutions used in the party walls were analyzed.

The party walls are made in all the cases by air brick of variable thickness, with plaster in both faces. 85.5% of those walls have a thickness greater than 8 cm, therefore, their fire resistance is at least 180 minutes. The remaining 14.5% are made by a double-air brick (7 cm), with plaster in both sides. According to the regulation, these walls have a fire resistance of 90 minutes, so they do not comply with the requirement. There is no pattern in the distribution by typology or construction period of the buildings with this kind of party walls.

Table 5. Fire reaction of the facade construction solutions of the case studies

Facade Construction Solution	Reaction to Fire
F1: Perforated or Solid Brick Wall (24 cm) + Plaster	A1
F2: Brick cladding or cement mortar + Double-air brick wall (24 cm) + plaster	A1
F3: Double-air brick wall (12 cm) + Air chamber + Single-air brick partition + plaster	A1
F4: Perforated or solid brick wall (12 cm) + air chamber + single-air brick partition + plaster	A1
F5: Brick cladding or cement mortar + double-air brick wall (12 cm) + Air chamber + single-air brick partition + plaster	A1
F6: Brick cladding or cement mortar + double-air brick wall (24 cm) + Air chamber + single-air brick partition + plaster	A1
F7: Perforated or Solid Brick Wall (12 cm) + mortar + fiberglass insulation (4 cm) + double-air brick partition (10 cm) + Plaster	A1
F8: "Granulite" type Cladding + Cement mortar + double-air brick wall (12 cm) + Air chamber + single-air brick partition + plaster	A2-s1.do
F9: "Granulite" type Cladding + Cement mortar + double-air brick wall (12 cm) + plaster	A2-s1.do
F10: Ventilated façade (Ceramic panel + Mineral wool + Metal Framing) + Cement mortar + double-air brick wal (12 cm) + Air chamber + single-air brick partition + Plaster	A1
F11: Ventilated facade, no data	Test Needed

Table 6. Percentage of blocks that comply with the required number of floor exit, by typology

The block complies with the number of floor exits		Percentage of blocks									
		T1.1	**T1.2**	**T1.3**	**T1.4**	**T2.1**	**T2.2**	**T2.3**	**T3**	**T4**	**Total**
Yes:	The floor exit can be and is a non-protected stair	78%	86%	67%	23%	100%		29%		10%	65%
No	The stair should be protected. The evacuation route is longer than allowed.	22%	12%		62%			29%	70%	50%	22%
	The stair should be especially protected. There should be more floor exits		2%	33%	15%		100%		30%	20%	10%
	The hole of the non-protected stair is > 1.30 m^2. It is not a floor exit; the evacuation route is longer than allowed							43%		20%	3%

All blocks comply with the requirements for distances between elements with a fire resistance lower than 60 minutes, for the limitation of the horizontal and vertical spread through the facade.

Regarding the reaction to fire, the facade construction systems are shown in Table 5. Solutions F1 to F6 are different variations of a brick facade, which is always made of materials that belong to class A1. Solution F7, present in a block built in 1976, is the first with thermal insulation: 4 centimeters of fiberglass, covered on both sides by layers with a fire resistance higher than 30 minutes.

Facades F8 and F9 have a synthetic-mineral coating, composed of marble granules and acrylic resins. They are present in 6 grouped linear blocks, built in the 60s, and 2 towers built in the 70s. Similar materials used today belong to class A2-s1,d0; but in order to accurately determine their reaction to fire, it would be necessary to perform the necessary test.

Facades F10 and F11 are the result of later interventions to improve the quality of the thermal envelope of the building. In both cases, exterior thermal insulation has been added, through the installation of a ventilated facade over the exiting one. Data about F10 was obtained from the plans of three T3 blocks, built in the 1970s. However, it was not possible in the case of the six blocks with a facade F11: Ventilated facade, no data. The modification introduced in the last version of the DB-SI document alludes specifically to the insulation placed inside ventilated chambers. F10 complies with the regulation since mineral wool is class A1. For facades F11 it would be necessary to obtain detailed data to know if they meet the current standard.

The requirements for the limitation of the exterior spread through the roof are met in all the cases, and the covering materials -ceramic tile, fiber cement and gravel- always belong to class B_{ROOF} (Table 5).

Evacuation of Occupants

The most conflicting element when assessing the adequacy of the case studies to the regulation about the occupant evacuation is the type of staircase. As mentioned above, all blocks have a single open staircase. Therefore, those that exceed 28 meters of descending evacuation height will not comply with the regulations because they require more than one floor exit. On the other hand, blocks exceeding 14 meters of evacuation height should have a protected staircase to be considered as a floor exit. If this

Table 7. Percentage of blocks that comply with the number of required floor exits, by construction period

The block complies with the number of floor exits		Percentage of blocks				
		1940	1950	1960	1970	Total
Yes	The floor exit can be and is a non-protected stair	88%	53%	77%	15%	65%
No	The stair should be protected. The evacuation route is longer than allowed.	12%	37%	15%	42%	22%
	The stair should be especially protected. There should be more floor exits		11%	7%	30%	10%
	The hole of the non-protected stair is > 1.30 m2. It is not a floor exit; the evacuation route is longer than allowed			1%	12%	3%

is not the case, the evacuation route extends from the door of the dwelling to the exit of the building, exceeding the maximum permitted length (25 m). The same occurs in those buildings lower than 14 meters, in which the floor area of the staircase hole is greater than 1.30 m^2.

35% of case studies do not comply with the number of floor exits and type of stair (Tables 6 and 7). In order to verify whether the width of the means of escape was adequate, the occupancy of each block has been calculated. Then, the formulas contained in DB-SI-3 for the dimensioning of the evacuation elements were used. For non-protected descending stairs, the formula is $A \geq P/160$, being A the element width (in meters) and P the total number of persons whose passage is expected. In the case of dimensioning doors and passages, the formula is $A \geq P/200 \geq 1.00$ m.

10 buildings have a stair narrower than 1.00 m (between 0.75 and 0.90 meters), so they do not reach the minimum dimension established in the Basic Document "Safety of Use and Accessibility", DB-SUA (Ministerio de Fomento, 2019a). 7 of them belonging to the H-Block typology (T2), and the remaining three to the T-Block typology (T4). In addition, eleven blocks do not have the appropriate dimension according to the formula. All of these are high-rise buildings, so they have high occupancies. Nine of them, built in the 70s, do not comply with the minimum width of the main door.

In all the cases, the building exit doors are hinged with a vertical rotation axis by means of an easy and quick opening locking system. In "Housing" use, the exit doors must open in the direction of evacuation if the occupancy exceeds 200 people. This condition occurs in nine blocks analyzed, of which only three meet the requirement.

Those residential buildings that exceed 28 meters height must dispose means for the evacuation of occupants with disabilities, that may be: a pass to an alternative fire compartment through an accessible floor exit, or a refuge area where they can wait safely to the rescue services. None of the studied blocks, of which 16 are higher than 28 meters, had any of these means. This is an important issue, since data shows that the elderly and people with mobility impairments are the most common fatal victims in fires in Spain (Fernández-Vigil & Echeverría, 2019).

Fire Protection Systems

The analyzed residential buildings must dispose the following fire protection systems: portable extinguishers of 21A-113 effectiveness, located at less than 15 meters of any point of the evacuation route (excluding the interior of the dwelling), and in the special risk zones; exterior hydrants when the descendent evacuation height exceeds 28 meters, and dry column when it exceeds 24 meters. None of the analyzed blocks is higher than 33 meters, so the detection and alarm systems and automatic extinguisher systems are not compulsory for any building.

It should be noted that 63% of the case studies do not have any fire protection system. As it can be seen in Figure 3, the linear typology and those blocks built in the 40s are those with the highest percentage of buildings without protection systems. 54.8% of the blocks that exceeds 24 meters have a dry column, and only 12.5% of those higher than 28 m have an exterior hydrant.

The Section 4 of the DB-SI document also states that the manual fire protection systems must be correctly signaled, according to the Standard "Fire Protection Systems Regulation", RIPCI, published in 2017 ("Real Decreto 513/2017, de 22 de Mayo, Por El Que Se Aprueba El Reglamento de Instalaciones de Protección Contra Incendios.," 2017). However, 34.5% of the buildings that have these systems lack signage.

Figure 3. Percentage of blocks with fire protection systems, by typology and decade of construction

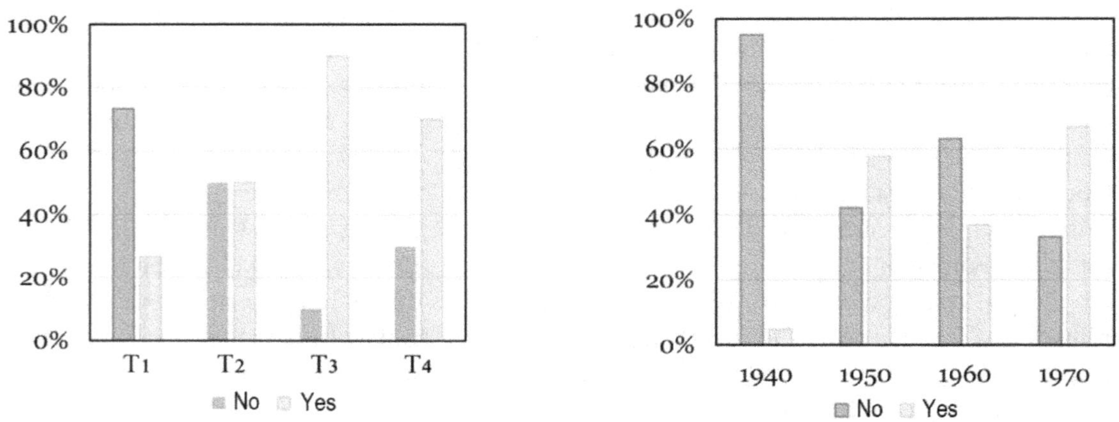

Fire Department Operation

When the descendant evacuation height exceeds 9 meters, it is necessary to have a maneuvering space that allows for the adequate intervention of the rescue services in case of fire. The conditions of such space are included in Section 5 of the DB-SI document, and they refer to the width, height, slope and resistance to puncturing of the operating space, the separation of the vehicle to the facade and accesses; and the width, height and bearing capacity of the approach roads.

45% of the case studies infringe some of the described conditions. There is not a pattern in the distribution of the cases that comply with the requirements and those that do not by typology or construction period, but it depends on the urban location of the block. On the other hand, the accessibility conditions of the facade to allow the access of the fire department are guaranteed in all the case studies.

Structure Fire Resistance

The analysis of the fire resistance of structural elements is a very complex task since most of the projects lack structural drawings. Those that do have these plans, do not provide enough information to estimate the fire resistance. This would imply to analyze element by element of each building in order to provide reliable results, which is an unfeasible task in a study of these characteristics.

All the buildings have a structure made of reinforced concrete columns and beams, in some cases with load-bearing brick walls. The first Instruction for the Design and Execution of Concrete Projects was approved in 1939 (Ministerio de Obras Públicas, 1939), and it required a minimum steel reinforcement cover of "more than one diameter and more than one centimeter". This standard underwent a constant process of review and modifications along the period under study. The last modification was in 1973, with the publication of the Instruction EH-73 ("Real Decreto 3062/1973, de 19 de Octubre, Por El Que Se Aprueba La Instrucción Para El Proyecto y La Ejecución de Obras de Hormigón En Masa o Armado," 1973). This regulation maintains the same minimum coverings.

The first mandatory fire protection regulation in Spain (although it was later declared voluntary) is the Basic Standard of Fire Protection, NBE-CPI-81, approved in 1981, which is a date after the period covered by this research. This standard establishes that the steel reinforcements of the concrete structures should have a minimum cover of 2 centimeters, in order to ensure a fire resistance of 60 minutes ("Real Decreto 2059/1981 Por El Que Se Aprueba La Norma Básica de La Edificación," 1981).

Therefore, it is not possible to guarantee the structure fire resistance of the case studies, since the reinforced concrete regulations applicable at the period when they were built required minimum coverings lower than those stated in the fire protection regulation.

Table 8. Percentage of blocks that comply with the prescription of the DB-SI document, by typology

Number of requirements met	Percentage of blocks				
	T1	T2	T3	T4	Total
Less than 6	3%	27%	30%	60%	13%
7 – 8	77%	40%	70%	30%	67%
9 – 10	20%	33%	0%	10%	20%

Overall Assessment of the Case Studies

Once each of the sections of the DB-SI document had been analyzed, an overall rating was performed to identify which morphologies and construction periods have the highest level of risk. In order to carry out this assessment, the number of requirements -10 in total- met by each block was quantified as follows:

Interior Spread (3): adequate compartmentalization in fire sectors, adequate compartmentalization of special risk locals and adequate reaction to fire of the materials.

Exterior Spread (2): The exterior spread is properly limited through dividing walls and facades, and through roofs.

Evacuation of Occupants (3): The building has the necessary number of floor exits, adequate dimensions and opening of the means of evacuation, there are means for the evacuation of occupants with disabilities in buildings over 28 m high.

Fire Protection Systems (1): The building has all the fire protection systems required by the regulation, according to its characteristics.

Fire Department Operation (1): The approach conditions and the environment of the building comply with the requirements of the regulation.

The typologies that comply less with the prescriptions of the Building Code, and therefore they have the highest level of risk, are T3 Tower, and T4 T-block (Table 8). When the comparison is made according to the construction decade, the buildings with the highest level of risk are those built between 1970 and 1979 (Table 9).

Table 9. Percentage of blocks that comply with the prescription of the DB-SI document, by construction decade

Number of requirements met	Percentage of blocks				
	1940	1950	1960	1970	Total
Less than 6	10%	6%	5%	36%	13%
7 – 8	90%	72%	58%	55%	67%
9 – 10		22%	36%	9%	20%

As it was detailed in the previous sections, this is since most of the high-rise buildings belong to these typologies and decade. The building height is one of the most determining factors when applying the prescriptions of the DB-SI document, since it affects several parameters: the requirements for the interior spread (the necessity for compartmentalization for large surface areas), exterior spread (it is needed a better reaction to fire of facade material at higher height), evacuation of occupants (number of floor exits, protection of the stairs) and fire protection systems (the necessity of a dry column, external hydrant…).

On the other hand, T1 buildings, linear block, and those built in the 40s, are the cases that must comply with the less demanding prescriptions: their small size and low height imply that it is not necessary to divide the building into fire sectors, nor to dispose more than one floor exit or protected evacuation routes, nor sophisticated fire protection systems. However, the fact that they meet many requirements of the regulation and that they do not have the hazards associated with high-rise buildings, does not mean that they are exempt of risks. Fire statistics show that older residential buildings have more fatal fires (Fernández-Vigil & Echeverría, 2019). This may be due to the poor condition of the electrical installation in many cases, which is the second most frequent cause of this type of accidents. Or this may be due to the fact that the low quality of the thermal envelope causes many occupants use portable heating devices, the leading cause of fatal fires (Fernández-Vigil & Echeverría, 2019).

In addition, none of the analyzed buildings had a detection and alarm system. However, several international studies confirm the effectiveness of smoke detectors in the reduction of mortality and morbidity rates from domestic fires (Bruck, 2001; Marshall et al., 1998; Stevenson & Lee, 2003; Warda et al., 1999) and they even conclude that these systems reduce the probability of death in a fire by about half (Ahrens, 2019; Echeverría Trueba et al., 2020; Rohde et al., 2016). A study conducted in the University of Navarra revealed that 96% of fatal victims in residential fires between 2014 and 2016 died inside their dwelling (Fernández-Vigil & Echeverría, 2019). This does not mean that the safety of the evacuation routes should be underestimated, and even less in high-rise buildings. But it does mean that focus should be placed on the protection measures that can be implemented inside dwellings, which are currently almost non-existent.

It is necessary to consider these circumstances when assessing the fire risk level of the building stock, as well as other factors such as the age of the occupants, the presence of people with disabilities or mobility impairments… All of them are elements that increase the risk of fire in residential buildings. It would be desirable to enforce risk evaluation methods that allow a more accurate study of fire safety in residential buildings (Echeverría Trueba et al., 2020).

50% of building stock in Spain is previous to the first mandatory fire regulation. The urgent need in this country to refurbish residential buildings in order to comply with the energy efficiency standards are a great opportunity to also improve their fire safety. In those cases when it is not possible to meet the prescriptions of the regulation, the performance-based design is the appropriate tool to reduce the level of risk, relying on active protection systems (Echeverría Trueba et al., 2020).

Results Obtained after Application of the MEREDICTE© Method

All the cases studies were assessed with the MEREDICTE© Method (Pérez-Martín, 2014). The results show that none of the building have a Global Risk Level below 1. In other words, all the residential buildings have an unacceptable fire risk, according to MEREDICTE©. Within the "unacceptable risk" group, the method presents 5 categories: high (1 < NRG ≤ 1.5), very high (1.5 < NRG ≤ 2), severe (2 < NRG ≤ 5), very severe (5 < NRG ≤ 10) or catastrophic (NRG > 10).

Figure 4. Global risk level of the analyzed buildings

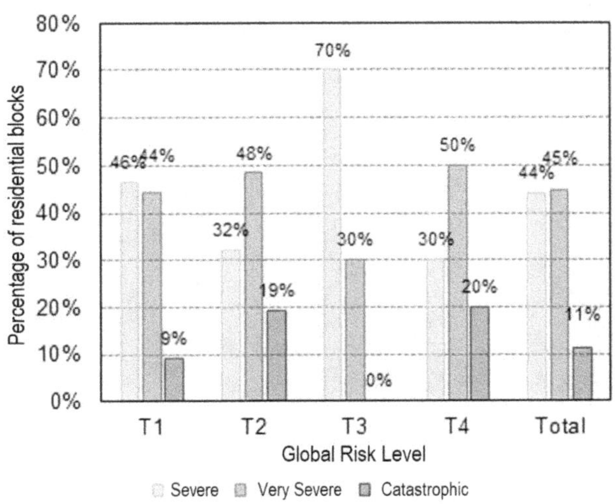

Figure 5. Global risk level, by sub-typology

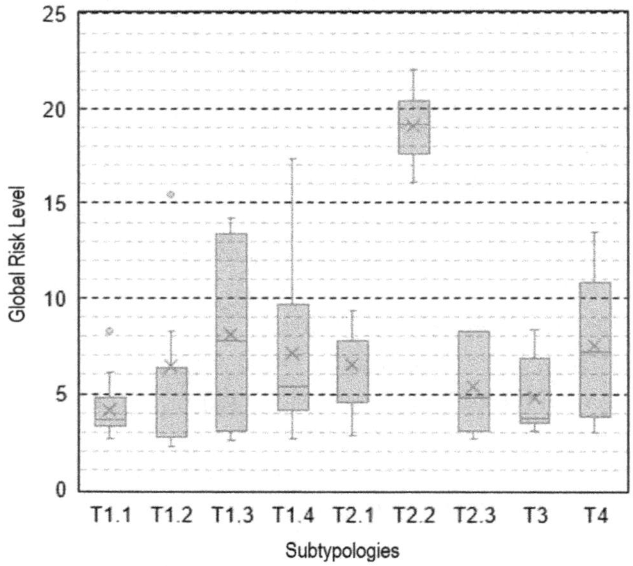

The average Global Risk Level obtained in the case studies was 6.43, very severe risk. The lowest average Global Risk Level was obtained in typology T1.1, linear block with nearby building environment, and the highest in typology T2.2, H-block, height above Ground + 4. Table 11 shows a summary of the results (see Appendix 2).

All the blocks have a risk over 2, that is, all of them have at least a sever fire risk. 11% of the buildings had a catastrophic Global Risk Level (Figure 4). Figure 5 shows the Global Risk Level in a boxplot, by sub-typology.

Figure 6. Relation between building height and global risk level

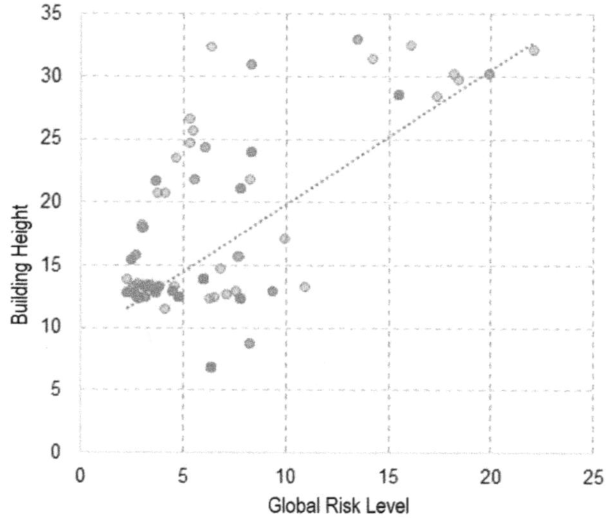

The results confirm that the high-rise buildings have the highest fire risk, and that buildings belonging to T1 typology have the lowest fire risk. As it was detailed in the previous sections, this is due to all the requirements that must be complied by the highest buildings, according to the Spanish regulations.

In fact, there is a positive moderate correlation between the building height and their global risk level, with a Pearson correlation coefficient of 0.62 (Figure 6).

One of the great challenges of urban sustainability in Spain is the regeneration of social neighborhoods built in the second half of the 20th century. There is a consensus on the importance of their energy restoration, but less importance has been given to other aspects, such as fire safety. This chapter discusses which factors and typologies increase the risk of fire and shows how safety is currently very weak. In addition, the social and demographic characteristics of the occupants further increase the risk of fire. For this reason, safety should be considered when addressing the regeneration of these neighborhoods.

CONCLUSION

- The systematic study of the buildings fire risk by typologies allows affording big urban ensembles. The results can be of interest for regulators and fire fighters. The first can inform their future decisions and the latter can develop risk maps to optimize their action. The methodology created for this analysis can be easily replicated in any other city to obtain an approach to the fire risk level of its residential buildings.
- To evaluate the level of risk, it is taken as a model of "safe building" that which complies with all the requirements of the CTE DB-SI. It would be interesting to implement a specific fire risk assessment method to analyze dwellings, which would allow a more precise study of specific factors of this type of buildings.
- The results for Pamplona show that the highest blocks built before the implementation of fire regulations present the highest risk level. The systematic application of the MEREDICTE© Method confirms this result.
- Some of the conflictive elements in the analyzed buildings are: the non-watertight concealed spaces generated under the roof, when it rests on a wooden substructure; the open staircases in high-rise buildings, the absence of fire protection systems and the presence of unprotected special risk locals.
- Unless some of the studied buildings have been renovated, none of them has had an intervention in the evacuation routes. The most significative improvements consisted in the installation of extinguishers, lighting and signing.
- As part of urban sustainability, it is necessary to address the regeneration of social neighborhoods built in the second half of the 20th century. This regeneration must be comprehensive, and it must include the aspect of fire safety.
- The future architectural interventions that may be performed in these buildings are a great opportunity to increase their fire safety. In those cases where it is not possible to comply with the prescriptions of the regulation, the performance-based design is the appropriate tool to reduce the level of risk, based on a correct evaluation of each case study.

REFERENCES

Ahrens, M. (2019). *Smoke Alarms in U.S. Home Fire.* https://www.nfpa.org/-/media/Files/News-and-Research/Fire-statistics-and-reports/Detection-and-signaling/ossmokealarms.pdf

Bruck, D. (2001). The who, what, where and why of waking to fire alarms: A review. *Fire Safety Journal, 36*(7), 623–639. doi:10.1016/S0379-7112(01)00025-X

Domínguez Arribas, L. E., Serrano Lanzarote, B., García-Prieto Ruiz, A., Ortega Madrigal, L., & Soto Francés, L. (2015). *Estudio del Potencial de ahorro energético y reducción de emisiones de CO2 en la Comunitat Valenciana.* www.five.es

Echeverría Trueba, J. B., Rubio Poves, N., Rodríguez Oyarbide, I., Jönsson, J., Vigne, G., & Sánchez Peña, J. (2020). Retos de la seguridad en caso de incendio de levantes en edificios de vivienda existentes. *Informes de la Construcción, 559.*

Eurostat. (n.d.). *Housing statistics.* Retrieved December 18, 2019, from https://ec.europa.eu/eurostat/statistics-explained/index.php?title=Housing_statistics#Type_of_dwelling

Fernández-Vigil, M., & Echeverría, J. B. (2019). Elderly at Home: A Case for the Systematic Collection and Analysis of Fire Statistics in Spain. *Fire Technology, 55*(6), 2215–2244. doi:10.100710694-019-00852-6

Fundación Mapfre & APTB. (2017). *Víctimas de Incendios en España en 2016.* doi:10.2865/5721

Fundación Mapfre & APTB. (2010). *Víctimas de incendios en España 2010.* https://www.fundacionmapfre.org/fundacion/es_es/images/informe-victimas-incendios-en-espana-2010-actualizado_tcm1069-211526.pdf

Fundación Mapfre & APTB. (2011). *Víctimas de Incendios en España 2011.* https://www.fundacionmapfre.org/fundacion/es_es/images/informe-victimas-incendios-en-espana-2011_tcm1069-211527.pdf

Fundación Mapfre & APTB. (2014). *Víctimas de incendios en España 2012 y 2013.* https://www.fundacionmapfre.org/fundacion/es_es/images/informe-victimas-incendios-en-espana-2012-2013_tcm1069-211528.pdf

Fundación Mapfre & APTB. (2015). *Víctimas de incendios en España en 2014.* https://www.fundacionmapfre.org/fundacion/es_es/images/informe-victimas-incendios-en-espana-2014_tcm1069-211529.pdf

Fundación Mapfre & APTB. (2016). *Víctimas de incendios en españa en 2015.* https://www.fundacionmapfre.org/fundacion/es_es/images/informe-victimas-incendios-espana-2015_tcm1069-388743.pdf

Fundación Mapfre & APTB. (2018). *Estudio de víctimas de incendios en España en 2017.* https://www.fundacionmapfre.org/fundacion/es_es/images/informe-victimas-incendios-espana-2017_tcm1069-522308.pdf

Instituto Nacional de Estadística. (2011). *Censos de Población y Viviendas 2011.* Cifras de Población y Censos Demográficos. https://www.ine.es/censos2011_datos/cen11_datos_inicio.htm

Instituto Nacional de Estadística. (2014). *Proyección de la Población de España 2014–2064.* Author.

Marshall, S. W., Runyan, C. W., Bangdiwala, S. I., Linzer, M. A., Sacks, J. J., & Butts, J. D. (1998). Fatal Residential Fires: Who dies and who survives? *Journal of the American Medical Association*, *279*(20), 1633–1637. doi:10.1001/jama.279.20.1633 PMID:9613913

Ministerio de Fomento. (2019a). *Código Técnico de la Edificación: Documento Básico de Seguridad de utilización y accesibilidad.* https://www.codigotecnico.org/pdf/Documentos/SUA/DBSUA.pdf

Ministerio de Fomento. (2019b). *Código Técnico de la Edificación: Documento Básico de Seguridad en caso de incendio.* https://www.codigotecnico.org/images/stories/pdf/seguridadIncendio/DccSI.pdf

Ministerio de Fomento. (2019c). *Código Técnico de la Edificación. Parte I.* https://www.codigotecnico.org/pdf/Documentos/Parte1/Parte_I_20dic2019.pdf

Ministerio de Obras Públicas. (1939). *Instrucción para el Proyecto y Ejecución de Obras de Hormigón.*

Ministerio de transportes movilidad y agenda urbana. (2020). *Nota sobre el Documento Reconocido CTE-DR/048/14.* https://www.mitma.es/carreteras/trafico-velocidades-y-accidentes-mapa-estimacion-y-evolucion/mapas-de-trafico/2019

Ortega Madrigal, L., García-Prieto Ruiz, A., & Serrano Lanzarote, B. (2013). *Catálogo de tipología edificatoria residencial.* Ámbito. https://episcope.eu/fileadmin/tabula/public/docs/brochure/ES_TABULA_TypologyBrochure_IVE.pdf

Pérez-Martín, J. C. (2014). *MEREDICTE Método de evaluación de riesgo en caso de incendio en el marco del Código Técnico de la Edificación.* Academic Press.

Pérez-Martín, J. C., Díaz-Díaz, R., & Santos García, R. (2010). Método de evaluación del riesgo de incendio en el marco del Código Técnico de la Edificación. *DYNA Ingeniería e Industria*, *85*(4), 303–314. doi:10.6036/3437

Real Decreto 1053/1985, de 25 de mayo, sobre ordenación de la estadísticas de las actuaciones de los Servicios contra Incendios y de Salvamento. (1985). In *Boletín Oficial del Estado* (Vol. 158, Issue 12771, pp. 20843–20844).

Real Decreto 2059/1981 por el que se aprueba la Norma Básica de la Edificación. (1981). In *Boletín Oficial del Estado* (Vol. 224, pp. 21707–21727).

Real Decreto 3062/1973, de 19 de octubre, por el que se aprueba la instrucción para el proyecto y la ejecución de obras de hormigón en masa o armado. (1973). In *Boletín Oficial del Estado* (Vol. 2993, pp. 23701–24204). https://www.boe.es/boe/dias/1973/12/07/pdfs/R23701-24204.pdf

Real Decreto 513/2017, de 22 de mayo, por el que se aprueba el Reglamento de instalaciones de protección contra incendios. (2017). In *Boletín Oficial del Estado* (Vol. 139, Issue I, pp. 48349–48386). http://www.boe.es

Real Decreto 842/2013, de 31 de octubre, por el que se aprueba la clasificación de los productos de construcción y de los elementos constructivos en función de sus propiedades de reacción y de resistencia frente al fuego. (2013). In *Boletín Oficial del Estado* (Vol. 281, Issue I, pp. 93493–93527).

Rohde, D., Corcoran, J., Sydes, M., & Higginson, A. (2016). The association between smoke alarm presence and injury and death rates: A systematic review and meta-analysis. *Fire Safety Journal, 81*, 58–63. doi:10.1016/j.firesaf.2016.01.008

Rubio del Val, J. (2011). Rehabilitación Urbana en España (1989-2010). Barreras actuales y sugerencias para su eliminación. *Informes de La Construccion, 63*, 5–20. doi:10.3989/ic.11.060

SAVIArquitectura. (2016). *Proyecto prestaRener. Capítulo 3: Tipologías*. Author.

SAVIArquitectura & Worcester Polythecnic Institute. (2016). *Proyecto prestaRener: Protocolo de Actuación en la Rehabilitación de la Envolvente de Edificios basado en el Diseño Prestacional*. Author.

Stevenson, M. R., & Lee, A. H. (2003). Smoke alarms and residential fire mortality in the United States: An ecologic study. *Fire Safety Journal, 38*(1), 43–52. doi:10.1016/S0379-7112(02)00048-6

Warda, L., Tenenbein, M., & Moffatt, M. E. K. (1999). House fire injury prevention update. Part II. A review of the efectiveness of preventive interventions. *Injury Prevention, 5*(3), 217–225. doi:10.1136/ip.5.3.217 PMID:10518271

KEY TERMS AND DEFINITIONS

Combustion: Exothermic reaction of a combustible substance with an oxidizer.

Compartmentation: The technique of dividing a building into a number of compartments.

Escape: The fire safety tactic of ensuring that the occupants of the building are able to move to places of safety before they are threatened by heat and smoke.

Fire Engineering: Design which considers the building as a complex system, and fire safety as one of the many interrelated subsystems which can be achieved through a variety of equivalent strategies.

Fire Resistance: The properties that establish the maintenance of the capacity of sustenance of the work for a determined time in case of fire.

Fire Risk: The product of 1) probability of occurrence of a fire to be expected in a given state, 2) consequence or extent of damage to be expected on occurrence of a fire.

Reaction to Fire: The properties that limit the initiation and propagation of the fire and the smoke inside the work.

APPENDIX 1

Table 10. List of codes and abbreviations used

Abbreviation	Spanish Meaning	English Meaning
	STANDARDS AND CODES	
CTE	Código Técnico de la Edificación	Spanish Building Code
DB-SI	Documento Básico: Seguridad en caso de Incendio	Basic Document: Fire Safety
DB-SI-1	Documento Básico: Seguridad en caso de Incendio – Propagación interior	Basic Document: Fire Safety – Interior Fire Spread
DB-SI-2	Documento Básico: Seguridad en caso de Incendio – Propagación exterior	Basic Document: Fire Safety – Exterior Fire Spread
DB-SI-3	Documento Básico: Seguridad en caso de Incendio – Evacuación de ocupantes	Basic Document: Fire Safety – Evacuation of occupants
DB-SI-4	Documento Básico: Seguridad en caso de Incendio – Instalaciones de protección contra incendios	Basic Document: Fire Safety – Fire protection Systems
DB-SI-5	Documento Básico: Seguridad en caso de Incendio – Intervención de bomberos	Basic Document: Fire Safety – Intervention of extinction services
DB-SI-6	Documento Básico: Seguridad en caso de Incendio – Resistencia al fuego de la estructura	Basic Document: Fire Safety – Fire resistance of the load-bearing structure
DB-SUA	Documento Básico: Seguridad de utilización y accesibilidad	Basic Document: Safety of Use and Accesibility
RIPCI	Reglamento de Instalaciones de Protección contra Incendios	Fire Protection Systems Regulation
-	Instrucción para el Proyecto y Ejecución de Obras de Hormigón (1939)	Instruction for the Design and Execution of Concrete Projects (1939)
EH-73	Instrucción para el Proyecto y la ejecución de obras de hormigón en masa o armado (1973)	Instruction for the Design and Execution of Concrete Projects (1973)
NBE-CPI-81	Norma Básica de la Edificación - Condiciones de protección contra incendio en los edificios (1981)	Basic Standard of Fire Protection
	BUILDING TYPOLOGIES	
T1	Bloque lineal	Linear block
T1.1	Bloque lineal en manzana	Linear block with nearby building environment
T1.2	Bloque lineal no agrupado	Non-clustered linear block
T1.3	Bloque lineal con una orientación	Linear block with one orientation
T1.4	Bloque lineal con patio mínimo	Linear block with minimum courtyard
T2	Bloque en H	H-Block
T2.1	Bloque en H	H-block
T2.2	Bloque en H, altura mayor que B+A	H-block, height above Ground + 4
T2.3	Bloque en H con patio mínimo	H-block with minimum courtyard
T3	Torre	Tower
T3.1	Torre en H	H-tower
T3.2	Torre en cruz	Cross Tower
T4	Bloque en T	T-Block
T4.1	Bloque en T, 3 viviendas por planta	T-block, 3 dwellings per floor
T4.2	Bloque en T, 2 viviendas por planta	T-block, 2 dwellings per floor
T5	Unifamiliar	Single-family houses
T5.1	Unifamiliar, Planta baja	Single-family house, ground floor
T5.2	Unifamiliar, Baja + 1	Single-family house, ground floor + 1
	MEREDICTE	
NRG	Nivel de Riesgo Global	Global Risk Level
PP	Peligro Potencial	Potential Hazard
AS	Aforo del sector	Sector Capacity
T	Tetraedro del fuego	Fire tetrahedron
CO	Características de los ocupantes	Occupants Characteristics
CA	Características arquitectónicas	Architectural characteristics
NPG	Nivel de protección global	Level pf Global Protection
NPA	Plan de Autoprotección	Self-protection Plan
RFE	Resistencia al fuego de la estructura	Fire Resistance of the structure
P	Propagación interior y exterior	Interior and exterior fire spread
EO	Evacuación de ocupantes	Evacuation of the occupants
IPCI	Instalaciones de protección contra incendios	Fire Protection Systems
IB	Intervención de bomberos	Intervention of the fire services

APPENDIX 2

Table 11. Global risk level, potential hazard and global protection level obtained in the case studies, by sub-typology

Typology	Global Risk Level			Potential Hazard Average	Global Protection Level Average
	Average	Minimum	Maximum		
T1.1	4.14	2.63	8.25	2.3	0.56
T1.2	6.44	2.26	15.48	4.1	0.68
T1.3	8.07	2.53	14.20	4.7	0.62
T1.4	7.09	2.68	17.38	3.7	0.54
T1	5.75	2.26	17.38	3.5	0.63
T2.1	6.55	2.78	9.34	4.3	0.65
T2.2	19.07	16.08	22.06	8.6	0.45
T2.3	5.34	2.68	8.27	3.6	0.68
T2	8.74	2.68	22.06	5.0	0.62
T3	4.79	3.04	8.32	3.4	0.73
T4	7.55	3.01	13.44	5.5	0.73
Total	6.43	2.26	22.06	3.9	0.64

Chapter 9
Research of Alternative Ecological Waste Materials Used in Geopolymers for Sustainable Built Environments

mahmoud Ziada
Istanbul University-Cerrahpasa, Turkey

Yosra Tammam
Istanbul Gelisim University, Turkey

Savaş Erdem
https://orcid.org/0000-0003-0906-5233
Istanbul University-Cerrahpasa, Turkey

ABSTRACT

Infrastructure and industrial projects continue to expand, resulting in a steady rise in demand for cement. However, it is claimed that nature and the environment suffer significant harm throughout the cement manufacturing process due to the cement factory's greenhouse gas emissions. Additionally, economic difficulties have emerged as a result of energy usage during manufacturing. As a result, waste materials have begun to be utilized in geopolymer concrete in place of cement. Therefore, environmentally friendly materials with a low carbon footprint have been in high demand across the world's building sector. Activators are used with waste materials as binders of geopolymers without using cement. Thus, geopolymers have gained importance due to their environmental compatibility, sustainability, and durability. This chapter comprehensively reviews the environmental suitability of geopolymers produced using waste materials. In addition, the authors tried to collect most geopolymer physical, mechanical, microstructural, and durability properties.

DOI: 10.4018/978-1-6684-4030-8.ch009

INTRODUCTION

The construction sector is a significant user of natural assets. Construction consumes more than 40% of all resources utilized globally (Horvath, 2004), and specific resources, such as aggregate or sand, are becoming rare in particular areas (Ioannidou et al., 2017; Peduzzi, 2014). Reduced resource usage must thus be matched with reduced greenhouse gas emissions. Attempts have been devoted in recent decades to design more energy-efficient buildings, which has lowered the energy required to operate the structures. Unfortunately, the embodied energy involved with the manufacture of building materials was not adequately examined (Sartori & Hestnes, 2007). Also, cement, which is the principal element of concrete, is perhaps the biggest obstacle in combating climate change, and the world's demand for it seems limitless. Cement has been extensively used as an efficient binder in mortar and concrete, and other building materials in the construction engineering business. Cement manufacturing is well acknowledged as a significant greenhouse gas emission source (Du et al., 2017; Leung et al., 2014; Ziada et al., 2021), accounting for around 6% of total emissions of CO_2, as recorded by the International Energy Agency (Á. Palomo et al., 2011).

Nonetheless, worldwide demand for OPC will almost double by 2050 (Pacheco-Torgal et al., 2011). A novel form of green and sustainable materials known as geopolymer was developed (Xie & Ozbakkaloglu, 2015). In general, geopolymer is significantly more environmentally friendly than cement in terms of CO_2 emissions and energy consumption (Duxson, Provis, et al., 2007; McLellan et al., 2011; Tammam et al., 2021). Joseph Davidovits (Davidovits, 1991) presented the concept of geopolymers as a novel material in 1978. The previous study (S. Zhang et al., 2004) exhibited several intriguing properties, including high strength, corrosion resistance, water resistance, high-temperature resistance, and the presence of contained metal ions. Blast furnace slag, red mud, fly ash, glass waste, brick powder, and other waste materials are often employed as aluminosilicates in geopolymer concrete applications. The ultimate attributes of geopolymers, such as strength and setting time, are substantially influenced by the silica, alumina, sodium, and water ratios, which must be considered in the mix design (Rowles & O'connor, 2003). Geopolymers have a wide array of uses in the nuclear waste disposal industries, membrane materials, coating, metallurgy, repair, and transportation (He et al., 2013; H. Y. Zhang et al., 2015; J. Zhang et al., 2014). As aggregates typically account for 60 to 80 percent of the volume of concrete, designers may explore utilizing green aggregates and substituting regular cement with green binders such as geopolymer in an effort to make concrete more ecologically friendly. Coarse or fine aggregates are crucial features in concrete because they promote mechanical strength, volume stability, and cost-effectiveness (Tammam et al., 2021). Waste materials such as fly ash are also used in engineered cement composites (ECC). Ziada et al. (Ziada et al., 2022) produced a high-performance, sustainable cementitious composite by using a fly ash/cement ratio of 1.25.

In industrialized nations, emphasis has been placed on fire and heat resistance and radioactive and hazardous waste management, resulting in advanced geopolymer resins and binders. Green-House and global warming concerns are the main forces for sustainable development in developing nations. Geopolymer Green chemistry creates new forms of low-CO_2 cement in construction and infrastructure, using geological and industrial waste resources. Ukrainian scientist Glukhovsky discovered the potential of synthesizing binders using aluminosilicates (clays, rocks, and slags) and alkali metal solutions. He coined the term "soil cement" for the binder and "soil silicates" for the resulting concrete. This material was used to construct two nine-story residential structures in Mariupol, Ukraine, in the 1960s. These structures stand more than 50 years later (Hua et al., 2008). There were a few further structures constructed using this approach. However, the first residential structure constructed entirely of alkali-activated concrete without Portland cement was constructed in 1989 in Lieptsk, Russian Federation (Almutairi et al., 2021).

Recent urbanization, particularly in developing nations, has exacerbated the negative environmental effect of cement manufacturing (Alaloul et al., 2020). As a result, it is critical that sustainable alternatives to cement be utilized in building applications to preserve the environment's sustainability (Hamada et al., 2020). Numerous waste products created by different sectors may be utilized as sustainable substitutes for the traditional resources used in the cement manufacturing process. Consequently, the utilization of such wastes in manufacturing a sustainable alternative to cement would result in a considerable decrease in GHG emissions, the cost of raw materials, and the use of natural raw resources connected with cement (Das et al., 2020). Alkali-activated materials, which result in geopolymer concrete, are a potential option that may be utilized entirely instead of cement in concrete (Tayeh et al., 2021).

This chapter explores the geopolymer potential as a building material with more significant energy savings, sustainability, and decreased carbon footprints for application in severe locations. The appropriate utilization of industrial waste recycling and the prevention of diminishing natural resources are also emphasized. The performance of geopolymers in terms of durability and mechanical properties can be emphasized. The widespread use of environmentally friendly geopolymers has been essential to reducing carbon emissions.

GEOPOLYMERIZATION

Geopolymers are formed when the materials containing aluminosilicates are mixed with the activator solutions and dissolved in them at various temperatures, leading to a 3-dimensional network structure of silicoaluminate and amorphous (De Silva et al., 2007; A. Palomo et al., 1999). Although experts have differing views on the reaction mechanism behind geopolymerization, the majority believe that the process may be separated into three various steps (De Silva et al., 2007; Duxson, Fernández-Jiménez, et al., 2007; Prud'Homme et al., 2011). Firstly aluminosilicate materials are dissolved in the activator solutions. They create the alumina tetrahedron unit and free silica. The second is material transfer, solidification/gelation, which occurs when silica hydroxyl and alumina react to produce a gel. Water is expelled from the structure at this step due to the hydrolysis process. The third step is that once the gel phase condenses, it results in forming a 3-dimensional network of silicoaluminate called a geopolymer.

GEOPOLYMER MATERIALS

Binder Materials

Fly ash (FA), metakaolin (Mk), and granulated blast furnace slag (GBFS) are often employed as geopolymer source materials. They are owing to their ease of availability and attractive mechanical qualities (Şahin et al., 2021; Tammam et al., 2021; Ziada et al., 2021). The calcium reaction at room temperature may cure a high calcium FA geopolymer combination. However, the poor binding strength of high calcium FA without additions is due to their very slow geopolymerization at ambient conditions (Somna et al., 2011).

Huseien et al. (Ghasan F Huseien et al., 2018) investigated the effect of the metakaolin on the mechanical properties of the produced samples. Metakaolin was utilized to partially substitute for 5%, 10%, and 15% of GBFS in the geopolymer. Ternary blended geopolymer with different amounts of plum oil fuel ash, FA, and GBFS was created. These materials influenced the flexural strength of the geopolymer mortars (Ghasan Fahim Huseien et al., 2016). Numerous alkaline solutions were utilized, including a classic alkaline solution, sodium silicate, and sodium hydroxide composed of sodium silicate and sodium hydroxide. The curing period's effect on the composites' mechanical properties was also examined. Al-Majidi et al. (Al-Majidi et al., 2016) investigated the impact of GBFS substitution with fly ash at 10, 20, 30, 40, and 50% ratios on the mechanical properties of geopolymer mortars. Hani et al. (Alanazi et al., 2016) investigated the effect of Mk on the strengths and long-term qualities of geopolymer mortars.

Moura et al. (Moura et al., 2011) examined the impact of replacing metakaolin-based geopolymer with calcium hydroxide at 5% and 10% under various alkali solution concentrations. Hu et al. (Hu et al., 2008) investigated the bonding and abrasion resistance of geopolymer mortar when GGBFS was substituted for MK at 0% and 20%. Torgal et al. (Pacheco-Torgal et al., 2006) investigated using a geopolymeric binder composed of tungsten mine waste and calcium hydroxide at a 10% substitution rate.

Aggregates

Sands like river sand are often utilized as a filler material in the geopolymers. In addition, industrial wastes and recycled concrete aggregates are also used as aggregates in geopolymer mortars and concretes. Ziada et al. (Ziada et al., 2021) used basalt waste as a filler in fly ash-based geopolymer. Also, Tammam et al. (Tammam et al., 2021) used the powders of basalt waste, marble waste, and limestone waste with various ratios in fly ash-based geopolymer mortar.

Activators

Alkali Activators

Several studies have demonstrated that alkali activators are commonly used to activate geopolymer materials. Alkaline activators are available in solid and liquid form. Activators such as Na_2SiO_3 and NaOH have been frequently employed in prior investigations. The activator sodium water glass was used to create a geopolymer based on MK with an approximately 60 MPa compressive strength result (Tchakouté & Rüscher, 2017). Using hydrated lime and solid Na_2CO_3 as activators in the geopolymer samples cured at 25 and 85 °C was able to achieve a strength of 50 and 85 MPa (Kovtun et al., 2015). Sodium sulfate as an activator substantially affects the early strength of FA-based geopolymers, but not on those with a high Fe_2O_3 concentration (Velandia et al., 2016).

Alkali activators based on sodium and potassium have been the most frequently utilized activators. Earlier research has demonstrated that sodium-based alkali activators are more effective in activating FFA than potassium-based activators (Helmy, 2016). When potassium compounds were used in geopolymer systems, they demonstrated greater alkalinity than sodium hydroxide. Potassium carbonate solid was an excellent activator (Askarian et al., 2018). Lithium carbonate, calcium aluminate, sodium aluminate, calcium oxide, and lithium hydroxide activators are combined to form the geopolymer. In addition, the 28-day strength of these FA-slag-based geopolymers reached 38 MPa (Askarian et al., 2019). The researchers showed that lithium hydroxide solution might be used as an alkali initiator to limit active silica dissolution (Taha & Nounu, 2008). $NaAlO_2$ would be an intriguing raw material since it contains active aluminum (Al) (Helmboldt et al., 2000). As a result, it is critical to choose activators that are compatible with the properties of raw materials. Olive oil biomass ash has the potential to be employed as an activator of geopolymers contained blast furnace slag (BFS) (Alonso et al., 2019).

Acidic Activators

While the majority of geopolymers are alkali-base composites, a few are acid-base composites. Phosphoric acid solutions were utilized as an activator solution to produce metakaolin-based geopolymers. The compressive strength of produced geopolymers was about 90 MPa (Shuai et al., 2020; Tchakouté & Rüscher, 2017). The consolidation of phosphate base geopolymer samples with varying Si/Al and Al/P ratios demonstrates that Al-rich, Si-rich, and P-poor components consolidate more effectively (H Celerier et al., 2018). Additionally, another research showed that acid-base geopolymers have superior thermal endurance and mechanical qualities than alkali-activated composites (Hélène Celerier et al., 2019).

GEOPOLYMER PREPARATION PROCEDURES

According to the literature, three primary processes are used to create geopolymer materials: The first one is the pouring technique. This technique is the most often utilized way of preparation, and it is identical to the procedure used to prepare standard cement concrete. Since this type of preparation demands a fluid paste, more water is required in the preparation, which typically amounts to 20%–40% of the total mass of the cementitious material. Pouring is often used to create products with intricate forms. Pouring is a promising approach for in situ applications, particularly for items with complicated forms. The compressive strength of the materials, on the other hand, is often less than 100 MPa (Barbosa et al., 2000; H. Ma et al., 2002; Ziada et al., 2021). The second technique is the compression molding technique. This approach creates a viscous colloidal slurry from the aluminosilicate solid and alkaline activator solutions. Following stirring, the mixture is pressed and molded at a 5–10 MPa to get reasonably high strength, up to 100 MPa (Wang et al., 2005). Compression modeling looks to be a viable option for precast construction, although managing alkali solutions might be challenging. The third method is the hot-pressing method, this technology combines heat and pressure to provide significantly increased mechanical strength in a shorter period. This allows for forming a better-developed geopolymer matrix and a nearly poreless structure (Ranjbar et al., 2017). Also, some researchers employ auxiliary techniques such as microwave-assisted method, ultrasonic aided method, and others to improve the production of geopolymer materials (David Feng et al., 2004). The study of Tammam et al. (Tammam et al., 2021) is presented as an example of geopolymer preparation. In their study, firstly, alkaline activator solution (a combination of NaOH and Na_2SiO_3) was mixed with fly ash, and then slag was added and stirred until homogenous. The required quantity of filler was then combined into fly ash paste. The resultant mixture was then cast into molds to prevent entrapped air and voids created by the vibrating sample molds. After an hour of casting heat, all specimens were cured in an oven at 80C for 24 hours; the samples were then stored in laboratory conditions until the planned testing. Similarly mixing procedure was performed by Al-mashhadani et al. (Al-mashhadani et al., 2018) and their mixing procedure is shown in Figure 1.

Figure 1. The mixing procedure of the fly ash-based geopolymer mortar
(Al-mashhadani et al., 2018)

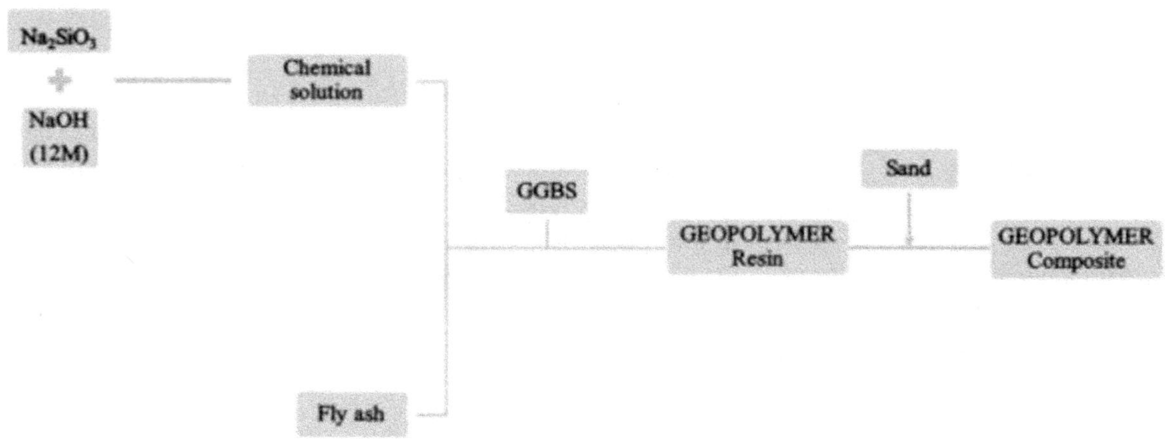

GENERAL APPLICATIONS OF GEOPOLYMER MATERIALS

Geopolymers, in general, have significant compressive strength and low thermal conductivity. Thus, the geopolymers can be used as structural materials, especially given their advantageous properties, for instance, remaining devoid of fibers and keeping structural stability after and during fire exposure (Z. Zhang et al., 2014). Also, porous geopolymers have the potential to be used as adsorbent components. Geopolymers are used as sorbents, and owing to geopolymer cation exchange characteristics, they may be used as practical water treatment elements (Cheng et al., 2012). Porous monoliths (such as honeycombs and foams) outperform typical powdered or packed-bed adsorbents because they increase contact duration, minimize pressure reduction, and are readily collected and recycled (Grosse et al., 2009). In addition, Zhang et al. (Z. Zhang et al., 2015) examined the sound absorption of porous geopolymers utilizing the technique of impedance tube, and the sound absorption capacity was evaluated utilizing the coefficient of sound absorption. They found that the composition of slag and fly ash geopolymers and their thickness influenced their coefficient of sound absorption.

Moreover, Ge et al. (Ge et al., 2015) examined the Ni^{2+} removal using a membrane of metakaolin-based geopolymer with porosity up to 62% (volume). Additionally, they demonstrated an excellent prospective use for ammonium ion reduction and possible recovery from municipal wastewaters. Rice husk ash, Mk, and foamed with water vapor were utilized to prepare porous geopolymer to eliminate the cesium ions (López et al., 2014). Duan et al. (Duan et al., 2016) investigated the removal of Cu ions from porous geopolymers made from fly ash and iron ore tailings and foamed with H_2O_2. The data indicate that geopolymer dose, the starting Cu^{2+} concentration, pH, temperature, and duration of contact all had a significant effect on adsorption capacity. After exchanging the copper ions, specimens demonstrated a high degree of antibacterial activity (Hashimoto et al., 2015). Minelli et al. (Minelli et al., 2016) studied the selectivity, and the CO_2 adsorption of porous geopolymers generated utilizing fume silica. At atmospheric pressure, the test results exhibited a promising capacity of adsorption.

PROPERTIES OF GEOPOLYMER MATERIALS

Mechanical Properties of Geopolymers

Various precursors and activators have been used to manufacture different geopolymers in the literature. Also, the durability, mechanical and microstructural properties, and other properties were investigated by many researchers. Table 1 is shown the precursor and activators used in the geopolymers and their compressive strength.

Corrosion and Durability Properties of Geopolymer Composites

This section contains various studies conducted on the durability of geopolymers that are reviewed. Tammam et al. (Tammam et al., 2021) investigated the effect of high temperature on fly ash-based geopolymer's mechanical properties. Their study exposed fly ash-based geopolymer samples to 200, 400, 600, and 800°C. They found that the strength characteristics of the geopolymer matrix deteriorated significantly as the temperature increased between 600 and 800°C. As the temperature increased to 600°C, the UPV readings decreased significantly. Also, cracks began to form when the material's compressive strength decreased significantly between 600 and 800°C, and cracks began to form. When the temperature hit 800 C, a visible color shift occurred in the geopolymer specimen. However, the fissures remained at a slower pace, indicating that the geopolymer samples maintained stable conditions when exposed to high temperatures. Ziada et al. (Ziada et al., 2021) examined the high-temperature properties of fly ash-based and basalt powder waste-filled geopolymer mortars. They also investigated the effect of basalt fiber (BF) on the high-temperature exposed geopolymer samples. They found that the basalt fiber enhanced the strengths of the geopolymer samples. Also, 12BF samples treated to 800°C increased by 15.79 percent for the compressive strength and 44.69 percent for the flexural strength compared to 0BF. Özcan Karakoç (Özcan & Karakoç, 2019) examined the resistance of slag based geopolymer produced to 5% hydrochloric, hydrofluoric, phosphoric, and sulfuric acids during 90 days. The samples produced using Elazığ ferrochrome slag demonstrated greater resistance to all acids than those manufactured using blast furnace slag. Zhang et al. (P. Zhang et al., 2020) provided an overview of geopolymer's fresh qualities, including its endurance. This research established that geopolymer is more resistant to acid attack than cement composites, as strength decreases were much lower when these samples were subjected to 2% H_2SO_4 (Okoye et al., 2017) and 5% H_2SO_4 (Manjunath et al., 2019). Aliabdo et al. (Aliabdo et al., 2018) studied geopolymer's resistance to corrosion. Numerous experiments were conducted to determine the resistance to corrosion, the amount of water absorbed, the depth of water penetration, and the void ratio. Temperature, cement addition, binder ratio, and the molarity of the activator were all evaluated in this study. Based on the test's findings, corrosion resistance was more significant in geopolymer composites than cement composites.

Additionally, by increasing the binder concentration and adding cement to the geopolymer composites, they noticed a decrease in water penetration depth, voids, water absorption, and corrosion resistance. The MK-based geopolymer was shown to be more resistant to corrosion than the FA-based geopolymer. Albitar et al. (Albitar et al., 2017) investigated the corrosion resistance of geopolymer composites in various corrosive conditions. Magnesium sulfate solutions, sodium sulfate, and sodium chloride at a concentration of 5% were used to treat samples of the composites. Additionally, they were exposed to heating cycles, cooling, drying, alternating wetting, and 3% sulphuric acid. The results demonstrated

Table 1. The precursor and activators used in the geopolymers and their compressive strength

Activators	Precursor	Compressive strength range (MPa)	References
Sodium sulfate and Na_2SiO_3-anhydrous	Fly ash sinking beads GGBFS	50–70	(C. Ma et al., 2020)
NaOH and wood biomass ash	Diatomite	28–48	(Hassan et al., 2019)
Sodium carbonate	Feldspar, cement kiln dust	13–52	(Abdel-Gawwad & Khalil, 2018)
Na_2CO_3, dolomite	Bentonite	10–38	(Peng et al., 2017)
Na_2CO_3, NaOH	Albite	42–45	(Dingwu Feng et al., 2012)
NaOH	Feldspar, air cooled slag	55–85	(Abdel-Gawwad et al., 2018)
Na_2SiO_3, NaOH	Fly ash, GGBFS	47–53	(Ziada et al., 2021)
Sodium hydroxide, sodium carbonate	Low-quality kaolin	1–63	(Peng et al., 2015)
Sodium hydroxide, hydrous Na_2SiO_3, anhydrous Na_2SiO_3	Metakaolin, low calcium fly ash	20–46	(H.-Y. Zhang et al., 2021)
calcium oxide, sodium carbonate, Sodium silicate,	Kaolin	7–19	(Ababneh et al., 2020)
Sodium carbonate, potassium hydroxide	Rice husk ash, municipal solid waste ash, high calcium fly ash	14–50	(Almalkawi et al., 2019)
Sodium aluminate, chlorosilane production residue,	Microsilica, rice husk ash, GGBFS	30–58	(Sturm et al., 2018)
Desulphurization dust	Microsilica, GGBFS	32–35	(Adesanya et al., 2020)
NaOH, solid sodium silicate,	Metakaolin	38–70	(Hajimohammadi et al., 2017)
Sodium carbonate, lime, sodium sulfate	Natural pumice	5–22	(Almalkawi et al., 2017)
Anhydrous Na_2SiO_3, Na_2CO_3, NaOH	Nickel slag	78–82	(Xu et al., 2020)
Na_2SiO_3, NaOH	Fly ash, GGBFS	51–68	(Tammam et al., 2021)
NaOH, Red mud	Silica fume	10–30	(Ye et al., 2017)
Sodium aluminate	Rice hull ash	10–22	(Hajimohammadi & van Deventer, 2017)

that geopolymer composites were better in corrosion resistance to all corrosive solutions and conditions. Furthermore, the loss in compressive and tensile strengths was less pronounced in geopolymers than in cement composites. Additionally, after exposure to corrosive conditions, the flexural tensile and splitting strengths of geopolymer composites were more significant than the cement composites.

Also, Bellum et al. (Bellum et al., 2020) examined the effect of different FA and slag concentrations on the durability of geopolymer composites. They evaluated the chloride and acid attack resistance and the porosity, water absorption, and sorptivity. The results suggested that incorporating slag into FA-based geopolymer improved its mechanical characteristics. Also, slag addition to geopolymer samples resulted in decreased porosity and water absorption and increased compressive strength due to C–A–S–H gel development. This resulted in a more uniform microstructure of the samples. Valencia-Saavedra et al. (Valencia-Saavedra et al., 2020) investigated the resistance of FA/OPC and FA/GBFS composites to sulfuric and acetic acid solutions and compared the obtained results to cement composites. They discovered that alkali-activated composites degraded less in compressive strength and mass after 12 months, while cement composites degraded more. Law et al. (Law et al., 2015) examined the long-term durability of geopolymer composites mixtures. Numerous tests, including sorptivity, chloride diffusion, carbonation, and strength, were conducted on specimens composed of various mixtures, including varying amounts of blended cement, Ordinary Portland, and fly ash. According to the test findings, the strength increased as the dissolution of the FA grains and the heat used to cure the geopolymer specimens increased. Chloride diffusion was observed to be equivalent in geopolymer specimens to those of blended cement and Ordinary Portland cement. Another research used the Taguchi approach to optimize geopolymer blends using FA to increase their endurance in saltwater (Olivia & Nikraz, 2012). The mechanical qualities of geopolymer samples were observed to be higher compared to cement composites samples. Kupwade and Allouche (Kupwade-Patil & Allouche, 2013) investigated the durability of reinforced geopolymer composites for one year. Chemical analysis, mechanical analysis, ocular observations, element analysis, chloride penetration, and electrochemical analysis were performed on geopolymer samples. For comparison, similar experiments were done on the cement composites. According to the test findings, geopolymer samples had a lower chloride concentration, porosity, and diffusion coefficient than cement composites samples. Fly ash (class C) containing geopolymer composites exhibited greater corrosion resistance due to a denser matrix than fly ash (class F). Three different sources of fly ash (class F) were used, and none of them demonstrated similar corrosion resistance to fly ash (class C).

MICROSTRUCTURE ANALYSIS OF GEOPOLYMERS

Various analyzes as Fourier transform infrared spectroscopy (FTIR), X-ray powder diffraction (XRD), energy-dispersive X-ray spectroscopy (EDS), and scanning electron microscopy (SEM) are carried out to examine the properties of geopolymers. Scanning electron microscope (SEM) is the most widely used of these analyzes. Luo et al. (Luo et al., 2022) examined the microstructure of silty clay stabilized by metakaolin. They examined the microstructures of plain soil, lime stabilized soil, conventional Portland cement stabilized soil, and MKG stabilized soil using a scanning electron microscope. Figure 2 illustrates the usual properties at magnifications of 500 and 3000. MKG stabilized soil has a denser and more regular structure than standard Portland cement stabilized soil, lime stabilized soil, and untreated plain soil. Phoo-ngernkham, Maegawa, et al. (Phoo-ngernkham, Maegawa, et al., 2015) evaluated the shear bond and compressive strengths of GBFS-FA geopolymer in the presence of sodium hydroxide and sodium silicate solutions. Figure 3 illustrates the findings of SEM studies of the fracture surfaces of geopolymer pastes acquired in their investigation. They discovered that the matrix of the FA pastes was rather loose, as seen in Figures 3a, d, and g. The FA pastes, including sodium silicate (NS) and sodium hydroxide (NH), presented a more porous, but the FA-based paste containing NS and NH exhibited greater partly reacted and unreacted fly ash particles in the matrix of the paste.

THE ADVANTAGES AND DISADVANTAGES OF USING GEOPOLYMERS

Geopolymers have superior early mechanical characteristics over other materials (Phoo-ngernkham, Sata, et al., 2015). Geopolymer is unquestionably better than OPC for multiple reasons: less cracking, a

Figure 2. A typical microstructure of plain soil and different stabilizer
(Luo et al., 2022)

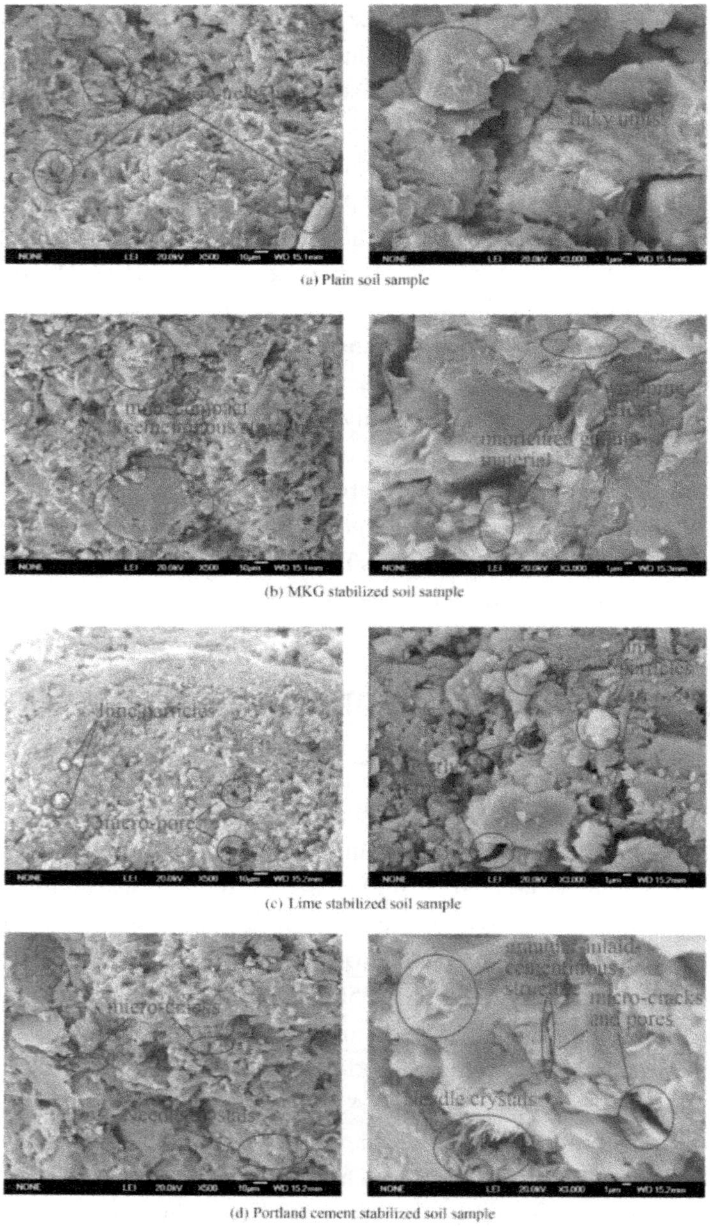

much smaller carbon footprint, greater resilience to sea salt, and endurance in cold areas. Bondar (Bondar, 2009) evaluated the carbon footprint and cost of mortars built with alternative binders and alkali activators. Given the increased demand for the concrete sector to promote sustainability, these issues are critical. The carbon footprint and the cost and compositions of geopolymers are approximately quantified compared to cement composites. Geopolymer has superior endurance to cement and repair materials and superior characteristics in acid and salt conditions (Ariffin et al., 2013). Currently, geopolymeric

Figure 3. SEM of geopolymer pastes
(Phoo-ngernkham, Maegawa, et al., 2015)

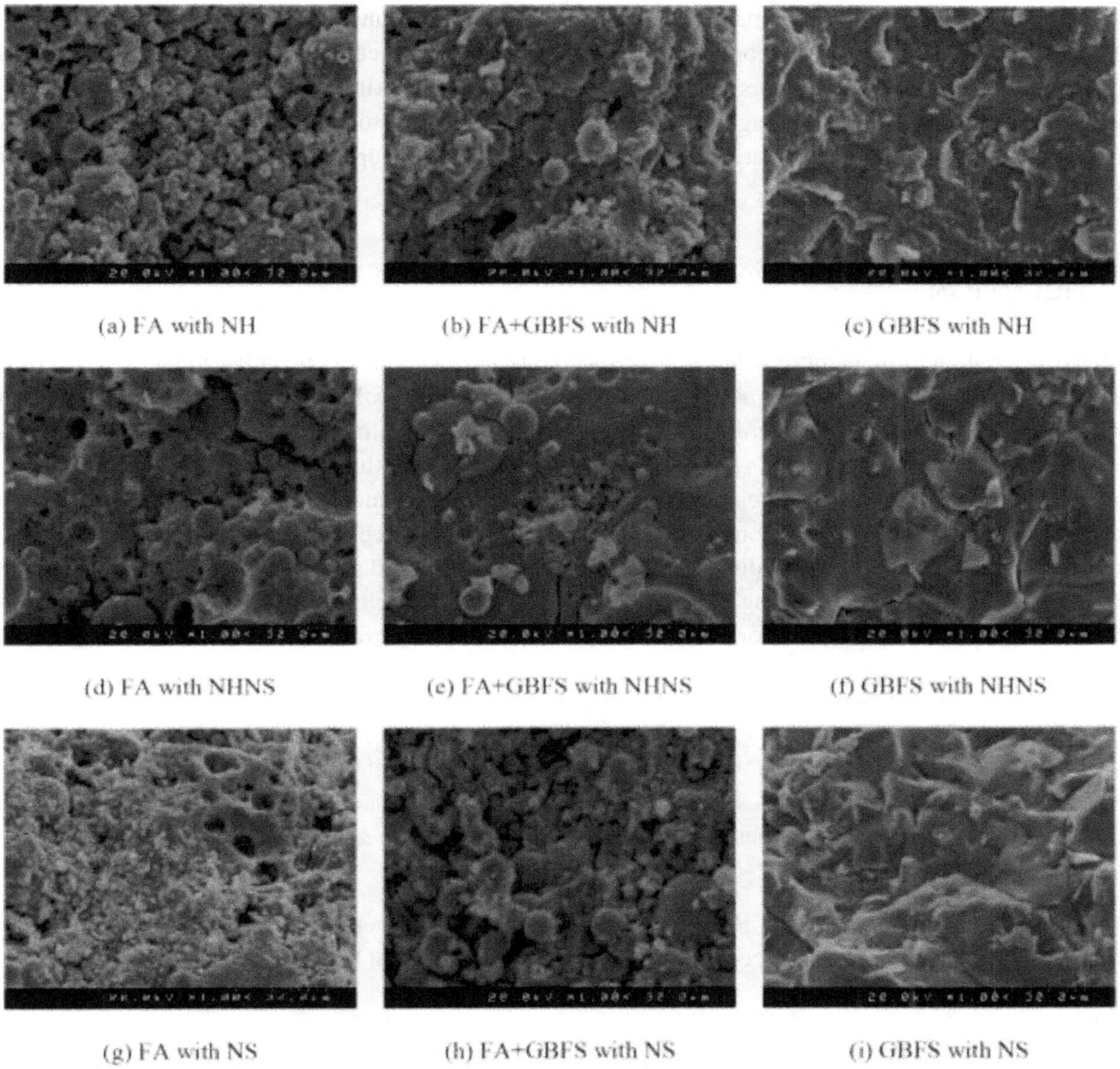

binders are economically viable only for structural applications with high performance. Geopolymers are up to seven times more affordable than currently available commercial repair mortars (Pacheco-Torgal et al., 2006). However, the cheapest commercial repair mortar costs approximately 14 times as much as geopolymers. Due to viscous components such as sodium silicate and sodium hydroxide, the geopolymers exhibit stiff workability. Numerous studies indicated placement challenges as a result of geopolymers' limited workability. Another researcher (Chindaprasirt et al., 2007) discovered that using a superplasticizer improves the workability of geopolymers. According to the sodium hydroxide to sodium silicate ratio, it may also result in a decrease in strengths.

FUTURE RESEARCH DIRECTIONS

This study discussed the environmental contributions, advantages, and disadvantages of geopolymers. Despite some disadvantages of geopolymers, it has gained importance in many sectors. Therefore, it is recommended to use various fibers, nanoparticles, new waste materials (binders and fillers), and new activators to reduce the disadvantages of geopolymers. In addition, studies on the recycling of geopolymers and the autonomous and autogenous recovery of cracked geopolymers could be important for future research direction.

CONCLUSION

Industrial waste and incinerator ashes degrade valuable land and have a negative impact on ecology. They may be recycled and utilized as raw materials in producing geopolymers. Geopolymer is a potential environmentally friendly material, and its development will help minimize carbon dioxide emissions associated with the cement industry's growth. Geopolymer materials exhibit superior mechanical qualities and exhibit various superior features such as corrosion and fire resistance. Also, geopolymers' expansion ratio, weight loss, and thermal conductivity are much lower than cement composites. Geopolymer's microstructural, mechanical, and durability qualities are proportional to the amount of each constituent in the formulation. When designing the mix, all factors affecting mechanical qualities should be considered, including the ratio of activator to the binder, the fraction of hydroxide silicate, and binder quantity.

REFERENCES

Ababneh, A., Matalkah, F., & Aqel, R. (2020). Synthesis of kaolin-based alkali-activated cement: Carbon footprint, cost and energy assessment. *Journal of Materials Research and Technology*, *9*(4), 8367–8378. doi:10.1016/j.jmrt.2020.05.116

Abdel-Gawwad, H. A., García, S. R. V., & Hassan, H. S. (2018). Thermal activation of air cooled slag to create one-part alkali activated cement. *Ceramics International*, *44*(12), 14935–14939. doi:10.1016/j.ceramint.2018.05.089

Abdel-Gawwad, H. A., & Khalil, K. A. (2018). Application of thermal treatment on cement kiln dust and feldspar to create one-part geopolymer cement. *Construction & Building Materials*, *187*, 231–237. doi:10.1016/j.conbuildmat.2018.07.161

Adesanya, E., Ohenoja, K., Di Maria, A., Kinnunen, P., & Illikainen, M. (2020). Alternative alkali-activator from steel-making waste for one-part alkali-activated slag. *Journal of Cleaner Production*, *274*, 123020. doi:10.1016/j.jclepro.2020.123020

Al-Majidi, M. H., Lampropoulos, A., Cundy, A., & Meikle, S. (2016). Development of geopolymer mortar under ambient temperature for in situ applications. *Construction & Building Materials*, *120*, 198–211. doi:10.1016/j.conbuildmat.2016.05.085

Al-mashhadani, M. M., Canpolat, O., Aygörmez, Y., Uysal, M., & Erdem, S. (2018). Mechanical and microstructural characterization of fiber reinforced fly ash based geopolymer composites. *Construction & Building Materials*, *167*, 505–513. doi:10.1016/j.conbuildmat.2018.02.061

Alaloul, W. S., Musarat, M. A., Tayeh, A. B., Sivalingam, S., Rosli, M. F. Bin, Haruna, S., & Khan, M. I. (2020). Mechanical and deformation properties of rubberized engineered cementitious composite (ECC). *Case Studies in Construction Materials, 13*.

Alanazi, H., Yang, M., Zhang, D., & Gao, Z. (2016). Bond strength of PCC pavement repairs using metakaolin-based geopolymer mortar. *Cement and Concrete Composites, 65*, 75–82.

Albitar, M., Mohamed Ali, M. S., Visintin, P., & Drechsler, M. (2017). Durability evaluation of geopolymer and conventional concretes. *Construction & Building Materials*, *136*, 374–385. doi:10.1016/j.conbuildmat.2017.01.056

Aliabdo, A. A., Abd Elmoaty, A. E. M., & Mohamed, M. F. (2018). Permeability indices and corrosion resistance of geopolymer and Portland cement concretes. *Magazine of Concrete Research*, *70*(12), 595–609. doi:10.1680/jmacr.17.00151

Almalkawi, A. T., Balchandra, A., & Soroushian, P. (2019). Potential of Using Industrial Wastes for Production of Geopolymer Binder as Green Construction Materials. *Construction & Building Materials*, *220*, 516–524. doi:10.1016/j.conbuildmat.2019.06.054

Almalkawi, A. T., Hamadna, S., & Soroushian, P. (2017). One-part alkali activated cement based volcanic pumice. *Construction & Building Materials*, *152*, 367–374. doi:10.1016/j.conbuildmat.2017.06.139

Almutairi, A. L., Tayeh, B. A., Adesina, A., Isleem, H. F., & Zeyad, A. M. (2021). Potential applications of geopolymer concrete in construction: A review. *Case Studies in Construction Materials, 15*, e00733.

Alonso, M. M., Gascó, C., Morales, M. M., Suárez-Navarro, J. A., Zamorano, M., & Puertas, F. (2019). Olive biomass ash as an alternative activator in geopolymer formation: A study of strength, radiology and leaching behaviour. *Cement and Concrete Composites*, *104*, 103384. doi:10.1016/j.cemconcomp.2019.103384

Ariffin, M. A. M., Bhutta, M. A. R., Hussin, M. W., Mohd Tahir, M., & Aziah, N. (2013). Sulfuric acid resistance of blended ash geopolymer concrete. *Construction & Building Materials*, *43*, 80–86. doi:10.1016/j.conbuildmat.2013.01.018

Askarian, M., Tao, Z., Adam, G., & Samali, B. (2018). Mechanical properties of ambient cured one-part hybrid OPC-geopolymer concrete. *Construction & Building Materials*, *186*, 330–337. doi:10.1016/j.conbuildmat.2018.07.160

Askarian, M., Tao, Z., Samali, B., Adam, G., & Shuaibu, R. (2019). Mix composition and characterisation of one-part geopolymers with different activators. *Construction & Building Materials*, *225*, 526–537. doi:10.1016/j.conbuildmat.2019.07.083

Barbosa, V. F. F., MacKenzie, K. J. D., & Thaumaturgo, C. (2000). Synthesis and characterisation of materials based on inorganic polymers of alumina and silica: Sodium polysialate polymers. *International Journal of Inorganic Materials*, *2*(4), 309–317. doi:10.1016/S1466-6049(00)00041-6

Bellum, R. R., Muniraj, K., & Madduru, S. R. C. (2020). Influence of slag on mechanical and durability properties of fly ash-based geopolymer concrete. *Journal of the Korean Ceramic Society*, *57*(5), 530–545. doi:10.100743207-020-00056-7

Bondar, D. (2009). *Alkali activation of Iranian natural pozzolans for producing geopolymer cement and concrete*. University of Sheffield.

Celerier, H., Jouin, J., Gharzouni, A., Mathivet, V., Sobrados, I., Tessier-Doyen, N., & Rossignol, S. (2019). Relation between working properties and structural properties from 27Al, 29Si and 31P NMR and XRD of acid-based geopolymers from 25 to 1000°C. *Materials Chemistry and Physics, 228*, 293–302.

Celerier, H., Jouin, J., Mathivet, V., Tessier-Doyen, N., & Rossignol, S. (2018). Composition and properties of phosphoric acid-based geopolymers. *Journal of Non-Crystalline Solids*, *493*, 94–98. doi:10.1016/j.jnoncrysol.2018.04.044

Cheng, T. W., Lee, M. L., Ko, M. S., Ueng, T. H., & Yang, S. F. (2012). The heavy metal adsorption characteristics on metakaolin-based geopolymer. *Applied Clay Science*, *56*, 90–96. doi:10.1016/j.clay.2011.11.027

Chindaprasirt, P., Chareerat, T., & Sirivivatnanon, V. (2007). Workability and strength of coarse high calcium fly ash geopolymer. *Cement and Concrete Composites*, *29*(3), 224–229. doi:10.1016/j.cemconcomp.2006.11.002

Das, S. K., Mishra, J., Singh, S. K., Mustakim, S. M., Patel, A., Das, S. K., & Behera, U. (2020). Characterization and utilization of rice husk ash (RHA) in fly ash – Blast furnace slag based geopolymer concrete for sustainable future. *Materials Today: Proceedings*, *33*, 5162–5167. doi:10.1016/j.matpr.2020.02.870

Davidovits, J. (1991). Geopolymers: Inorganic polymeric new materials. *Journal of Thermal Analysis and Calorimetry*, *37*(8), 1633–1656. doi:10.1007/BF01912193

De Silva, P., Sagoe-Crenstil, K., & Sirivivatnanon, V. (2007). Kinetics of geopolymerization: Role of Al2O3 and SiO2. *Cement and Concrete Research*, *37*(4), 512–518. doi:10.1016/j.cemconres.2007.01.003

Du, K., Xie, C., & Ouyang, X. (2017). A comparison of carbon dioxide (CO2) emission trends among provinces in China. *Renewable & Sustainable Energy Reviews*, *73*, 19–25. doi:10.1016/j.rser.2017.01.102

Duan, P., Yan, C., Zhou, W., & Ren, D. (2016). Development of fly ash and iron ore tailing based porous geopolymer for removal of Cu(II) from wastewater. *Ceramics International*, *42*(12), 13507–13518. doi:10.1016/j.ceramint.2016.05.143

Duxson, P., Fernández-Jiménez, A., Provis, J. L., Lukey, G. C., Palomo, A., & van Deventer, J. S. J. (2007). Geopolymer technology: The current state of the art. *Journal of Materials Science*, *42*(9), 2917–2933. doi:10.100710853-006-0637-z

Duxson, P., Provis, J. L., Lukey, G. C., & van Deventer, J. S. J. (2007). The role of inorganic polymer technology in the development of 'green concrete.'. *Cement and Concrete Research*, *37*(12), 1590–1597. doi:10.1016/j.cemconres.2007.08.018

Feng, D., Provis, J. L., & Deventer, J. S. J. (2012). Thermal activation of albite for the synthesis of one-part mix geopolymers. *Journal of the American Ceramic Society*, *95*(2), 565–572. doi:10.1111/j.1551-2916.2011.04925.x

Feng, D., Tan, H., & van Deventer, J. S. J. (2004). Ultrasound enhanced geopolymerisation. *Journal of Materials Science*, *39*(2), 571–580. doi:10.1023/B:JMSC.0000011513.87316.5c

Ge, Y., Yuan, Y., Wang, K., He, Y., & Cui, X. (2015). Preparation of geopolymer-based inorganic membrane for removing Ni2+ from wastewater. *Journal of Hazardous Materials*, *299*, 711–718. doi:10.1016/j.jhazmat.2015.08.006 PMID:26295686

Grosse, J., Dietrich, B., Garrido, G. I., Habisreuther, P., Zarzalis, N., Martin, H., Kind, M., & Kraushaar-Czarnetzki, B. (2009). Morphological characterization of ceramic sponges for applications in chemical engineering. *Industrial & Engineering Chemistry Research*, *48*(23), 10395–10401. doi:10.1021/ie900651c

Hajimohammadi, A., Ngo, T., Mendis, P., Nguyen, T., Kashani, A., & van Deventer, J. S. J. (2017). Pore characteristics in one-part mix geopolymers foamed by H2O2: The impact of mix design. *Materials & Design*, *130*, 381–391. doi:10.1016/j.matdes.2017.05.084

Hajimohammadi, A., & van Deventer, J. S. J. (2017). Solid reactant-based geopolymers from rice hull ash and sodium aluminate. *Waste and Biomass Valorization*, *8*(6), 2131–2140. doi:10.100712649-016-9735-6

Hamada, H. M., Al-attar, A. A., Yahaya, F. M., Muthusamy, K., Tayeh, B. A., & Humada, A. M. (2020). Effect of high-volume ultrafine palm oil fuel ash on the engineering and transport properties of concrete. *Case Studies in Construction Materials, 12*.

Hashimoto, S., Machino, T., Takeda, H., Daiko, Y., Honda, S., & Iwamoto, Y. (2015). Antimicrobial activity of geopolymers ion-exchanged with copper ions. *Ceramics International*, *41*(10, 10, Part A), 13788–13792. doi:10.1016/j.ceramint.2015.08.061

Hassan, H. S., Abdel-Gawwad, H. A., Vásquez-García, S. R., Israde-Alcántara, I., Flores-Ramirez, N., Rico, J. L., & Mohammed, M. S. (2019). Cleaner production of one-part white geopolymer cement using pre-treated wood biomass ash and diatomite. *Journal of Cleaner Production*, *209*, 1420–1428. doi:10.1016/j.jclepro.2018.11.137

He, J., Jie, Y., Zhang, J., Yu, Y., & Zhang, G. (2013). Synthesis and characterization of red mud and rice husk ash-based geopolymer composites. *Cement and Concrete Composites*, *37*, 108–118. doi:10.1016/j.cemconcomp.2012.11.010

Helmboldt, O., Keith Hudson, L., Misra, C., Wefers, K., Heck, W., Stark, H., Danner, M., & Rösch, N. (2000). Aluminum compounds, inorganic. Ullmann's Encyclopedia of Industrial Chemistry, 1–17.

Helmy, A. I. I. (2016). Intermittent curing of fly ash geopolymer mortar. *Construction & Building Materials*, *110*, 54–64. doi:10.1016/j.conbuildmat.2016.02.007

Horvath, A. (2004). Construction materials and the environment. *Annual Review of Environment and Resources*, *29*(1), 181–204. doi:10.1146/annurev.energy.29.062403.102215

Hu, S., Wang, H., Zhang, G., & Ding, Q. (2008). Bonding and abrasion resistance of geopolymeric repair material made with steel slag. *Cement and Concrete Composites*, *30*(3), 239–244. doi:10.1016/j.cemconcomp.2007.04.004

Hua, X., Provis, J., Van Deventer, J., & Krivenko, P. (2008). Characterization of Aged Slag Concretes. *ACI Materials Journal*, *105*, 131–139.

Huseien, G. F., Mirza, J., Ismail, M., & Hussin, M. W. (2016). Influence of different curing temperatures and alkali activators on properties of GBFS geopolymer mortars containing fly ash and palm-oil fuel ash. *Construction and Building Materials*, *125*, 1229–1240.

Huseien, G. F., Mirza, J., Ismail, M., Ghoshal, S. K., & Ariffin, M. A. M. (2018). Effect of metakaolin replaced granulated blast furnace slag on fresh and early strength properties of geopolymer mortar. *Ain Shams Engineering Journal*, *9*(4), 1557–1566. doi:10.1016/j.asej.2016.11.011

Ioannidou, D., Meylan, G., Sonnemann, G., & Habert, G. (2017). Is gravel becoming scarce? Evaluating the local criticality of construction aggregates. *Resources, Conservation and Recycling*, *126*, 25–33. doi:10.1016/j.resconrec.2017.07.016

Kovtun, M., Kearsley, E. P., & Shekhovtsova, J. (2015). Dry powder alkali-activated slag cements. *Advances in Cement Research*, *27*(8), 447–456. doi:10.1680/jadcr.14.00078

Kupwade-Patil, K., & Allouche, E. N. (2013). Examination of chloride-induced corrosion in reinforced geopolymer concretes. *Journal of Materials in Civil Engineering*, *25*(10), 1465–1476. doi:10.1061/(ASCE)MT.1943-5533.0000672

Law, D. W., Adam, A. A., Molyneaux, T. K., Patnaikuni, I., & Wardhono, A. (2015). Long term durability properties of class F fly ash geopolymer concrete. *Materials and Structures*, *48*(3), 721–731. doi:10.161711527-014-0268-9

Leung, D. Y. C., Caramanna, G., & Maroto-Valer, M. M. (2014). An overview of current status of carbon dioxide capture and storage technologies. *Renewable & Sustainable Energy Reviews*, *39*, 426–443. doi:10.1016/j.rser.2014.07.093

López, F. J., Sugita, S., & Kobayashi, T. (2014). Cesium-adsorbent geopolymer foams based on silica from rice husk and metakaolin. *Chemistry Letters*, *43*(1), 128–130. doi:10.1246/cl.130851

Luo, Y., Meng, J., Wang, D., Jiao, L., & Xue, G. (2022). Experimental study on mechanical properties and microstructure of metakaolin based geopolymer stabilized silty clay. *Construction & Building Materials*, *316*, 125662. doi:10.1016/j.conbuildmat.2021.125662

Ma, C., Zhao, B., Wang, L., Long, G., & Xie, Y. (2020). Clean and low-alkalinity one-part geopolymeric cement: Effects of sodium sulfate on microstructure and properties. *Journal of Cleaner Production*, *252*, 119279. doi:10.1016/j.jclepro.2019.119279

Ma, H., Ling, F., Yang, J., & Wang, G. (2002). Preparation of mineral polymer from potassium feldspar wastes: An experimental study. *Earth Sciences (Paris)*, *27*(5), 576–583.

Manjunath, R., Narasimhan, M. C., & Umesha, K. M. (2019). Studies on high performance alkali activated slag concrete mixes subjected to aggressive environments and sustained elevated temperatures. *Construction & Building Materials, 229*, 116887. doi:10.1016/j.conbuildmat.2019.116887

McLellan, B. C., Williams, R. P., Lay, J., van Riessen, A., & Corder, G. D. (2011). Costs and carbon emissions for geopolymer pastes in comparison to ordinary portland cement. *Journal of Cleaner Production, 19*(9), 1080–1090. doi:10.1016/j.jclepro.2011.02.010

Minelli, M., Medri, V., Papa, E., Miccio, F., Landi, E., & Doghieri, F. (2016). Geopolymers as solid adsorbent for CO2 capture. *Chemical Engineering Science, 148*, 267–274. doi:10.1016/j.ces.2016.04.013

Moura, D., Vasconcelos, E., Pacheco-Torgal, F., & Ding, Y. (2011). *Concrete repair with geopolymeric mortars: Influence of mortars composition on their workability and mechanical strength*. Academic Press.

Okoye, F. N., Prakash, S., & Singh, N. B. (2017). Durability of fly ash based geopolymer concrete in the presence of silica fume. *Journal of Cleaner Production, 149*, 1062–1067. doi:10.1016/j.jclepro.2017.02.176

Olivia, M., & Nikraz, H. (2012). Properties of fly ash geopolymer concrete designed by Taguchi method. *Materials & Design, 36*, 191–198.

Özcan, A., & Karakoç, M. B. (2019). The resistance of blast furnace slag-and ferrochrome slag-based geopolymer concrete against acid attack. *International Journal of Civil Engineering, 17*(10), 1571–1583. doi:10.100740999-019-00425-2

Pacheco-Torgal, F., Gomes, J. P., & Jalali, S. (2006). *Bond strength between concrete substance and repair materials: comparisons between tungsten mine waste geopolymeric binder versus current commercial repair products.* Academic Press.

Pacheco-Torgal, F., Moura, D., Ding, Y., & Jalali, S. (2011). Composition, strength and workability of alkali-activated metakaolin based mortars. *Construction & Building Materials, 25*(9), 3732–3745. doi:10.1016/j.conbuildmat.2011.04.017

Palomo, Á., Fernández-Jiménez, A., López-Hombrados, C., & Lleyda, J. L. (2011). Railway sleepers made of alkali activated fly ash concrete. *Revista de Ingeniería de Construcción, 22*(2), 75–80.

Palomo, A., Grutzeck, M. W., & Blanco, M. T. (1999). Alkali-activated fly ashes: A cement for the future. *Cement and Concrete Research, 29*(8), 1323–1329. doi:10.1016/S0008-8846(98)00243-9

Peduzzi, P. (2014). Sand, rarer than one thinks. *Environmental Development, 11*, 208–218. doi:10.1016/j.envdev.2014.04.001

Peng, M. X., Wang, Z. H., Shen, S. H., & Xiao, Q. G. (2015). Synthesis, characterization and mechanisms of one-part geopolymeric cement by calcining low-quality kaolin with alkali. *Materials and Structures, 48*(3), 699–708. doi:10.161711527-014-0350-3

Peng, M. X., Wang, Z. H., Xiao, Q. G., Song, F., Xie, W., Yu, L. C., Huang, H. W., & Yi, S. J. (2017). Effects of alkali on one-part alkali-activated cement synthesized by calcining bentonite with dolomite and Na2CO3. *Applied Clay Science, 139*, 64–71. doi:10.1016/j.clay.2017.01.020

Phoo-ngernkham, T., Maegawa, A., Mishima, N., Hatanaka, S., & Chindaprasirt, P. (2015). Effects of sodium hydroxide and sodium silicate solutions on compressive and shear bond strengths of FA–GBFS geopolymer. *Construction & Building Materials*, *91*, 1–8. doi:10.1016/j.conbuildmat.2015.05.001

Phoo-ngernkham, T., Sata, V., Hanjitsuwan, S., Ridtirud, C., Hatanaka, S., & Chindaprasirt, P. (2015). High calcium fly ash geopolymer mortar containing Portland cement for use as repair material. *Construction & Building Materials*, *98*, 482–488. doi:10.1016/j.conbuildmat.2015.08.139

Prud'Homme, E., Michaud, P., Joussein, E., Peyratout, C., Smith, A., & Rossignol, S. (2011). In situ inorganic foams prepared from various clays at low temperature. *Applied Clay Science*, *51*(1–2), 15–22. doi:10.1016/j.clay.2010.10.016

Ranjbar, N., Mehrali, M., Maheri, M. R., & Mehrali, M. (2017). Hot-pressed geopolymer. *Cement and Concrete Research*, *100*, 14–22. doi:10.1016/j.cemconres.2017.05.010

Rowles, M., & O'connor, B. (2003). Chemical optimisation of the compressive strength of aluminosilicate geopolymers synthesised by sodium silicate activation of metakaolinite. *Journal of Materials Chemistry*, *13*(5), 1161–1165. doi:10.1039/b212629j

Şahin, F., Uysal, M., & Canpolat, O. (2021). Systematic evaluation of the aggregate types and properties on metakaolin based geopolymer composites. *Construction & Building Materials*, *278*, 122414. doi:10.1016/j.conbuildmat.2021.122414

Sartori, I., & Hestnes, A. G. (2007). Energy use in the life cycle of conventional and low-energy buildings: A review article. *Energy and Building*, *39*(3), 249–257. doi:10.1016/j.enbuild.2006.07.001

Shuai, Q., Xu, Z., Yao, Z., Chen, X., Jiang, Z., Peng, X., An, R., Li, Y., Jiang, X., & Li, H. (2020). Fire resistance of phosphoric acid-based geopolymer foams fabricated from metakaolin and hydrogen peroxide. *Materials Letters*, *263*, 127228. doi:10.1016/j.matlet.2019.127228

Somna, K., Jaturapitakkul, C., Kajitvichyanukul, P., & Chindaprasirt, P. (2011). NaOH-activated ground fly ash geopolymer cured at ambient temperature. *Fuel*, *90*(6), 2118–2124. doi:10.1016/j.fuel.2011.01.018

Sturm, P., Gluth, G. J. G., Jäger, C., Brouwers, H. J. H., & Kühne, H.-C. (2018). Sulfuric acid resistance of one-part alkali-activated mortars. *Cement and Concrete Research*, *109*, 54–63. doi:10.1016/j.cemconres.2018.04.009

Taha, B., & Nounu, G. (2008). Using lithium nitrate and pozzolanic glass powder in concrete as ASR suppressors. *Cement and Concrete Composites*, *30*(6), 497–505. doi:10.1016/j.cemconcomp.2007.08.010

Tammam, Y., Uysal, M., & Canpolat, O. (2021). Effects of alternative ecological fillers on the mechanical, durability, and microstructure of fly ash-based geopolymer mortar. *European Journal of Environmental and Civil Engineering*, 1–24. doi:10.1080/19648189.2021.1925157

Tayeh, B. A., Zeyad, A. M., Agwa, I. S., & Amin, M. (2021). Effect of elevated temperatures on mechanical properties of lightweight geopolymer concrete. *Case Studies in Construction Materials, 15*.

Tchakouté, H. K., & Rüscher, C. H. (2017). Mechanical and microstructural properties of metakaolin-based geopolymer cements from sodium waterglass and phosphoric acid solution as hardeners: A comparative study. *Applied Clay Science*, *140*, 81–87. doi:10.1016/j.clay.2017.02.002

Valencia-Saavedra, W. G., Mejía de Gutiérrez, R., & Puertas, F. (2020). Performance of FA-based geopolymer concretes exposed to acetic and sulfuric acids. *Construction & Building Materials, 257*, 119503. doi:10.1016/j.conbuildmat.2020.119503

Velandia, D. F., Lynsdale, C. J., Provis, J. L., Ramirez, F., & Gomez, A. C. (2016). Evaluation of activated high volume fly ash systems using Na2SO4, lime and quicklime in mortars with high loss on ignition fly ashes. *Construction & Building Materials, 128*, 248–255. doi:10.1016/j.conbuildmat.2016.10.076

Wang, H., Li, H., & Yan, F. (2005). Synthesis and tribological behavior of metakaolinite-based geopolymer composites. *Materials Letters, 59*(29–30), 3976–3981. doi:10.1016/j.matlet.2004.08.049

Xie, T., & Ozbakkaloglu, T. (2015). Behavior of low-calcium fly and bottom ash-based geopolymer concrete cured at ambient temperature. *Ceramics International, 41*(4), 5945–5958. doi:10.1016/j.ceramint.2015.01.031

Xu, L., Wang, X., Guan, C., Wu, W., & Zhang, L. (2020). The effect of activators on the mechanical properties and microstructure of alkali-activated nickel slag. *Advances in Civil Engineering, 2020*, 2020. doi:10.1155/2020/1764108

Ye, N., Chen, Y., Yang, J., Liang, S., Hu, Y., Hu, J., Zhu, S., Fan, W., & Xiao, B. (2017). Transformations of Na, Al, Si and Fe species in red mud during synthesis of one-part geopolymers. *Cement and Concrete Research, 101*, 123–130. doi:10.1016/j.cemconres.2017.08.027

Zhang, H. Y., Kodur, V., Qi, S. L., & Wu, B. (2015). Characterizing the bond strength of geopolymers at ambient and elevated temperatures. *Cement and Concrete Composites, 58*, 40–49. doi:10.1016/j.cemconcomp.2015.01.006

Zhang, H.-Y., Liu, J.-C., & Wu, B. (2021). Mechanical properties and reaction mechanism of one-part geopolymer mortars. *Construction & Building Materials, 273*, 121973. doi:10.1016/j.conbuildmat.2020.121973

Zhang, J., He, Y., Wang, Y., Mao, J., & Cui, X. (2014). Synthesis of a self-supporting faujasite zeolite membrane using geopolymer gel for separation of alcohol/water mixture. *Materials Letters, 116*, 167–170. doi:10.1016/j.matlet.2013.11.008

Zhang, P., Gao, Z., Wang, J., Guo, J., Hu, S., & Ling, Y. (2020). Properties of fresh and hardened fly ash/slag based geopolymer concrete: A review. *Journal of Cleaner Production, 270*, 122389. doi:10.1016/j.jclepro.2020.122389

Zhang, S., Gong, K., & Lu, J. (2004). Novel modification method for inorganic geopolymer by using water soluble organic polymers. *Materials Letters, 58*(7), 1292–1296. doi:10.1016/j.matlet.2003.07.051

Zhang, Z., Provis, J. L., Reid, A., & Wang, H. (2014). Geopolymer foam concrete: An emerging material for sustainable construction. *Construction & Building Materials, 56*, 113–127. doi:10.1016/j.conbuildmat.2014.01.081

Zhang, Z., Provis, J. L., Reid, A., & Wang, H. (2015). Mechanical, thermal insulation, thermal resistance and acoustic absorption properties of geopolymer foam concrete. *Cement and Concrete Composites, 62*, 97–105. doi:10.1016/j.cemconcomp.2015.03.013

Ziada, M., Erdem, S., Tammam, Y., Kara, S., & Lezcano, R. A. (2021). The Effect of Basalt Fiber on Mechanical, Microstructural, and High-Temperature Properties of Fly Ash-Based and Basalt Powder Waste-Filled Sustainable Geopolymer Mortar. In Sustainability (Vol. 13, Issue 22). doi:10.3390u132212610

Ziada, M., Tammam, Y., Erdem, S., & Lezcano, R. A. (2022). Investigation of the Mechanical, Microstructure and 3D Fractal Analysis of Nanocalcite-Modified Environmentally Friendly and Sustainable Cementitious Composites. In Buildings (Vol. 12, Issue 1). doi:10.3390/buildings12010036

Chapter 10
Smart City Buildings for a Resilient, Sustainable Future

Vinod Anand Bijlani
Independent Researcher, Singapore

ABSTRACT

We are living in rapidly changing times where the world is running low on its carbon budget in trying to rapidly transition to net zero. We are also living amidst transformations, innovation, and operational modernization to be able to provide for a Smart City, including climate health. Future smart cities demand a balanced blend of data and technology to create diversely inclusive and sustainable solutions. In this chapter, the author focuses on how smart buildings can be designed for a sustainable future, exploring the possibilities of converting existing buildings into interoperable spaces, powered by technology and data. While the author outlines the differences between brownfield and greenfield smart buildings, he makes a case for each and explains their potential in achieving productive sustainability. The chapter delineates some of the policy implementations required to align smart buildings with sustainable climate action and substantiates global practices with ROI data. The author concludes the chapter by offering the way forward in the urban sustainability journey.

INTRODUCTION

The need for flexibility, safety, and sustainability in our world has never been greater: Flexibility in how we live, work, or travel, safety in the environment around us, and sustainability in various aspects of life. Buildings generate 38% (UNEP, 2020) of the total greenhouse gases (GHGs), the highest in a single industry worldwide, justifying the demand to recast them into more sustainable, energy-efficient, spaces. Cities account for 60 to 80% of energy consumption, and generate as much as 70% of the human-induced GHG emissions (UN Habitat, 2016). Surprisingly, the World Green Building Council (WGBC) reports that there are currently only about 500 net zero commercial buildings across the world (IISD, 2017). The global push towards net zero carbon goals has accelerated the efforts to create Smart buildings that enable businesses and empower people. Adaptability and sustainability are the catchphrases, as buildings adapt to the people using them, constantly collecting data and making smart inferences.

DOI: 10.4018/978-1-6684-4030-8.ch010

In a recent global report, it was estimated that cutting emissions from cities, its buildings and infrastructure included, by about 90% was possible using proven technologies (CfUT, 2019). This could just be a clarion call for all new buildings to operate under net zero from 2030, if we are to win the 2050 race against our planet heating up.

The objective of this chapter is to study the current state of the buildings industry and understand the concept of sustainability in the construction sector. The author aims to delineate the definitions of, and differences between, greenfield and brownfield structures before diving into the concept of sustainable Smart buildings. The author explains the viability of technology-driven sustainability solutions that can help achieve smart civic growth along with public health and wellbeing.

Part of the solution includes enhancing the efficiency of renewable energy, water supply, and air quality, besides increasing environment-friendly practices. However, with such ambitious plans requiring large investments for long-term results, financing sustainable development goals (SDGs) is no mean feat. Given that each country's development paradigm is different, the world needs to ramp up aid to include private financing as well as unlocking national resources, based on a shared model where possible. To address concerns arising from policy implementation, performance systems can be put in place to set targets calibrated with local governments and national policies. The author describes some of the policy and financial implementations that can encourage sustainability and profitability to co-exist.

BACKGROUND AND LITERATURE REVIEW

The main objectives of sustainability are to check depletion of natural resources, preserve the environment while neutralizing any negative impact from human interference, and provide a safer urban environment – a secure community to help ease future generations into healthier and greener surroundings.

'Sustainability' as a concept goes as far back as the 1970s (Grevelman & Kluiwstra, 2010). However, most literature points to the Brudland Report (UN WCED, 1989) as being the first to spotlight global sustainability concerns, while establishing the relationship between social, economic, and environmental aspects of sustainable development. The Report defines sustainability as "meeting the needs of the present without compromising the ability of future generations to meet their own needs." Subsequently, there have been several reports and studies on the meaning and significance of sustainability in a modern world.

Most published works relating to sustainability in the buildings industry have been influenced by the concept of limiting resources and reducing the impact to the natural environment, with emphasis on technical issues, such as construction material, building components, construction and energy-related design and technologies (Mohd Isa et al., 2014). Blutstein and Rodger (2001) opine that a sustainable building, more than identifying solutions to specific problems, requires changes to attitudes, paradigms, processes, and systems. The United States Energy Protection Agency (EPA) lists efficiency in renewable energy, water usage, construction material, waste and toxic reduction, air quality, as well as smart sustainability to be the tenets of a 'sustainable building' (US EPA, n.d.). The WGBC (n.d.) defines a green building as one that "in its design, construction or operation, reduces or eliminates negative impacts, and can create positive impacts, on our climate and natural environment, preserve precious natural resources and improve our quality of life."

Sustainable buildings could be developed using brownfield assets or on greenfield sites. Brownfields are sites that were previously built and used, but underutilized or abandoned later due to economic, financial, or political reasons. Leaving them unused can contaminate air and soil. Brownfield regeneration is, therefore, critical to sustainability. Greenfield structures are based on the idea of promoting conservation of natural resources and exploiting local materials considering the climate and topography of the region (Wolfson, 2018). Sustainable greenfield buildings impact the environment less during construction, provide a healthier place for their occupants and are more cost-efficient in the long run than conventional structures.

Tizouiar and Boussoualim (2018) state that the relationship between a sustainable and comfortable built-environment with intelligent living can be ensured through adopted technologies, passive architectural devices, and materials used. The components of natural resources, data, and Smart technology could be interconnected, making for an interoperable building, where every component is able to seamlessly communicate with the other and act on instructions based on data. Such a networked intercommunicable system results in a sustainable building powered by Smart technology, including the latest in Artificial Intelligence (AI), the Internet of Things (IoT), and 5G. Gartner (2019) predicts that there will be over four billion connected IoT devices in commercial Smart buildings by 2028 (Deloitte, 2021). They will be powered by telecommunications infrastructures, with 5G and High Efficiency Wi-Fi 6 at the forefront, and will have smart utilities for power, waste, and water.

Energy-related carbon emissions from the construction industry include embodied carbon, jammed inside construction material, as well as operational carbon from energy-gorging resources such as Heating, Ventilation, and Air Conditioning (HVAC) and lighting systems required for building operations. Converting buildings into interoperable, energy-efficient spaces that are driven by data as well as by the constant exchange of information between the Building Management System (BMS) and the occupants, powered by data analytics can lead the way to a greener world. Although the preliminary costs of sustainable buildings that are developed on greenfield or brownfield sites may seem higher than traditional projects, it is observed that longer-term cost savings in operations and maintenance can help recover those costs within a few years (Mohd Isa et al., 2014).

SUSTAINABLE SMART BUILDINGS AT THE CORE OF SMART CITIES

Buildings that consume large amounts of energy and natural resources have a significant impact on climate change and air and water quality. Deterioration of livability concerns related to waste management, scarcity of resources, air pollution, traffic congestion, aging public infrastructure, among other problems, cause human health concerns (Washburn et al., 2010). Therefore, cities with Smart buildings are fast emerging as the solution to sustainability challenges that have cropped up from rapid urbanization. As Toli and Murtagh (2020) note, Smart cities are those that "perform well on six characteristics: environment, economy, mobility, people, living and governance."

Planning and design of internal spaces and the overall urban-scape are indicators of both positive building performance as well as positively redirected priorities to achieving a Smarter City. Careful, sustainability-centric, intelligent planning not only helps generate and restore sustainable spaces in an urban cityscape but also helps realize SDGs. The United Nations' SDGs are increasingly being recognized as the foundation for greater global sustainability as the world shifts its focus to the Environmental, Social, and Governance (ESG) criteria.

Figure 1. Summary of factors influencing Smart building technology investment, top-ranking regions with LEED certification buildings and building performance management apparatus
Source: Meadows (2021); Markets & Markets (2020); Urban Land Institute (2021); Fernandes (2021); USGBC (2022). Infographic by the author.

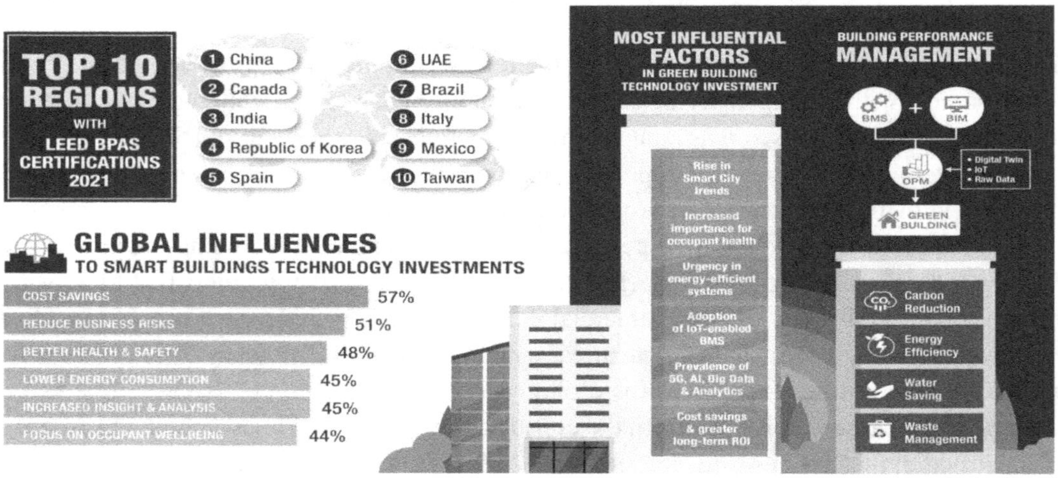

Performance Measurement and Management of Sustainable Smart Buildings

Besides the local regulations that are in place across the globe, the sustainability-related building standards and methodologies recognized worldwide spotlight evaluating environmental performance in buildings (Mohd Isa et al., 2014). These include the universally acknowledged LEED (Leadership in Energy and Environmental Design) and WELL (International Well Building Institute – IWBI) of the United States, BREEAM (Building Research Establishment Environmental Assessment Method) conceptualized in the United Kingdom, LBC (Living Building Challenge, the International Living Future Institute's advanced sustainability measure), CASBEE (Comprehensive Assessment System for Building Environmental Efficiency) from Japan, France's HQE (High Environmental Quality), and Australia's Green Star, among others. With every country having its own Building Performance Assessment System (BPAS), it is still rare to find any that incorporate the ESG aspect or its mutuality. Most of these BPASs include similar broad evaluation categories, including energy consumption, indoor air quality, site and waste management, water resourcing, building materials, and innovation.

BPASs such as LEED and LBC are leaders in certification programs that focus on "healthy" materials, while also allowing buildings to earn points as they use Smart building technology (Bowen, 2021). They rate new and pre-existing buildings per environmental attributes and sustainability features (Jacomit et al., 2009). For the LBC and LEED certification credits, Smart tech solutions can be customized to make a marked difference in overall on-site energy consumed by buildings. Figure 1 summarizes the factors that influence technology investments and trends in Smart buildings, the top-ranking regions with LEED certification buildings, and the performance measurement and management apparatus for a Smart building.

It is essential that building performance be measured quantitatively, in the context of optimizing environmental impact and lifecycle cost of the buildings. Using material that has a Life Cycle Assessment (LCA) certificate further supports energy goals. The LCA considers several vital factors, such as energy consumption, use of natural resources, GHG and pollution reduction, etc. (Wang & Adeli, 2014). Along with the LCA, a brownfield building's Smart Readiness Indicator (SRI) rates its capacity to use new technologies and systems (Fichte & Hollfelder, 2020) to reduce energy consumption and carbon emissions, and adapt to occupants' needs.

Sustainability measurement tools are imperative to assess compliance with sustainability principles and encourage reduction of environmental burden (Aktas & Ozorhon, 2015). The combination of sustainability measurement tools and technology such as Building Information Modeling (BIM), Building Management Systems (BMS), AI, and Big Data show great promise with further research serving as an anchorage.

BIM can help with complex analysis of building performance to certify an optimally sustainable design by visualizing the functional and physical aspects of a building. The BIM represents the process of developing and using a computer-generated model to stimulate planning, design, construction, and operation of a building. Data apposite to occupants' needs can be extracted and scrutinized from this virtual building model to obtain insights for better decision-making at the actual site. With research indicating the Return on Investment (ROI) from BIM to be a gigantic 634% to 1633% (Schueter & Thessling, 2008; Autodesk, 2008), linking the model to energy analysis tools allows for early and continual energy detection and modifications to boost sustainability.

BMS, on the other hand, is the overarching intelligent computer-based system that helps regulate and monitor a building's electrical and mechanical equipment. The system helps building managers exercise intelligent control over the power system, programmable lighting and illumination, plumbing, solar thermal units, fire alarm, HVAC, video surveillance, access control, and a host of other building operations Hossain, 2019), with the incorporation of smart control software. Having an extensive implementation stack, from condition monitoring to predictive maintenance using sensors, the BMS can be implemented across a range of devices. Its core functionality is within a specified range of the building, offering malfunction alarms, monitoring device failures, and controlling operations based on occupancy and need (Hossain).

The main intent of the integration of data from BIM and BMS using AI techniques is effective energy and power management, environmental conservation, automation, and keeping occupants satisfied (Fazli et al., 2019), using data from sensors, power control systems, actuators, processes, and occupants themselves. While data from internal lighting, temperature and air quality can serve as examples of AI algorithm inputs that help in building operation automation, the resulting optimized parameters help change the actuators per occupant usage and requirements. These parameters help in conceptualizing an efficient Operations Performance Management (OPM) strategy for the Smart building (Verma, 2018).

OPM is aimed at achieving optimal functioning of operations. OPM software brings together raw data to determine the best suited actions to be taken based on consistent and clear semantic information (ThoughtWire, 2019) thanks to ML and AI algorithms. It rests on the ability to draw out context-specific information from the data collected, which includes real-time information about occupants and operational systems. Based on predefined rules, processes can self-adjust for the ideal functioning under the given conditions, thereby minimizing or eliminating human intervention. In summary, it brings together BIM and BMS to optimize operational performance, making the case for digital twin technology.

Digital twin technology for operations management has the ability to intuitively store, organize, and access large amounts of data (Koerner, 2021) generated by the Smart, yet complex, building systems. Being a virtual model of a physical building, the digital twin supports simulations that can help accurately predict the operation and performance of the building based on the set parameters. Along with digital twin, IoT technologies act as the catalyst to continuous operational enhancements helping drive occupant comfort, optimize building performance, and drive savings, consequently bridging the gap between buildings facilities and operational efficacy.

Greenfield Smart Buildings: Pivoting to Purposeful Profitability

The structure of a greenfield building is one of the most important factors in sustainable design. The strategies, planning and design are primarily established based on the form of the structure (Wang & Adeli, 2014). The use of land and material, energy consumption (Pinto et al., 2013; Lee et al., 2013), GHG emissions, maintenance, risk management, life cycle costs (Elhakeem & Hegazy, 2012), and recycling of resources largely depend on the selection of the structural system and form. Collins et al. (2008) suggest various strategies for structural forms, including optimization of climatic conditions (such as solar gain, day lighting, and wind harvesting), wind load response-based structures, and using a combination of nature-based solutions (such as service design, façade design, and structural form for natural ventilation), among others.

Additionally, designing tall buildings for mixed-use and multi-function, providing for more open community spaces, insulating interiors from external temperature variations, using vegetation and hydroponic greenhouse and façade farms for urban and structural relief (Wood, 2008) as well as dual purpose exteriors with solar control skins doubling up as tuned mass damper could provide distinction in form and purpose. Studies show that creating healthier buildings is the chief trigger for designing 'green' globally, with occupant health topping the list of social reasons with 77% (WGBC, 2018). Designers may even carry out additional iterations on plans to satisfy the need for climate positivity, optimal building shape, as well as occupant comfort, health, and security.

Sustainability Measures in Greenfield Buildings

While designing a new building, there is great scope to thoughtfully consider the material that is being used in the construction. Concrete, steel, and aluminum used in the built environment are believed to be responsible for 23% of total global emissions (IEA & UNEP, 2018). Concrete, made from cement paste and aggregates, can release CO_2 due to chemical reactions over time. With a pound of gas released for every pound of cement, it is recommended that fly ash and slag be added to cement, along with other recycled material, to improve structural durability and mechanical properties while reducing carbon emissions (Wang & Adeli, 2014). Further, real-world studies reveal that aluminum structures have a higher ratio of load-bearing capacity, better corrosion resistance, and lower maintenance requirements compared to similar steel structures. In other words, if designed and executed suitably, aluminum could be eco-friendlier and more economical than steel (Wang & Adeli).

In dense and humid cities, it is proposed that a Combined Heat and Power (CHP) system be used for the building structure where power, heat, and chilled water for air-conditioning are produced simultaneously through highly energy-efficient technology (Ali & Armstrong, 2008). CHP demonstrates marked savings in cost and reduced carbon emissions. Another interesting structural solution for greenfield buildings is the building envelope, which acts as a skin, creating an intermediate layer between the building interiors and exteriors (Premier, 2012). The envelope acts as a dynamic façade, creating a filter between the interior and exterior, providing shade, sunlight, and ventilation. Leung and Weismantle (2008) explain that creating wind movement between the interior and exterior environment conditions can be captured for passive cooling and ventilation in the building, improving air quality and general occupant wellbeing.

Self-draining glass, made from nanomaterials and installed as part of the building exteriors absorb heat energy from the sun and convert it to electrical energy (Alsarraf & Alobaidi, 2020). Moreover, solar skin, created with nanoparticles called quantum dots, that pass electrons and generate electric current when exposed to solar energy, will also soon see the light of day on building exteriors. Developed by the engineers at the University of Queensland, the dots can be printed on any flexible surface and have the potential to be applied to windows and other surfaces to harvest natural energy even on cloudy days or in cities with weak sunlight (Cox, 2020). Other aspects of greenfield design include heat-reducing facades, circuit-embedded windows, electric curtains, and naturally ventilated roofs and tiles that reduce energy consumption and boost occupants' wellness.

When the building is operational, occupancy data and energy utilization become key to optimizing its energy usage. The data collected from the web of integrated IoT-based devices, including the security system, the BMS, elevators, lighting, HVAC, and other sensor-based utilities, collect data across occupant behavior and building performance to incorporate all of it into the real-time representation of the building (Verma, 2018) using digital twins. Such a collaboration is important because the alliance of OPM tools with a digital twin can provide building owners complete control over optimizing energy efficiency and tenant wellbeing.

Phases of Sustainable Greenfield Implementation

Phase One: Map requirements with availability: A cohesive plan on what is to be built, listing the requirements, identifying the features that are conducive to the location of the building as well as the climate of the city are top priority. Exploiting locally available resources while integrating the grey with the green should also be part of the initial stage of planning.

Phase Two: Strategy formulation and modeling: Having a sound strategy and creating the BIM to understand and evaluate the outcome of the plan ahead of execution can help save time, resources, and monetary losses.

Phase Three: Blueprint for implementation before production: While the BIM serves as a ready reckoner from the time of planning to implementation, creating a blueprint based on the observations from the BIM is essential.

Phase Four: Operations Management: The operations management of the complete structure will be a combination of the BMS, the OPM and digital twin technology. The digital twin of the greenfield building will provide a real-time, holistic view of what is happening inside the structure, even offering building owners remote access to build and deploy their own sustainable data strategy.

Figure 2 illustrates some real-world examples of greenfield Smart buildings.

Brownfield Smart Buildings

The phrase "brownfield development" was conceived by the American Environmental Protection Agency, and universally acknowledged to signify the opposite of greenfield development (Thornton et al., 2007; Popescu & Pătrăscoiu, 2012). Sustainable brownfield regeneration has been defined as the management, rehabilitation and return to beneficial use of brownfields ensuring the attainment and continued satisfaction of human needs for present and future generations in environmentally sensitive,

Figure 2. Greenfield Smart buildings - examples from the world over
Sources: Arch20 (n.d).; Smart Cities World (2017); PwC (2017); Roxburgh (2016); Smart Energy International (2021); Wong, 2019. Infographic by the author.

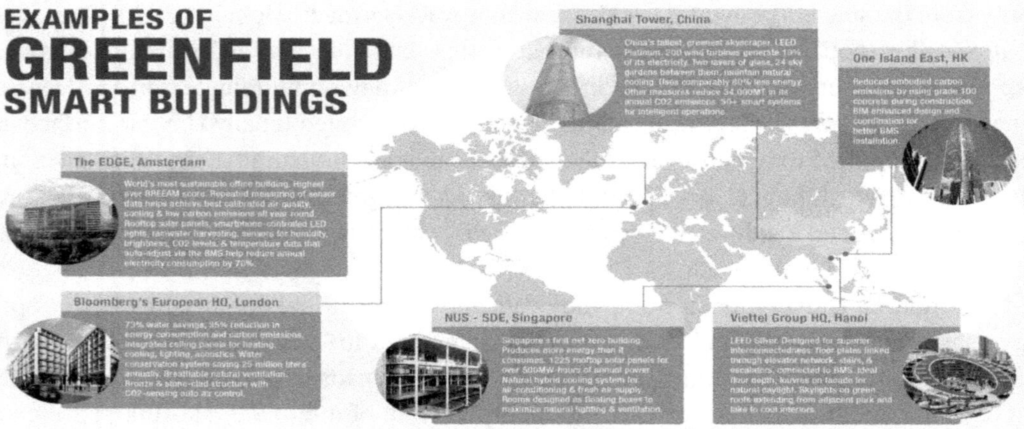

economically viable, institutionally robust and socially acceptable ways within the particular regional context (Franz et al., 2006).

Brownfield sites may come with outdated infrastructure, contaminated assets, and extant carbon footprint. However, as former president of the American Institute of Architects, Carl Elefante (2012), remarked, the greenest building is one that already exists. His report concluded from several case studies that reusing and upgrading an existing building for efficiency was better, regardless of the type of building or the climate around. Sarni (2010) agrees that the redevelopment of brownfield sites can create value from a liability via, what he terms, 'asset conversion', while emphasizing that incorporating sustainable land use and green building practices can create much greater value.

The adoption of nature-based and nature-identical solutions offers to decontaminate existing sites and promote brownfield redevelopment. These could result in a number of ESG benefits ranging from lesser energy consumption and greater material efficiency (O'Connor, 2020) to nurturing increased resilience. Connectivity is important to unbridling the potential of Smart buildings. Power over Ethernet (POE), coupled with 5G, run on an effective BMS can become the key enabler of brownfield smart buildings.

Implementing Sustainability Measures in Brownfield Buildings

The simplest way to improve energy efficiency in brownfield buildings is to replace old equipment with newer, higher efficiency models, increasing data observability. Upgrading lighting to automation capabilities, fitting existing HVAC with economizers, or maximizing on variable speed drivers and refrigeration systems can yield an average 10-15% energy savings (Hatcher, 2021). And with a BMS being easy to install, maintain, and integrate with existing platforms, it becomes the go-to for brownfield redevelopment.

Smart space management can be achieved through occupancy sensors installed at appropriate locations within the structure. Occupancy thresholds can be set across different floors or areas of the building to optimize operations, prevent overcrowding, and conserve energy. Sensor data can also be used to track and ensure that humidity levels are per the standards set by the American Society of Heating, Refrigerating and Air-Conditioning Engineers (ASHRAE) and the EPA (Hedayat, 2020). The BMS can also be configured to auto-ensure air quality standards and adequate air circulation. Further, it can compare HVAC performance against the set specifications from the BIM to optimize operation, reduce energy consumption, and adjust temperature where and when needed, creating a healthy building environment.

While the BMS and OPM can be combined to create a digital twin for the brownfield structure, renovating brownfields with Restorative Environmental Design (RED) can help reverse damage caused by its past. Biophilic design, with special attention to environmental stimulation, coherence, and visual contact with natural elements, coupled with the presence of biomorphic forms and structures (Pasini et al., 2021) can not only promote health but also a stress-free building environment. In fact, views of greenery in office buildings can increase job satisfaction and reduce absenteeism (Gibbs & Peiser, 2022). RED also emphasizes maximizing daylighting, thoughtfully configuring spatial plans, natural ventilation strategies, and interior greenery (Browning & Ryan, 2020). Several studies have demonstrated that natural daylight boosts occupant health, improves quality of sleep, enhances overall wellbeing, and increases employee productivity. It is important that key considerations be factored prior to retrofitting a brownfield site with sustainable measures: (i) health and environmental benefits from redevelopment (ii) direct and indirect benefits (iii) comprehensive benefits (Dong et al., 2020).

Phases of Brownfield Retrofit Considerations

Phase One: Evaluation for fit and identification of constraints: Several legacy sites may not be upgrade-ready, having been earlier designed in silos with little provision beyond their original facilities. The consideration in such a situation would be the requirements that the project should meet to be deemed sustainable, and whether it would fit the rip-and-replace approach.

Phase Two: Evaluation of cost factor: Evaluating the lifespan of the brownfield building could help determine if reconstruction could pose cost-prohibitive (Thieme, 2019). This would include the building's place in its life cycle since construction as well as the expected technical lifetime going forward.

Phase Three: Possible areas of implementation: There are several avenues for sustainably refurbishing a brownfield asset. Broadly, these would include occupant wellbeing and engagement, occupancy management, and building operations (Hedayat, 2020). Bi-facial as well as rooftop solar panels, natural ventilation, cooling, and shade using slatted facades, energy-recovery ventilators with air purging options, biophilic green exteriors to reduce heat, recycling condensed water from cooling systems, are some of them.

Phase Four: Digital readiness: A brownfield asset has to be evaluated for digital readiness as well as for its resilience to the remedial solutions being considered. The strategies for renewable energy and incorporating remote sensing capabilities (EPA, 2021) need to be assessed to bridge the gap between the existing site and the retrofitted structure. Besides conceiving the seamless accommodation of interoperable technologies, this stage will also include charting the course of the BMS based on the SRI to be able to build an OPM system with a digital twin. Generating a digital twin for a brownfield building can be challenging, considering most pre-existing sites may not have a machine-readable design in place (Sierla et al., 2021) at the time of renovation. However, some added effort in conducting a fit-gap analysis and creating the digital twin can go a long way in simulating the recommendations before implementing them, remote-surveying the operations, and predicting through modelling (Digital Europe, 2020). Considering the existing assets of brownfields, this would also help better conduct what-if analyses, improve communication, and successfully integrate existent disparate systems.

Figure 3. Brownfield Smart building implementations from around the world
Sources: Solxenergy (n.d.); Sayadi et al. (2016); Beavis (2014); AIA (2021); Tay (2021); Fix Development (2013). Infographic by the author.

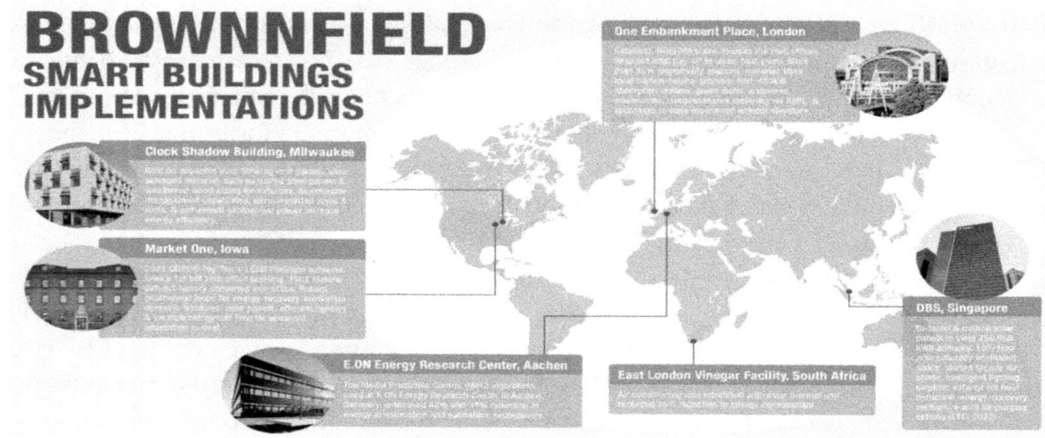

Figure 3 lists some of the real-world implementations of brownfield Smart buildings. Figure 4 compares the phases of implementing brownfield and greenfield projects.

Figure 4. Phases of implementation in greenfield and brownfield projects
Sources: Wang & Adeli, 2014; Collins et al., 2008; Sarni, 2010; O'Connor, 2020; Pasini et al., 2021; Thieme, 2019; Hedayat, 2020; EPA, 2021; Sierla et al., 2021; Digital Europe, 2020. Image created by the author.

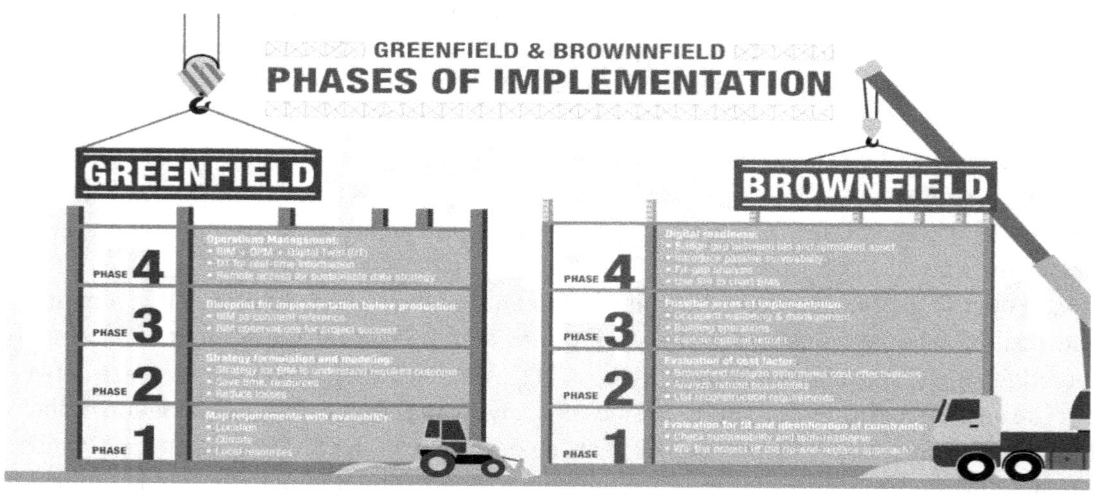

Tables 1 lists the differences between implementations of sustainability measures in greenfield and brownfield smart buildings.

Table 1. A comparison of greenfield and brownfield smart buildings

	Greenfield Smart Buildings	Brownfield Smart Buildings
Design	Provides flexibility in design to meet project requirements	Existing structure already in place, reducing scope for changes to design
Maintenance	Lower, planned maintenance	Higher maintenance, based on the additional fit-outs, modifications, and upgrades
Costs	Lower costs, using judicious strategies	Higher costs from decontamination, demolitions, and retrofit accommodations
Duration	Can be completed faster once the design is in place	Potential to take longer to complete projects as it involves deep revival and regeneration
Operations	BMS and digital twins selection possible	Need to conduct a fit-gap analysis with existing BMS and find operations optimization solutions to support OPM
Challenges	Additional developmental costs, council approvals, difficult terrain to build on are sometimes overlooked	Design and operational efficiency could be compromised due to current constraints

Source: Compiled and explained by the author

SOLUTIONS AND RECOMMENDATIONS: IMPLEMENTING RESILIENT SUSTAINABILITY

Energy usage by buildings is expected to rise by 32% by 2040 (US EIA, 2017). However, of the total carbon emissions from this increase, 57% is expected to come from embodied carbon and 43% from operational carbon (Architecture 2030, n.d.). In a bid to reducing carbon and achieving neutrality, planners must remember that building-related emissions can be separated into scope-1 emissions (such as those direct emissions from fossil fuel combustion), scope-2 (such as the indirect emissions from the generation of power and heat), and scope-3 (such as embodied carbon from the use of building materials). Paying adequate attention to emissions in every form if we are to create a more sustainable planet is, therefore, in order.

As seen in the examples in this chapter, several innovations are being shaped and deployed to accomplish the decarbonization objectives of sustainable buildings. These innovations withstand the test of time as well as incremental changes in weather and environment, making the buildings structurally sound, energy efficient, and therefore, resilient.

Industry Trends in Resilient, Sustainable Buildings

Digital twins (Burton, 2021) integrated with design and operational data are being used by developers and property management organizations to track and manage energy consumption. Digital twins help map data gathered from all sources within the building onto a single digital model that is identical to the original structure. A single interface connected to the digital model offers an accurate view of the actual power being utilized in and around the site. With the combination of BIM, BMS, IoT sensors, and other compatible tools, run over high-speed low-latency 5G networks, building planners can receive useful insights to strategize efficient energy management (IIoT-World, 2021). With sensors installed in new or retrofitted structures, data can be conveyed in real time to a centralized dashboard from where appropriate control over various power generation devices can be exercised instantaneously.

Smart strategies for energy management can ensure power consumption is restricted to when and where it is needed most, thereby optimizing efficiency. Studies (Minier & Tsai, 2021) have shown that carbon reduction is possible through upgrades to fabric, heating and cooling systems, lighting, and other power sources. The author classifies the solutions that can help advance the rollout of sustainable energy technologies in buildings as follows:

Heating, Cooling, and Ventilation Technologies: Fossil fuel and conventional electric equipment account for more than 80% of the world's building equipment (UNEP and IEA, 2017). Coal, oil and conventional gas boilers are being replaced by condensing gas boilers, with at least 15% higher efficiencies (Enerdata, 2017). There is an evident rise in air source heat pump sales in many markets in Europe, offering an energy-efficient alternative to furnaces and air conditioners. Renewable, they produce greater heating and cooling than the energy traditionally consumed to yield similar performance.

Fuel cells of hydrogen and oxygen are being used for sustainable design (Ali & Armstrong, 2008) although they are still expensive. However, hydrogen fuel cells present several advantages over other power sources, in that they are renewable, readily available through electrolysis, more efficient than fossil fuels, charge faster, are more durable, and provide clean and flexible energy to support net zero strategies.

While the use of Smart controls and Smart grids can help improve operational efficiency and lower consumption, passive techniques for heating, cooling, and ventilation can also help alleviate the load of power consumption. Evaporative cooling, radiative cooling, integrated solar panels and skin on roofs, walls and other sunlit surfaces, solar shading and small-scale wind generation, earth-air heat exchange, roof ponds, among others, are examples of clean, passive techniques and form a part of building enveloping technologies.

Building Enveloping Technologies: Building envelope performance is expected to play an important role in meeting net zero targets. It forms the primary thermal barrier between the internal and external environment. If the building envelope, comprising the doors, windows, roof, foundation, siding, and other components of structural masonry, is not in good shape, none of the energy-efficient enhancements will show expected results. The new concept of "daylightophil" architecture (Bazazzadeh et al., 2021), where daylight and energy performance are considered optimal, holds promise for further research.

Lighting Technologies: Market trends indicate that LEDs are on track to meet climate targets. However, it is important that natural light be harnessed when possible, using passive techniques and powered lighting be resorted to only when needed. Dimmers, occupancy-based light sensors, white interiors with open floor plans for greater natural light – these are just some of the simple lighting solutions that a building can use to save on carbon emissions.

Water and Waste Management: Studies have acknowledged that the relationship between sustainability, health, and water and waste management has seen new complexities from urban inequities, population density, and climate change. This has added to the challenges of sustainable health in large cities, because a burgeoning population increases both waste generation as well as water consumption.

IoT-enabled Smart water management solutions can provide real-time meaningful and actionable data about the flow, pressure, and distribution of water to enhance usage efficiency and reduce wastage. The gathered data can be used to optimize water distribution, and detect and repair leaks. Predictive analysis can be used to prevent water system failures. With outdated water systems resulting in millions of gallons of water lost each year, research shows that global investment in water management technology for Smart buildings is expected to grow from USD 2.0 billion in 2016 to USD 2.8 billion in 2025 (Hardcastle, 2016). The global Smart water management market size is expected to almost double from USD 13.8 billion in 2021 to USD 22.4 billion by 2026 (Research and Markets, 2021). Industry trends in Smart water management include water-efficient plumbing and water monitoring software. Cloud-connected submetering of Smart buildings helps monitor water usage closely using granular data.

Similar cloud-based analytics can also be used to optimize waste management. Waste bin sensors installed across all floors of a building can use AI solutions acting on gathered data to understand when bins need to be emptied, the kind of waste being collected, and the optimal disposal methods. Smart waste sorting solutions can also reduce environmental footprint by helping segregate waste based on disposal method. Certain types of waste can also be recycled to produce energy that can be redistributed to the building in the form of electricity or heating. Sensors at the central waste collection area of the building can also provide a schedule to waste disposal companies to clear out the waste to prevent diseases and contamination. This can also streamline the operations of the garbage trucks, reducing emissions and saving fuel.

Policy Implementations towards Net Zero

Achieving positive climate ambitions require net zero-intensive policies. A good balance of building codes, carbon credits and other incentives, sound financing, and governance support are vital. It will also likely require changes in behavioral and social practices (Edenhofer, 2015). From building energy codes and incentive programs to rebates and financing, net zero policies apply to the entire building and construction value chain working towards reducing its collective carbon footprint. It is imperative that governments work with stakeholders to leverage their expertise and seek their perspectives while formulating policies that are optimal and feasible to implement.

Policies for Smart Cities should be framed to create a demand for Smart buildings. These could include tax deductions on financing costs for green projects as well as income derived from rental (Beng, 2019) or sale of such properties. Property tax rebates could also be introduced to owners of green properties in the form of tax allowances on capital expenditure and lowered tax rates on income from such properties. These rebates could be extended to property buyers as well (Marpakwar, 2018). Several countries are already offering tax breaks and introducing tax reforms as part of their national construction policy. France, for instance, offers tax credits for the adoption and installation of energy-efficient technology (Hendrickson, 2018). The US, Denmark and Switzerland, among others, also offer similar breaks for eco-friendly upgrades. Tax reforms should be recognized as offering long-term solutions (Kelly, n.d.) to adopting sustainable green solutions in Smart Cities.

Becqué et al. (2016) make the following policy proposition:

- Work in alignment with national policies and local governments.
- Set targets to improve efficiency of buildings.
- Lead by example with public sector buildings.
- Implement technologies to optimum potential – most deliver positive results in relatively short payback periods.
- Set performance systems, provide financial and non-financial incentives, and support stakeholders to pursue energy efficiency.
- Enable companies to purchase carbon credits that can be used to meet part of their emission reduction requirements.

Governments can include "green conditionality" for low-carbon buildings and construction. They should also be able to collaborate with private global initiatives in reducing energy demand while increasing awareness about tools and technologies that can make a case for sustainable construction.

It is not just about the tools and technologies used, but digital transformation fostered by Smart technology (Deloitte, 2021) that will bring value to all stakeholders. Establishing technological guidelines and creating a roadmap to prioritize sustainability through policies, regulations, penalties, and taxes on the one hand, and incentives and rebates on the other, could be the way forward. Extracting value from data should come before deciding to invest in sensor technologies running on 5G to identify exactly what is required. To further enhance interoperability, data sharing standards should be established ahead of implementation.

Sustainability and the Pivot to 'Profit through Purpose'

Environmentally friendly neighborhoods and Smart cities are among the top choices to live and work. New sustainable buildings are set to represent a USD 24.7 trillion investment opportunity by 2030 (WEF, 2021). Energy efficient improvements to existing assets can expect attractive ROI. US LEED-certified projects command 3.7% more in rent, have 4% higher occupancy rates, and 5.6% higher tenant renewal rates than non-certified buildings (Wesco, 2020). While there are various studies in the US market that have demonstrated premiums of up to 26% in the commercial sector (IEA, 2019), some of London's energy-efficient offices have managed up to 12% rental premiums (Knight Frank, 2021).

However, the ROI of green buildings is not just financial but also social, environmental, and operational. Smart buildings with integrated interoperable systems can realize 30-50% energy savings in existing buildings that are otherwise inefficient (King & Perry, 2017). PoE systems can reduce AC-to-DC conversion power loss by 15% compared to poorly designed systems (Tuenge et al., 2019). Sustainably designed buildings release 36% lesser CO_2 emissions compared to the US national average (US GSA, 2011). Sustainable buildings noted a reduction in capital expenses due to 20% faster lease-up rates as well as a 3% rent premium and a 2% increase in occupancy rates. Also, sale premiums rose between 10% and 30% on buildings with sustainability certifications (EY.com, 2017). New green buildings have a 14% lower average operating cost, with a 7% increase in asset value for green buildings over traditional buildings (Dodge Data and Analytics, 2016).

FUTURE RESEARCH DIRECTION

Energy management cannot be altered overnight. And yet, the world is at a critical phase requiring urgency in action. Seizing this opportunity to make every effort toward a new climate-positive decade can avoid the lock-in of wasteful building performance mid-century.

It is evident that the buildings sector is off-track with its climate goals. The scope of sustainability must be kept open to constant-evolution, for better ideas, combinations of policies, rebates and economical innovations, and transformative solutions aimed at profit through purpose. Therefore, there is scope for further research in these areas, specifically to study the most effective global solutions. Advancement in passive systems that require minimal intervention could be one of the key factors for further improvements. Studying the demographic, economic, social, climactic, as well as technological drivers can point to the direction the future needs to take. Energy efficiency analytics and predictive analytics studies can enable better-informed decisions.

Technically, a super-low energy (SLE) efficient building with 60% energy savings is feasible. However, further research in the area to push the limits and achieve 80% efficiency can be strived for. The kind of structures that can achieve net-zero targets more easily need to be studied and replicated for optimal design. Insights into cost-effective and large-scale Smart technology solutioning with optimal use of advanced computational technologies need to be deduced.

CONCLUSION

Urbanization having become a worldwide phenomenon, with the world's rural population falling from 66% to 45% in the last 50 years (Kemp, 2019), congested cities and enormous energy usage are a reality. The United Nations Intergovernmental Panel on Climate Change (UN - IPCC) reported a 26% rise in direct emissions from buildings between 1970 and 1990 (WBCSD - EEB, 2009). Since then, there has been a steady rise in energy consumption and GHG emissions from buildings. In fact, the challenge to 2050 is projected to be so stark that it would take a 94% reduction in energy use to restrict emissions to those reported in 2002 (EIA, 2016), assuming all improvement came from greenfield buildings. As the buildings industry looks toward a greener future, green development should be made the main criteria for Smart Cities of tomorrow.

A study of the energy use of 274 cities across the globe reveals that each city requires its own carbon emissions mitigation model to prevent energy consumption from increasing three-fold by 2050 (Thorpe, 2015). This is why performance measurement tools and accreditations must vary in their approach depending on regional requirements. While some regions may have a dire need to improve water efficiency, other cities may need to focus on power efficacy. Therefore, a building rating system that is unified, yet adaptable to the native environment, could be the ideal solution.

The emissions-saving potential of the building sector is said to be 84 gigatons of carbon by 2050 (WGBC, 2016). Sustainable buildings in Australia that have achieved the Green Star Certification have seen a 62% reduction (WGBC, n.d.) in GHGs. Other countries, including India, South Africa, and the US, have also reported between 25 and 50% energy savings (WGBC) from green construction. Buildings represent significant untapped possibilities for a sustainable future. Along with a shift in mindset at different levels, there is a need to drive a change in the belief that anything green is big-budget. The efforts towards going green does not mean additional costs. Quite the contrary, it promises exponential returns on long-term running costs such as utilities, operations, and maintenance through the building's entire lifespan. A substantial 50% reduction in operational carbon (WEF, 2021) can be achieved through changes in lighting, heating, cooling, and ventilation. However, for global, large-scale execution of low carbon technologies, there need to be policy-level implementations.

Policy implementations require concerted efforts and sustained participation at various levels: governmental backing through funding and incentives, public and private investments, cost-effective technologies, legal policies and regulatory frameworks, and citizen support. Government tax rebates, results-based incentives, and concessions could extend to both construction and the buyer who can be awarded tax breaks for investing in sustainable projects. Financers can do well by offering competitive rates on borrowings to encourage loans for development of, as well as investment in, green projects. While helping grow the economy, organizations also need to move their focus to becoming green in all aspects from adopting Smart technology to reduce operational costs to promoting environmental initiatives with increased participation.

Harmonizing sustainability efforts and goals across the globe, coordinating planning and policies within and between countries, sharing resources and best practices, facilitating and supporting networks aimed at sustainability and energy management, can light the way ahead. Along with these, supplementing private investments with public sector financing, and driving the transformation roadmap powered by sustainable digital solutions can contribute to a holistic, integrated path to the carbon-negative journey. With a pragmatic framework and dedicated commitment, we should be able to get there.

REFERENCES

AIA. (2021). *Market one*. The American Institute of Architects. Retrieved January 5, 2022, from https://www.aia.org/showcases/6388129-market-one

Aktas, B., & Ozorhon, B. (2015). Green building certification process of existing buildings in developing countries: Cases from Turkey. *Journal of Management Engineering*, *31*(6), 05015002. doi:10.1061/(ASCE)ME.1943-5479.0000358

Ali, M., & Armstrong, P. (2008). Overview of sustainable design factors in high-rise buildings. *Proc. of the CTBUH 8th World Congress*, 282–291.

Alsarraf, A. A., & Alobaidi, M. M. (2020). Advanced techniques in design openings of smart facades in buildings. *IOP Conference Series. Materials Science and Engineering*, *757*, 012003. doi:10.1088/1757-899X/757/1/012003

Arch2O.com. (n.d.). *10 most sustainable office buildings worldwide*. Arch20 - In Depth. Retrieved January 4, 2022, from https://www.arch2o.com/sustainable-office-buildings-worldwide/

Architecture 2030. (n.d.). *Why the building sector?* Architecture 2030. Retrieved January 5, 2022, from https://architecture2030.org/why-the-building-sector/

Autodesk, Inc. (2008). *Improving Building Industry Results through Integrated Project Delivery and Building Information Modeling*. White Paper. Available online at www.autodesk.com

Bazazzadeh, H., Świt-Jankowska, B., Fazeli, N., Nadolny, A., Safar ali najar, B., Hashemi safaei, S., & Mahdavinejad, M. (2021). Efficient shading device as an important part of Daylightophil architecture; a designerly framework of high-performance architecture for an office building in Tehran. *Energies*, *14*(24), 8272. doi:10.3390/en14248272

Beavis, L. (2014, May 15). PWC creates the most sustainable building in the world. *The Guardian*. Retrieved January 5, 2022, from https://www.theguardian.com/sustainable-business/sustainability-case-studies-pwc-one-embankment-place

Becqué, R., Mackres, E., Layke, J., Aden, N., Liu, S., Managan, K., Nesler, C., Mazur-Stommen, S., Petrichenko, K., & Graham, P. (2016). *Accelerating Building Efficiency: Eight actions for urban leaders*. Accelerating Building Efficiency. Retrieved December 13, 2021, from https://publications.wri.org/buildingefficiency/

Beng, T. H. (2019, January 24). *A greener Singapore: Proptech, tax incentives crucial*. Retrieved January 7, 2022, from https://www.businesstimes.com.sg/opinion/a-greener-singapore-proptech-tax-incentives-crucial

Blutstein, H., & Rodger, A. (2001). The sustainable tall building of the third millennium. *Proc. of the CTBUH 6th World Congress*, 131–142. 10.4324/NOE0415232418.ch6

Bowen, M. (2021, June 2). *Four ways smart building technology creates sustainable environments*. Intelligent CIO North America. Retrieved December 10, 2021, from https://www.intelligentcio.com/north-america/2021/05/28/four-ways-smart-building-technology-creates-sustainable-environments/

Browning, W. D., & Ryan, C. O. (2020). *Nature Inside: A Biophilic Design Guide* (1st ed.). RIBA Publishing., doi:10.4324/9781003033011

Burton, J. (2021, November 29). *3D Digital Twins, IOT, and the future of decarbonization*. IIoT World. Retrieved January 5, 2022, from https://www.iiot-world.com/industrial-iot/connected-industry/3d-digital-twins-iot-and-the-future-of-decarbonization/

CfUT. (2019, September 19). *Executive Summary | Climate Emergency, Urban Opportunity*. Coalition for Urban Transitions. Retrieved January 10, 2022, from https://urbantransitions.global/en/publication/climate-emergency-urban-opportunity/executive-summary

Collins, A., Watts, S., & McAlister, M. (2008). The economics of sustainable tall buildings. *Proc. of the CTBUH 8th World Congress*, 175–185.

Cox, L. (2020, February 17). Researchers claim solar efficiency breakthrough for flexible 'skin'. *The Guardian*. Retrieved January 4, 2022, from https://www.theguardian.com/environment/2020/feb/18/researchers-claim-solar-efficiency-breakthrough-for-flexible-skin

Deloitte. (2021). *Smart and sustainable buildings and infrastructure*. Deloitte. Retrieved January 5, 2022, from https://www2.deloitte.com/global/en/pages/public-sector/articles/urban-future-with-a-purpose/smart-and-sustainable-buildings-and-infrastructure.html

Digital Europe. (2020, December 4). *Digital Building Transformations for Europe's green renovations: Sustainable Living and working*. DIGITALEUROPE. Retrieved January 5, 2022, from https://www.digitaleurope.org/resources/digital-building-transformations-for-europes-green-renovations-sustainable-living-and-working/

Dodge Data and Analytics. (2016). *World Green Building Trends 2016 | Developing Markets Accelerate Global Green Growth*. Smart Market Report. Retrieved January 7, 2022, from https://www.infrastructureusa.org/world-green-building-trends-developing-markets-accelerate-global-green-growth/

Dong, J., Zhang, H., Klapare, S., Wang, J., & Luo, Y. (2020). 16: Socioeconomic benefit of contaminated site remediation. In *Sustainable Remediation of Contaminated Soil and Groundwater* (pp. 433–450). Butterworth-Heinemann. doi:10.1016/B978-0-12-817982-6.00016-1

Edenhofer, O. (2015). *Climate change 2014: Mitigation of climate change: Working Group III Contribution to the fifth assessment report of the intergovernmental panel on climate change*. Cambridge University Press.

EIA. (2016, May). *International Energy Outlook 2016*. International Energy Outlook 2016 - With Projections to 2040. Retrieved February 28, 2022, from https://www.eia.gov/outlooks/ieo/pdf/0484(2016).pdf

Elefante, C. (2012). The Greenest Building Is... One That Is Already Built. *Forum Journal, 27*(1), 62-72. https://www.muse.jhu.edu/article/494514

Elhakeem, A., & Hegazy, T. (2012). Building Asset Management with deficiency tracking and integrated life cycle optimisation. *Structure and Infrastructure Engineering*, *8*(8), 729–738. doi:10.1080/15732471003777071

Enerdata. (2017). *Share of new dwellings built according to national nZEB definition or better than nZEB*. ZEBRA2020. www.zebramonitoring.enerdata.eu/overall-building-activities/share-of-new-dwellings-built-according-to-national-nzeb-definition-or-better-than-nzeb.html

EPA. (2021, June 10). *Climate Smart Brownfields Manual*. Retrieved January 5, 2022, from https://www.epa.gov/sites/default/files/2021-06/documents/final_climate_smart_brownfields_manual_6-10-21_508_complaint.pdf

EY.com. (2017, September 1). *Smart buildings: how sustainability strategies can impact investor ROI*. EY UK. Retrieved December 13, 2021, from https://www.ey.com/en_uk/real-estate-hospitality-construction/smart-buildings-how-sustainability-strategies-can-impact-investor-roi

Fazli, W., Hakim, I. L., Rozaida, G., & Muhammad, A. (2019). *An Efficient Artificial Intelligence Hybrid Approach for Energy Management in Intelligent Buildings. KSII Transactions on Internet and Information Systems.* doi:10.3837/tiis.2019.12.007

Fernandes, L. (2021, July 8). *Smart buildings and IOT: Supporting a safe return to the Office*. Quocirca. Retrieved March 4, 2022, from https://quocirca.com/content/smart-buildings-and-iot-supporting-a-safe-return-to-the-office/

Fichte, J., & Hollfelder, M. (2020, April 13). *Smart buildings: Making buildings smarter, greener, and more energy-efficient - page 2 of 3*. EE Times Europe. Retrieved January 3, 2022, from https://www.eetimes.eu/smart-buildings-making-buildings-smarter-greener-and-more-energy-efficient/2/

Fix Development. (2013). *Clock shadow building*. Fix Development. Retrieved January 5, 2022, from https://www.fixdevelopment.com/clock-shadow-building/

Franz, M., Pahlen, G., Nathanail, P., Okuniek, N., & Koj, A. (2006). Sustainable development and brownfield regeneration. What defines the quality of derelict land recycling? *Environmental Sciences*, *3*(2), 135–151. doi:10.1080/15693430600800873

Gartner Inc. (2019). *Hype cycle for Smart City Technologies and Solutions, 2019*. Gartner. Retrieved January 3, 2022, from https://www.gartner.com/en/documents/3955915/hype-cycle-for-smart-city-technologies-and-solutions-201

Gibbs, A., & Peiser, R. (2022, January 25). ResilientHub: A study in getting office buildings to net zero. *Urban Land Magazine*. Retrieved March 3, 2022, from https://urbanland.uli.org/sustainability/resilienthub-a-study-in-getting-office-buildings-to-net-zero/?Site=ULI2015

Grevelman, L., & Kluiwstra, M. (2010). Sustainability in project management: A case study on enexis. *PM World Today, 12*(7).

Hardcastle, J. L. (2016, April 14). *Why water management will take smart buildings to the next level.* Environment + Energy Leader. Retrieved March 6, 2022, from https://www.environmentalleader.com/2016/04/why-water-management-will-take-smart-buildings-to-the-next-level/

Hatcher, J. (2021, June 30). The making of a smart building: Revamping current structures. *Smart Buildings Magazine*. Retrieved January 4, 2022, from https://smartbuildingsmagazine.com/features/the-making-of-a-smart-building-revamping-current-structures

Hedayat, K. (2020, June 23). *The return to workplace is happening. Is your building ready?* Schneider Electric Blog. Retrieved January 5, 2022, from https://blog.se.com/building-management/2020/06/23/the-return-to-workplace-is-happening-is-your-building-ready/

Hendrickson, V. L. (2018, April 26). *Where are the best property tax breaks for Green Homes Around the world?* Mansion Global. Retrieved January 7, 2022, from https://www.mansionglobal.com/articles/where-are-the-best-property-tax-breaks-for-green-homes-around-the-world-95175

Hossain, F. (2019). Chapter Seven - Best Management Practices. In Sustainable design and build: Building, energy, roads, bridges, water and Sewer Systems (pp. 419–431). Butterworth-Heinemann.

IEA. (2017). *Energy Technology Perspectives 2017.* IEA/OECD. www.iea.org/et

IEA. (2019). *Multiple Benefits of Energy Efficiency*. IEA. https://www.iea.org/reports/multiple-benefits-of-energy-efficiency

IEA & UNEP. (2018). *2018 Global Status Report: towards a zero-emission, efficient and resilient buildings and construction sector.* Author.

IIoT-World. (2021, October 27). *The role of 3D Digital Twins and IIOT in decarbonization of buildings.* IIoT World. Retrieved January 5, 2022, from https://www.iiot-world.com/industrial-iot/connected-industry/the-role-of-3d-digital-twins-and-iiot-on-the-road-to-decarbonization-of-buildings/

IISD. (2017, June 7). *Mitigation update: Building Sector Must Reach Net Zero Carbon by 2050.* IISD - SDG. Retrieved January 3, 2022, from https://sdg.iisd.org/news/mitigation-update-building-sector-must-reach-net-zero-carbon-by-2050/

Jacomit, A. M., Granja, A. D., & da Silva, V. G. (2009). Can Sustainable Buildings Cost the Same as Conventional Buildings? In *SASBE 2009 proceedings of the 3rd CIB International Conference on Smart and Sustainable Built Environments:* June 15-19, 2009, *Delft, the Netherlands* (pp. 1–2). Delft University of Technology.

Kelly, R. (n.d.). Implementing sustainable property tax reform in developing countries. *Taxation and Development: The Weakest Link?*, 326–363. doi:10.4337/9781783474332.00018

Kemp, J. (2019). *Urbanisation - Selected Indicators.* Thomson Reuters. Retrieved February 28, 2022, from https://fingfx.thomsonreuters.com/gfx/ce/7/7352/7334/URBANISATION.pdf

King, J., & Perry, C. (2017, February). *Smart buildings: Using smart technology to save energy in ...* American Council for an Energy-Efficient Economy. Retrieved January 7, 2022, from https://www.aceee.org/sites/default/files/publications/researchreports/a1701.pdf

Knight Frank Research Reports. (2021, September). *The Sustainability Series*. Knight Frank. Retrieved December 13, 2021, from https://www.knightfrank.com.hk/research/the-sustainability-series-september-2021-8395.aspx

Koerner, D. (2021, October 22). *How are digital twins used in building automation?* Smart Building Blog. Retrieved January 4, 2022, from https://www.75f.io/blog/how-are-digital-twins-used-in-building-automation

Lee, S. H., Augenbroe, G., Lee, J.-K., & Zhao, F. (2013). A design methodology for energy infrastructures at the campus scale. *Computer-Aided Civil and Infrastructure Engineering*, *28*(10), 753–768. doi:10.1111/mice.12050

Leung, L., & Weismantle, P. (2008). Sky-sourced sustainability – how super tall buildings can benefit from height. *Proc. of the CTBUH 8th World Congress*, 328–335.

Markets & Markets. (2020, February). *Smart buildings market*. Markets and Markets. Retrieved March 4, 2022, from https://www.marketsandmarkets.com/Market-Reports/smart-building-market-1169.html

Marpakwar, C. (2018, December 1). *Home buyers to get property tax rebates in Green Buildings*. Mumbai Mirror. Retrieved January 7, 2022, from https://mumbaimirror.indiatimes.com/mumbai/civic/home-buyers-to-get-property-tax-rebates-in-green-buildings/articleshow/66893701.cms

Meadows, C. (2021, November 20). What is Smart Building Technology Worth to the market? *LEDs Magazine*. Retrieved March 4, 2022, from https://www.ledsmagazine.com/blogs/article/14214312/what-is-smart-building-technology-worth-to-the-market

Minier, V., & Tsai, O. (2021, December 6). *4 expert views on how to accelerate investment in green buildings and decarbonize cities*. World Economic Forum. Retrieved January 5, 2022, from https://www.weforum.org/agenda/2021/12/4-expert-views-how-to-accelerate-investment-green-buildings/

Mohd Isa, N. K., Abdul Samad, Z., & Alias, A. (2014). A review on sustainability principles of building: Formulation of a theoretical framework. *Journal of Surveying, Construction & Property*, *5*(1), 1–16. doi:10.22452/jscp/vol5no1.5

O'Connor, D., & Hou, D. (2020). 1: Green and Sustainable Remediation: concepts, principles, and pertaining research. In *Sustainable remediation of contaminated soil and groundwater: Materials, processes, and assessment* (pp. 43–73). Butterworth-Heinemann. doi:10.1016/B978-0-12-817982-6.00003-3

Pasini, M., Brondino, M., Trombin, R., & Filippi, Z. (2021). A participatory interior design approach for a restorative work environment: A research-intervention. *Frontiers in Psychology*, *12*, 718446. Advance online publication. doi:10.3389/fpsyg.2021.718446 PMID:34603143

Pinto, T., Praça, I., Vale, Z., Morais, H., & Sousa, T. M. (2013). Strategic bidding in electricity markets: An agent-based simulator with game theory for scenario analysis. *Integrated Computer-Aided Engineering*, *20*(4), 335–346. doi:10.3233/ICA-130438

Popescu, G., & Patrascoiu, R. (2012). Brownfield sites - Between abandonment and redevelopment case study: Craiova city. *Human Geographies*, *6*(1), 91–97. doi:10.5719/hgeo.2012.61.91

Premier, A. (2012). Dynamic facades and smart technologies for building envelope requalification. *Screencity-International Academic Journal*, *1*, 65–69.

PwC. (2017). *Building smart - How digital technology can help construction companies achieve more value*. Orange Business Services. Retrieved January 4, 2022, from https://www.orange-business.com/en/library/white-paper/building-smart

Research and Markets. (2021, December 13). *Global Smart Water Management Market Forecast to reach USD 22.4 billion by 2026*. Cision PR Newswire. Retrieved March 6, 2022, from https://www.prnewswire.com/news-releases/global-smart-water-management-market-forecast-to-reach-usd-22-4-billion-by-2026--301443307.html

Roxburgh, H. (2016, August 23). Inside Shanghai Tower: China's tallest skyscraper claims to be world's Greenest. *The Guardian*. Retrieved January 4, 2022, from https://www.theguardian.com/cities/2016/aug/23/inside-shanghai-tower-china-tallest-building-green-skyscrapers

Sarni, W. (2010). *Greening brownfields: Remediation through sustainable development*. McGraw-Hill.

Sayadi, S., Tsatsaronis, G., & Morozyuk, T. (2016). CLIMA 2016 - 12th REHVA World Congress. In *Exergy-based Control Strategies for HVAC Systems* (Vol. 8). Academic Press.

Schlueter, A., & Thesseling, F. (2009). Building Information Model Based Energy/Exergy Performance Assessment in early design stages. *Automation in Construction*, *18*(2), 153–163. doi:10.1016/j.autcon.2008.07.003

Sierla, S., Azangoo, M., Rainio, K., Papakonstantinou, N., Fay, A., Honkamaa, P., & Vyatkin, V. (2021). Roadmap to semi-automatic generation of Digital Twins for Brownfield Process Plants. *Journal of Industrial Information Integration*, *100282*. Advance online publication. doi:10.1016/j.jii.2021.100282

Smart Cities World. (2017, October 2). *The world's most sustainable office*. Smart Cities World. Retrieved January 4, 2022, from https://www.smartcitiesworld.net/news/the-worlds-most-sustainable-office-2167

Smart Energy International. (2021, March 22). *The role of smart buildings in the development of Smart Cities*. Smart Energy International. Retrieved January 4, 2022, from https://www.smart-energy.com/regional-news/asia/the-role-of-smart-buildings-in-the-development-of-smart-cities/

Solxenergy. (n.d.). *Solxenergy*. Downloads. Retrieved January 5, 2022, from https://www.solxenergy.com/index.php/downloads

Tay, V. (2021, July 9). *DBS to turn Newton office building into net zero development*. Businesstimes.com.sg. Retrieved January 5, 2022, from https://www.businesstimes.com.sg/banking-finance/dbs-to-turn-newton-office-building-into-net-zero-development

Thieme, W. (2019, April 3). *3 practical ways to transform brownfield plants into digital factories*. IIoT World. Retrieved January 4, 2022, from https://iiot-world.com/connected-industry/3-practical-ways-to-transform-brownfield-plants-into-digital-factories/

Thornton, G., Franz, M., Edwards, D., Pahlen, G., & Nathanail, P. (2007). The challenge of sustainability: Incentives for brownfield regeneration in Europe. *Environmental Science & Policy*, *10*(2), 116–134. doi:10.1016/j.envsci.2006.08.008

Thorpe, D. (2015, February 2). *How can cities cut energy use by 50%.* World Economic Forum. Retrieved February 28, 2022, from https://www.weforum.org/agenda/2015/02/how-can-cities-cut-energy-use-by-50/

ThoughtWire. (2019, February 14). *What is OPM and How is it Transforming Buildings?* Thoughts Over the Wire. Retrieved January 4, 2022, from https://blog.thoughtwire.com/what-is-opm-and-how-is-it-transforming-buildings

Tizouiar, O., & Boussoualim, A. (2018). The approach used for comfortable and high efficient buildings in sustainable cities by different actors. *Renewable Energy for Smart and Sustainable Cities*, 3–13. doi:10.1007/978-3-030-04789-4_1

Toli, A. M., & Murtagh, N. (2020). The concept of sustainability in Smart City definitions. *Frontiers in Built Environment - Sustainable Design and Construction, 6*. doi:10.3389/fbuil.2020.00077

Tuenge, J., Kelly, K., & Poplawski, M. (2019, January). *Connected Lighting Systems Efficiency Study -- POE cable Energy Losses, Part 1.* US Dept. of Energy. Retrieved January 7, 2022, from https://www.energy.gov/sites/prod/files/2019/01/f58/cls_poe-cable-pt1.pdf

UN. (2016, May 18). *Habitat launches the World Cities Report 2016, Urbanization and Development: Emerging Futures: UN-Habitat.* UN Habitat For a Better Urban Future. Retrieved January 3, 2022, from https://unhabitat.org/un-habitat-launches-the-world-cities-report-2016

UN Environment Programme and International Energy Agency. (2017). *Towards a zero-emission, efficient, and resilient buildings and construction sector.* Global Status Report 2017.

UN WCED. (1989). *Our common future: Brundtland report.* Oxford University Press.

UNEP. (2020, December 16). *Building sector emissions hit record high, but low-carbon pandemic recovery can help transform sector – UN report.* Retrieved January 3, 2022, from https://www.unep.org/news-and-stories/press-release/building-sector-emissions-hit-record-high-low-carbon-pandemic

Urban Land Institute. (2021). *State of Green: Greenprint Performance Report* (Vol. 12). Urban Land Institute.

US EIA. (2017, September 14). *International Energy Outlook 2017.* Retrieved January 5, 2022, from https://www.eia.gov/outlooks/ieo/pdf/0484(2017).pdf

US EPA. (n.d.). *Sustainability.* EPA. Retrieved January 3, 2022, from https://www.epa.gov/sustainability/

US GSA. (2011). *Green building performance - general services administration.* GSA Public Buildings Service. Retrieved January 7, 2022, from https://www.gsa.gov/cdnstatic/Green_Building_Performance2.pdf

Verma, U. (2018, November 21). *PropTech: How digital twins impact OPM in smart buildings.* In-Building Tech. Retrieved January 4, 2022, from https://inbuildingtech.com/uncategorized/digital-twins-proptech/

Verma, U. (2018, December 14). *How digital twins unlock value beyond BMS in commercial office buildings*. In-Building Tech. Retrieved January 4, 2022, from https://inbuildingtech.com/smart-buildings/digital-twins-opm-thoughtwire/

Wang, N., & Adeli, H. (2014). Sustainable building design. *Journal of Civil Engineering and Management, 20*(1), 1–10. doi:10.3846/13923730.2013.871330

Washburn, D., Sindhu, U., Balaoura, S., Dines, R. A., Hayes, N. M., & Nelson, L. E. (2010). *Helping CIOs Understand "Smart City" Initiatives*. Forrester Research, Inc.

WBCSD - EEB. (2009). *Energy Efficiency in Buildings*. World Business Council for Sustainable Development. Retrieved February 28, 2022, from https://docs.wbcsd.org/2008/07/EEB-FactsTrends-FullReport.pdf

WEF. (2021, January). *Net Zero Carbon Cities: An Integrated Approach - Insight Report*. World Economic Forum. Retrieved January 5, 2022, from https://www3.weforum.org/docs/WEF_Net_Zero_Carbon_Cities_An_Integrated_Approach_2021.pdf

Wesco. (2020, February 10). *The ROI of sustainable intelligent buildings*. WESCO International, Inc. Retrieved January 11, 2022, from https://blog.wesco.com/the-roi-of-sustainable-intelligent-buildings

WGBC. (2016). *Towards zero-emission efficient and resilient buildings - Global Status Report 2016*. Global Alliance for Buildings and Construction. Retrieved February 28, 2022, from https://www.worldgbc.org/sites/default/files/GABC_Global_Status_Report_V09_november_FINAL.pdf

WGBC. (2018). *World Green Building Trends in 2018: Green keeps growing*. U.S. Green Building Council. Retrieved March 3, 2022, from https://www.usgbc.org/articles/world-green-building-trends-2018-green-keeps-growing

WGBC. (n.d.a). *What is green building?* World Green Building Council. Retrieved January 3, 2022, from https://www.worldgbc.org/what-green-building

WGBC. (n.d.b). *The benefits of Green Buildings*. World Green Building Council. Retrieved February 28, 2022, from https://www.worldgbc.org/benefits-green-buildings#:~:text=The%20world%20over%2C%20evidence%20is,communities%2C%20and%20driving%20economic%20growth

Wolfson, A. (2018). *In Sustainability and the City: The Service Approach*. Business Expert Press.

Wong, D. (2019, January 30). NUS launches Singapore's first net-zero energy building to be built from scratch. *The Straits Times*. Retrieved January 4, 2022, from https://www.straitstimes.com/singapore/environment/nus-launches-singapores-first-net-zero-energy-building-to-be-built-from

Wood, A. (2008). Green or grey? The aesthetics of tall building sustainability. *Proc. of the CTBUH 8th World Congress*, 194–202.

ADDITIONAL READING

Clark, J. (2020). *Uneven innovation: The work of Smart Cities*. Columbia University Press. doi:10.7312/clar18496

Halegoua, G. R. (2020). *Smart cities*. The MIT Press. doi:10.7551/mitpress/11426.001.0001

Muller, H. (2020). *Future state 2025: How top technology executives disrupt and drive success in the Digital Economy*. John Wiley & Sons, Inc.

Susantono, B., & Guild, R. (2021). *Creating Livable Asian cities*. Asian Development Bank. doi:10.22617/SGP210110

Thornbush, M. J., & Golubchikov, O. (2020). *Sustainable urbanism in digital transitions: From low carbon to smart sustainable cities*. Springer International Publishing. doi:10.1007/978-3-030-25947-1

KEY TERMS AND DEFINITIONS

BIM: Building Information Modeling or BIM is a process that draws support from various tools, techniques, and technologies to digitally represent a physical space.

BMS: Building Management System or BMS is an all-encompassing control system that intelligently auto-controls and regulates the various interoperable aspects of a building for the purpose of active monitoring.

Digital Twin: A digital twin is a virtual representation of any object that offers real-time information about the object.

OPM: Operations Performance Management or OPM is used across industries and is the alignment of all operations within a structure to support them to work together to achieve a common objective.

Chapter 11
Sustainability and Health in a Smart City:
Health at the Heart of a Smart City

Vinod Anand Bijlani
Independent Researcher, Singapore

ABSTRACT

Smart cities have so far focused on improving citizens' standards of living, with emphasis on better transportation, improved housing spaces and workplaces, and energy conservation. A pandemic-stricken world emerging to adopt a new normal has shifted the focus to health and hastened the need for leveraging technology, including but not limited to Blockchain, artificial intelligence (AI), and the internet of things (IoT) to enable Smart Cities to move to the next level of wellbeing. With public health and sustainability in the spotlight, civic agencies and government bodies are now alert to potential future contingencies that cannot be ruled out. This chapter emphasizes the need for a future-proofed healthcare system within sustainable Smart Cities that not only foresees and contains public health emergencies but also provides for its citizens' general wellbeing.

INTRODUCTION

It has taken a century's worth of pandemics to emphasize that healthcare is an essential part of city life. Cities that are going the sustainable way are posed with the challenges of judiciously meeting the rising demand on hospitals, medical care, and related resources, while balancing sustainability. A strong healthcare system involves a steady flow of patients and a matching supply of physicians and medical experts. A city that can provide real-time health condition monitoring, predict disease diagnosis ahead of time, and offer preventive treatment and rehabilitation, has met its goal of being a Smart City with health at its core.

More than half the world's population currently lives in cities (WHO, 2020), with the figures set to rise to 70% by 2050. With not even three-quarters of the infrastructure or resources required to meet this growth in place today, urban planners have a huge transformative challenge ahead of them. Their

DOI: 10.4018/978-1-6684-4030-8.ch011

test is to create carefully-planned urban spaces enabling planetary health and citizen wellbeing, with a conscious link between the two. In other words, it is important to reform cities to have environmental, social, and population health at their center.

Individuals create thousands of data points throughout their daily lives. These data points help draw up data-driven solutions that solve challenges to health and sustainability. This data can be utilized to perform predictive and prescriptive analyses of various situations using AI and machine learning (ML). Digital models can help create value-based solutions that can fundamentally change life and health in a Smart City, with a focus on predictive and preventive healthcare for all. Figure 1 provides a snapshot of an intelligent healthcare ecosystem in a Smart City.

The multidimensionality of Smart Cities simultaneously focusing on sustainability and population health can pose a concern mainly due to the complexity of the concept. Appropriate policy implementations and a good governance model are needed to improve the quality of citizens' health. The pace of implementation and the amount of funding could be ambitious, while also calling for changes in social and economic systems. The author deliberates on the maturity of Smart Cities to accelerate these actions and drive innovation, and attempts a proposition on reaching these goals.

BACKGROUND AND LITERATURE REVIEW

The World Health Organization (WHO, 2018) believes that *where* people live can affect their health and chances of leading flourishing lives. And yet, healthcare is often separated from urban infrastructure, although population health is a vital facet of city life.

Promoting population health and developing appropriate strategies rely on instantaneous connectivity. With a shift in paradigm from an 'information society' to a hyperconnected society, a healthy Smart City is a reality. Smart Cities comprise a hyperconnected digital network of devices producing big data (Wachowicz et al., 2013). With the arrival of 5G, terabytes of data collected from various sources across a Smart City's data microcosm are all interrelated to provide valuable insights at low latency. Highly permeable data combined with ubiquitous technology and well-planned hyperconnected infrastructure in a Smart City can offer a solution to health challenges. Besides just hospitals and care facilities, it can extend healthcare to home, workplaces, and community spaces, and create a sustainable and safe living environment.

Rapid growth in population density and steady rise in the aging population has led to an upsurge in the incidence of communicable and chronic illnesses in large cities. Plextek conducted a survey of future health needs (Johnson, 2016) that revealed 76% of the population to be concerned about the elderly in their family living alone in cities, without adequate health support, the main anxiety being around them experiencing a health emergency. It is not just the concern around the elderly, but also others who are at risk from various urban lifestyle-induced illnesses and noncommunicable diseases (NCDs) such as diabetes, hypertension, heart disease, etc., that are seeing an increasingly alarming incidence at a young age lately. Constantly monitoring changes to health status and regular medical follow-ups are, therefore, in order. For an at-risk population, there could be several limitations to visiting a doctor, from lack of transportation to discouragement from long waiting time. The same concerns could beset a city plagued by a highly communicable disease. Pandemics, such as COVID-19, could be the scourge of the future, just as much as they have tortured cities in the past.

Figure 1. Snapshot of an intelligent healthcare ecosystem in a Smart City
Infographic created by author

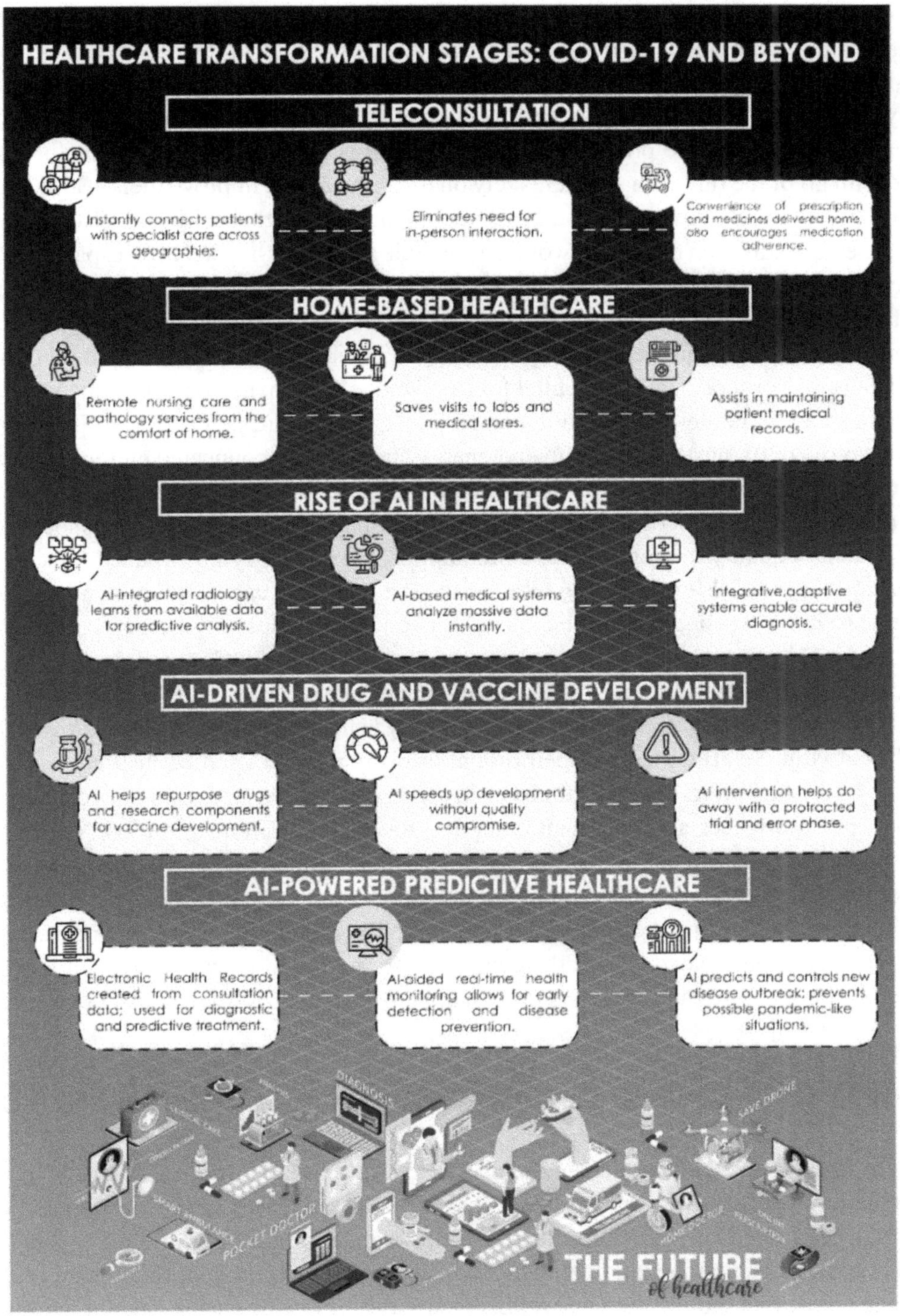

With the looming threat from population growth and inefficient patient management, there is an absolute need to build and adapt a more efficient and sustainable healthcare infrastructure. The need is made more dire by limited healthcare resources but increasing life expectancy, higher frequency of hospital stays and re-admission, as well as poor interlinked communication between civic sectors.

A Smart City is ideally a community "where residents can engage with smart services that are specifically designed to improve their health" (Gibbons, 2020). Such a city is made of multiple smart environments that are central to preventing health crises. It is comprised of a system "that can interact and engage with all of the things that patients rely on to optimize and improve their health… in between doctor visits" (Gibbons). This would include a networked mesh consisting of a healthy living environment, clean air and healthy habits on the way to work, sustainable and planet-friendly workspaces, safety for the elderly, secure community centers promoting social and physical wellness, besides a holistic approach to all of this bundled together. Complex interoperability across healthcare, human services, public safety, environmental health, social and emergency services, living spaces, workplaces, public transport, and the immediate neighborhood is most essential for a networked system to work to its maximum potential.

Smart Cities can facilitate collaboration and response as they are perceived as "living organisms" (Costa & Peixoto, 2020), bustling with life and energy, and are interconnected and interoperable, all at once. With comprehensive and systematic planning for good health, Smart Cities can address the issues of inequality in healthcare services and ensure greater participatory governance towards social, economic, and environmental equality among diverse communities. Participatory governance (Gao et al., 2020) is an important step towards empowering citizens to take proactive care of their own health as well as play an active role in assessing real public needs. It is one of the most compelling concepts for Smart Cities that are looking to meet the Sustainable Development Goals (SDGs), improve local government efficiency, and move towards a more collaborative form of living.

Driving this route to resilience in a healthy Smart City is innovation. Innovation is key to promoting sustainability of large Smart Cities and their digital infrastructures, including healthcare management. Innovation in healthcare offers the opportunity to lower costs, improve access to medical care, and revolutionize patient experiences and treatment outcomes for all. IoT for cities, run on AI algorithms and analyses, built on Blockchain technology, and fed by Big Data can be considered the biggest innovation for digitally-networked cities. Smart Cities employing embedded technologies that help with communication as well as monitoring could be the solution to preempting several health-related concerns. Healthcare that is continuous and automated can set the course for prevention of disease. The author discusses the possible healthcare strategies and solutions for sustainable and healthy Smart Cities in the next sections.

SMART HEALTHCARE MANAGEMENT FOR HEALTHY, SUSTAINABLE CITIES

In the modern world, preventable illnesses and diseases form a disproportionate chunk of the burden on a city's infrastructure (Kharas & Remes, 2018). Advancement in pervasive digitization and computing have resulted in networked infrastructures that can be managed to create healthier, safer spaces by detecting disease transmission, monitoring treatment, and managing rehabilitation. Healthcare ecosystems in Smart Cities are those that spotlight citizen wellbeing through preventive care and predictive intervention.

Powerful IoT-based Smart sensors and a range of wearables connected by wireless communication systems and integrated circuits can read health parameters to monitor patients remotely. By encouraging the use of wearables among citizens, the retrieved information can be managed to detect those who are unwell, make accurate predictions, and take preventive action. Besides mobile devices, digitized infrastructure fitted at every street corner and dotting every part of a Smart City can be leveraged to gather data to detect health concerns. Smart phones and mobile devices also play an important role in gathering data, connecting healthcare with other Smart City services, without having to spend additionally on sensor infrastructure (Cook et al., 2018), improving response type and time. Mobile citizen-sensing can inform city service providers of time-critical events from the data collected from individual devices (Totty, 2017).

Figure 2 illustrates the patient journey in a Smart City with robust healthcare infrastructure, where health prediction, prevention, and diagnosis are automated and speeded up. In the Smart healthcare model, diagnoses are proactive and preemptive, and treatments personalized and timely. This sets it apart from the conventional healthcare practices where prognosis is based solely on visible symptoms and historical records. The Smart healthcare paradigm encourages a broader view of citizen health. Remote patient monitoring with sensors can collect and record data from patients, report anomalies, and send result to doctors for analysis. Linking self-health monitoring to remote healthcare professionals, doctors can then analyze all the available data along with patient history to prescribe personalized medications. The process is streamlined, cost-effective, and convenient, all with minimal interference to daily routine. With this end-to-end journey, patients can receive advice and treatment before their condition worsens.

Figure 2. End-to-end patient journey in a Smart City with robust healthcare infrastructure
Infographic created by author.

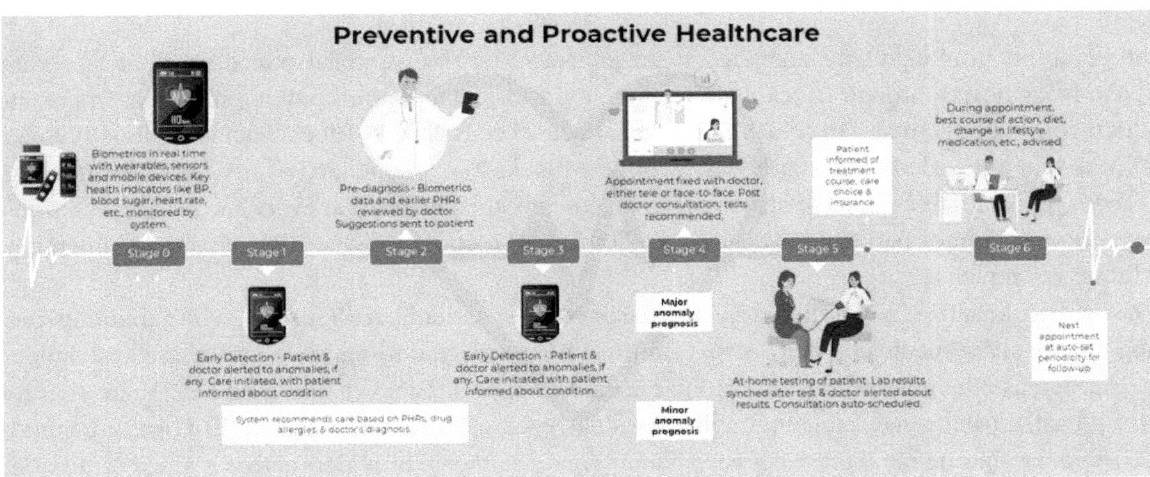

IoT can, therefore, also enable prediction and prevention of NCDs as well as communicable infections by catching signs and symptoms in groups and communities before there is widespread outbreak. For a Smart City, all this translates into healthier citizens, lower hospitalization rates, lesser waiting time in care facilities, awarding patients more control over their health (Uludag, 2016), evidence-based precautionary measures, well-targeted population health campaigns, and better citizen health.

DATA-DRIVEN PREEMPTIVE HEALTHCARE

Identifying the relevant data and analyzing it for impactful insights has the potential to markedly improve detection and mitigation of disease outbreaks. It can reduce the time taken for decision-making and action-taking to control critical situations. While data can be used to identify demographic groups with elevated health risks, it can also be used to curb disease incidence rates and further spread. The combination of AI and ML can help take a proactive approach through timely forecasts about how and where an outbreak is likely to spread – something that human effort cannot do with as much accuracy or speed (TechNative, 2018). Consider epidemiological investigators in South Korea who made use of their Smart City Data Hub tool (Millard, 2020) to trace COVID-19 transmissions based on citizen data pool to bring down infections by 90% without implementing lockdowns or self-isolation. The tool was part of several other smart projects being undertaken by the Korean government for citizen wellbeing.

Smart data can be used to find out where citizens are likely to need more medical assistance, and accordingly provide them with more relevant, localized healthcare facilities. For example, as part of its National AI Strategy to improve people's lives, Singapore's Multiple Readmission Predictive Model (Dhawan, 2021) uses AI and ML to create a list of potential high-risk NCD patients who may possibly need intervention. The intelligent solution auto-enrolls such patients into rehabilitation programs, mitigating serious illness and long-term hospitalization.

INTEROPERABILITY FOR HEALTH AND SUSTAINABILITY

Interoperability in a Smart City healthcare management system is essential to connect patient data points, and can be achieved through Blockchain technology. Blockchain helps combine multiple, fragmented health records about citizens to create a Smart healthcare ecosystem. A Blockchain is essentially a chain of blocks that are sealed in a secure way, with new blocks being appended as needed to form a larger and more holistic interconnected structure. In healthcare, this means that every medical transaction or every stop a citizen or patient makes along his journey is recorded on the Blockchain, documenting it for future reference.

Using Blockchain, it is possible to secure the privacy of patient health records while simultaneously being able to share medical history across multiple platforms and to multiple healthcare stakeholders. It helps increase the possibilities of available medical support. Blockchain-based healthcare solutions help with patient-managed data, longitudinal medical records, healthcare research (El-Gazzar & Stendal, 2020), and the sharing of all these between patients and healthcare providers across a safe, permissioned network. This complex interconnection uses encrypted data, retaining the sensitivity of medical records, anonymizing patient identity, and facilitating collaboration and data sharing on-demand and in real-time. Such instantaneous processing also leads to a more efficient and sustainable healthcare system, saving on time, resources and costs.

Piggybacking on IoT-based devices and wearables, and collaborating with cloud and Edge, Blockchain use cases can be extended to various other components of a Smart City infrastructure amplifying the benefits towards sustainability. The Centers for Disease Control in the US (CDC, n.d.) has been exploring Blockchain technology to preserve public health by identifying newer data sources to derive deeper insights. They have created a surveillance system for public agencies, accumulating and timestamping patient data, to research infections and predict possibility of disease outbreak.

Figure 3 depicts the interconnections in a sustainable Smart Healthcare Blockchain.

Figure 3. Interconnections in a sustainable Smart Healthcare Blockchain
Infographic created by author

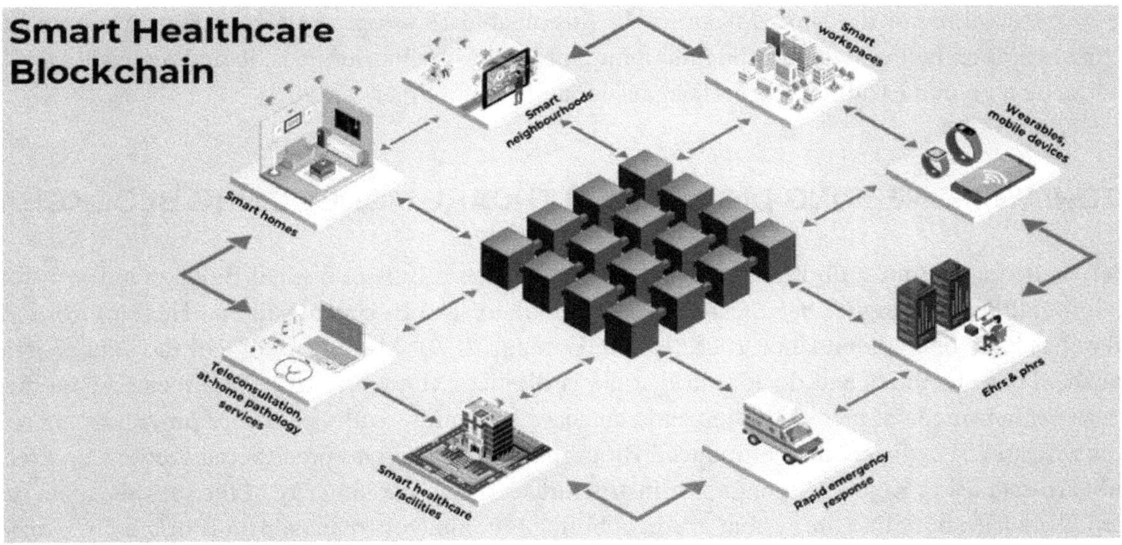

BLOCKCHAIN FOR PERSONAL HEALTH RECORDS

Medical practitioners typically do not, and practically cannot, monitor and study the real-time flow of patient health data. Studies show an estimated 55% of physicians (Fleming et al., 2013) have been making use of electronic health records (EHRs) and personal health records (PHRs), enabled by Blockchain, for quick and reliable diagnoses since close to a decade now. PHRs and EHRs gather data from different sources, including, but not limited to, patients' personal devices, clinics and medical facilities, care centers and care delivery organizations, and pharmacies (Kenney, 2021). PHRs are a form of EHR and help mine historical records for individual patient details as well as the common conditions that exist across a sub-population. This enables doctors to arrive at the best possible medical prognosis and understand health trends.

PHRs help obtain a unified view of health data across multiple healthcare providers (Roehrs et al., 2019) over a permissioned public network that can be accessed from any medical facility. Being inherently decentralized, the data ownership of PHRs lies with the patient (Lee et al., 2020), dismissing any doubts about privacy concerns. A PHR is completely paperless, saving resources and space, supporting sustainability. From medical records and DNA test data, all the way from birth to the end of life's stages can be stored, allowing for more holistic care and a closed-loop approach to health. With the PHR linked to an app, patients can receive personalized notifications when certain parameters are not in their normal range, encouraging self-service and a more patient-centric care model.

Blockchain-powered PHR allows patients the options of in-person or online consultations, or the support of a combination of both, while also being able to order medication online. Besides this, it can also be used for reminders for healthier lifestyles, early detection of illness and preventive care, and a more efficient risk management capability at a microlevel. At a broader governance level, this makes it possible to achieve superior clinical Big Data analytics capabilities for efficient planning of hospital facilities, patient care support features, and better preparation on urgent disease outbreak in the city. Collection of data, drawing meaningful insights, and ensuring data sharing trustworthiness can be listed among the top Key Performance Indicators (KPIs) of sustainable healthcare in a Smart City as they help uphold at least three of the United Nation's 17 Sustainable Development Goals. Blockchain technology that is optimally leveraged for the functioning of various public services, including healthcare, can therefore be a good measure of the performance maturity of a Smart city.

ESTONIA'S PIONEERING DIGITAL HEALTHCARE MODEL USING BLOCKCHAIN

Widely hailed as the most digitized nation on earth, Northern Europe's small Baltic country, Estonia, has a groundbreaking digital healthcare model to emulate. The Estonian National Health Information System has been operational since 2008 (Health Europa, 2019). More than 99% of the data generated by medical care facilities and doctors in Estonia is digitized (Lewin, 2020). This means that citizens can access their medical records online and can share it securely with healthcare providers as needed and at will. In 2012, the country introduced Blockchain technology to process transactions and secure healthcare data using Keyless Signature Infrastructure to ensure the integrity of the sensitive data being shared. With trust built into the system, medical billing is completely managed on Blockchain, with 95% of all health data being ledger-based and e-prescriptions being 99% digital.

For the people of Estonia, this not only means video consultations that save both time and a trip to the hospital but also medicine deliveries through e-pharmacies. Being digitized ensures reissuance of e-prescriptions, auto-checks for drug interactions or allergies, and instant drug counterindication support. Interconnectivity also helps registrations and applications to be done swiftly. Registering birth or death and applying for health, birth, or death certificates, which take time in most other countries, is done effortlessly online.

All this is possible thanks to a state-issued digital identity for every Estonian (PWC, 2019). The identity helps tie together all e-health services that are availed by the citizens. For example, the e-ambulance system is connected to the e-health services database, providing emergency care professionals with pre-filled health data forms even as they head out to save patients in dire need, improving overall efficiency and gaining time for resuscitation.

The e-health services also ensure sustainability of the overall healthcare system as more is achieved in lesser time with fewer resources. This is one of the most critical components in the shift from curative to preventive medicine, the backbone of a sustainable healthcare model in a Smart urban-scape. In fact, in the wake of COVID-19, the digital Republic proved that Smart healthcare solutions can help communities be crisis-ready. Estonia was able to continue all its public services uninterrupted by the pandemic.

As futuristic as its case sounds, Estonia has demonstrated the way forward to those countries that are seeking out radical ways to revolutionize their health systems towards a self-sustaining model. Estonia has managed its complete digital transformation journey at one percent (OECD, 2019) of its state budget to develop the technology to support digital ID, e-governance, e-health, e-school, i-voting, e-tax, and a host of other services. The pay-off in implementing most of these, step-by-step, has been seen in under four months of implementation of each. Moving further in preventive care, it started offering free gene tests at the Estonian Genome Centre in 2018, to advance personalized medicine studies in genomics.

The aim is to develop long-term AI and ML algorithms to detect high-risk groups of citizens prone to serious NCDs, including the more common cardiovascular disease and breast cancer. The gathered data can also be used in developing entire generations of healthier individuals through early diet planning and health predictions.

HEALTHCARE DEMAND PREDICTION USING DATA

While healthcare data is extremely powerful and insightful, it can be ginormous and unwieldly across heterogeneous platforms. Applying predictive analytics to this data and then sharing it securely back to hospitals and healthcare facilities can help with population health monitoring, healthcare demand forecasting, and disease prevention.

Predictive analytics using unsupervised learning can identify patterns in data using attribute-based algorithms. These algorithms can detect clusters and associations in data, creating dynamic models that can be used in developing responsive healthcare demand forecasts (Altmann-Richer, 2018). Conventionally, only structured data, such as medical records and patient admission records, was being used to make healthcare demand predictions. Today, the shift to semi-structured and unstructured data, such as data from wearables, sensors, patient notes, voice transcriptions, etc., is helping with quicker and more accurate predictions.

Prediction can be used to alert medical practitioners and clinicians of a likelihood of greater demand in healthcare and prevent resource shortage. Early signs and symptoms of illness or need for intervention can be modeled, whether in citizens connected to Smart technology or to warded patients in hospitals. Predictive analytics can help prevent acute challenges to healthcare preparedness by informing care facilities of the anticipated illnesses and rate of admissions. This can help in two ways: (1) Prevent the predicted outbreak of illnesses through enhanced city-level surveillance, health promotion, and early detection (2) Better manage the available manpower and resources to meet the demand surge when prevention fails, and take precautions to avert a worsening situation.

Meteorological factors such as changes in weather conditions, rain and wind patterns, and heat and cold waves often result in quicker transmission of communicable diseases, such as the flu, influenza, and dengue fever. Paired with socio-demographical factors, certain health emergencies can worsen population health. Healthcare demand prediction using Big Data analytics can help discover new biomarkers of emerging disease outbreaks and intelligent therapeutic intervention strategies aimed at preventing illness and sustaining good health in the city.

The four hospitals in Paris that form the *Assistance Publique-Hôpitaux de Paris* (AP-HP) successfully leverage internal and external data, including decade-long hospital admission records, to chart out daily and hourly patient admission predictions (Marr, 2016). This has helped the hospital chain deploy hospital resources effectively for more efficient patient care. Sweden's Sahlgrenska University Hospital and Södra Älvsborgs Hospital uses AI-modeling techniques with data-driven insights on predictive models to arrive at a reliable 14- to 21-day hospital admissions forecast that helped healthcare resources at the hospitals be prepared ahead for the demand during COVID-19. Internal crowd mobility data as well as external vaccination data were used, besides other available data to come up with the significantly accurate solution to prevent overburdening healthcare professionals.

The trends in usage of existing medical facilities, clinics, and hospitals, the cost of treatment involved, coupled with patient footfall, can help planners determine the need for more general physicians or medical specialists, or their lack thereof. This would ensure areas that are under-served receive the care and attention they seek out either by enhancing existing services or providing additional temporary access to health experts. Data can also help citizens connect with service providers and medical specialists when human interaction is either not possible or not required, as in the case of telehealth. Seat Pleasant, a small community in Maryland, launched a telehealth network integrated with its current Smart City platform called the Smart City Hub (Wicklund, 2020). This Nationwide Disease Surveillance and Chronic Disease Management Telehealth System uses telemedicine tools and mHealth platforms to monitor and aid local residents.

Besides, health experts can also use data to identify specific pockets of the city that are more prone to being affected by certain health issues. This can help them assist people of those areas with suitable health information and guidance to bring down mortality and promote preventive measures towards greater wellness. These statistics could also help urban planners incorporate certain design changes into the neighborhood blueprint. For instance, specific neighborhoods that show higher rates of obesity could have more parks and green spaces, walking and jogging paths, while encouraging cycling with the provision for dedicated lanes. Various earlier studies have shown that the combination of data and infrastructure are needed to optimize health outreach efforts to improve the population health of a city while also achieving sustainability targets of preventive care.

Analyzing data demand prediction data over a period of time can help recognize the demand troughs and peaks that a healthcare facility goes through along with its contributing factors. This data can be harnessed to devise preventive wellness strategies for targeted communities, promoting sustainable health. Among the most vulnerable in the community who require attention are the elderly.

ELDERLY CARE

With the world heading towards a longevity revolution, it is estimated that 1 in 6 people will be above 65 years of age by 2050 (UN, 2019). One of the key challenges, therefore, is to create a centralized Smart technology platform to coordinate all individual and community efforts to address the needs of elderly care from a strategic, long-term perspective. This would include a preventive and predictive living environment for them for holistic health using gerontechnology. Gerontechnology supports successful aging, and is a combination of the aging society and rapidly emerging new technologies (Bronswijk et al., 2009). The aim of such inclusive design would be to enable them to remain fairly independent and in close connection to their families and the community at large.

Under the purview of Smart City health, Singapore launched the SHINESeniors project (Zaobao, 2021) to help the elderly age in place independently, with healthcare professionals monitoring them remotely. In collaboration with the Singapore Management University, the solution makes use of IoT sensors that monitor the motion of the seniors inside their homes to detect any anomalies using AI algorithms (Gupta, 2017). This helps reach preventive medical assistance to the elderly in time and at home, reducing the pressure on nursing homes.

Several studies in the past (Tanner et al., 2008; Hwang et al., 2011) have considered various modifications in existing residences in order to support the elderly to age in place. The modifications are not just about changes in physical spaces to facilitate ease of safe movement but also enhancements driven by technology and greater connectivity (Cheek 2005, Noury 2000). Smart technology within homes can detect changes in the residents' health and offer insights on the effects of these health conditions on day-to-day life. Changes in cognitive performance when prescribed medication treatment is not regularly followed (Austin et al., 2017), constant gait analysis using ambient sensors to predict issues with balance and potential for falls (Mancini et al., 2016) as well as early onset of neuromusculature diseases (Huisinga, 2012), detection of visitors (Petersen et al., 2012), and recording of time spent outdoors or indoors, talking on the phone or using social media (Petersen et al., 2015), are all possible to study using ambient sensors.

If activities can be recognized, then their future occurrences can be predicted (Minor et al., 2017), and appropriate reminders could be triggered, such as those for medicines, having dinner on time, etc. Nice, France, with a third of its population over 60 years of age, has made at-home care for its elderly a priority. A Smart health project launched in the city (USE, 2016) brings together healthcare stakeholders to create tools and services to enable independent living among the elderly. The government has also set up an e-Health Business Innovation Centre and co-working spaces to support start-ups in the space of Smart elderly care.

While the creation of a social network can help the aging stay connected and forge bonds, the delivery of essentials, including medicines, and the use of telemedicine and video consultations (Kharas & Remes, 2018) can help them feel healthier and reassured when they are immobile and overwhelmed by the doubt of disease.

Along with the elderly, another neglected and oft-overlooked section of society are the homeless. With the homeless either not having access to quality care or accessing different emergency medical services each time leads to fragmentation of their medical history and records. Emergency health providers face challenges serving them with quality care due to the lack of longitudinal data about their health. In 2019, Dell Medical School at the University of Texas, Austin, and the Austin Blockchain Collective collaborated to improve health for the homeless in the city. Through their collective understanding of data science and health informatics (Gilkeson & Tilton, 2019), the team piloted an identifier system called the MyPass Initiative. The MyPass Initiative helped solve data fragmentation and medical care problems faced by the homeless by creating a unique identifier for each individual to store and access their health records from any computer that is within the network of the permissioned Blockchain run by the city. This structure allowed doctors and emergency personnel access the complete health records of the homeless without a physical identification and provide them with equitable and quality care. With data recorded and stored away, the homeless can be rounded up for periodical preventive health check-ups to ensure equitable quality care.

SOLUTIONS AND RECOMMENDATIONS: INTEGRATED, INTELLIGENT HEALTHCARE FOR SUSTAINABLE CITIES

Intelligent cities are about creating Smart healthcare ecosystems that can sustain themselves for the larger benefit of communities. They aim to diagnose and treat illness and disease in the best available way. More importantly, they are about leveraging digital technology to create an integrated system for early intervention and prevention.

A comprehensive combination of a wide range of computational systems in an integrated data environment will be able to support interoperability between the various components of urban life. Sanguinetti et al. (2016) propose a Smart City Information Model that integrates data from sensors in public and community spaces, Building Information Model (BIM), Building Management Systems (BMS), neighborhood-level space syntax analyses, and Geographic Information Systems (GIS) to support the development of Smart data translators. This would include workplaces, common spaces, neighborhood and community centers, and public transport, among others.

INTEGRATED CARE IN SHARED SPACES

With sustainability spotlighting preventive healthcare, Smart cities can create physical and virtual communities that can help encourage health and hygiene. However, the most basic shared unit of a Smart City, every building can also be an interoperable and fundamental part of healthcare.

Smart Workplaces

Creating and maintaining an environment that enables people to stay healthy and engaged at the same time can have a highly positive impact on their mental and physical wellness, making it a growing consideration in Smart building initiatives with embedded technology (The Economist, n.d.).Research indicates that chronic health conditions can be prevented or reversed with lifestyle changes, and that office designs that promote health can enhance productivity, reduce attrition, and improve wellness. Building developers are now globally creating workspaces that support a healthy lifestyle through Smart technology, considering employees spend over a third of their waking time working (McCormick, 2018). Office buildings can learn and enhance their services, become smarter and improve the experience for employees and occupants through the use of intelligent sensors, ML, AI, and advanced data analytics, all mapping back to cloud-based BMS. Smart controllers can support a range of sensor capabilities, including the heating, ventilation, and air conditioning (HVAC) system, air quality monitoring and CO_2 detection, air filtering to keep off airborne viruses, crowd monitoring, lighting modifications, and other energy-consuming devices (Stone, 2020) throughout office spaces. Healthy adjustments to workspaces for better wellness, including circadian lighting (CBRE, 2017), biophilic office design, and temperature moderation for a more relaxed environment can boost employee productivity during the day and restful sleep at night, ensuring a healthier body-clock rhythm.

WELLBEING IN THE NEIGHBORHOOD

The advantage of a Smart City neighborhood is that it can facilitate installing plug-and-play Smart objects (Jin et al., 2014) including sensors of different types and measuring numerous inputs, which can be deployed anywhere to blend into their surroundings. The IoT-powered devices, appliances, buildings, vehicles, etc., are seamlessly interconnected and interfaced with one another in real time, and can be relied on for an abundance of Smart data. They can not only help with health monitoring but also with conveying the demand for modifications to the neighborhood structures and facilities based on citizen requirements, predicting environmental and weather changes to send out alerts in the locality if citizens need to be forewarned of bad weather, keeping watch on safety and security in the area, as well as with informing citizens of arrival/departure timings and crowding in local public transport.

The hospitals and healthcare facilities at the coastal city of Cascais, Portugal, are world-class. And yet this Smart City endeavors to create an intelligent health ecosystem that is aimed at proactive healthcare management through a local-level decentralized system. Cascais uses an interconnected network of mobile apps and other Smart tools to connect its citizens (Deloitte, 2019) with national healthcare institutions, managing health, emergency and medical resources on a single platform. These apps also help maintain clear and constant connection within neighborhood communities. This platform was used to enlist all 200,000 residents of the country for mass testing of coronavirus antibodies during the COVID-19 outbreak, helping manage the crisis commendably. The city's robust network of neighborhood and community-based caregivers collect and monitor health indicators of the more vulnerable sections for preventive assistance.

Drawing from this example, we see that healthcare experts can rely on data collected from various sources to study which pockets of the neighborhood need to maximize on its public health outreach efforts. Such data can also be used to reclaim space by shutting down some roads and reducing automobile-centric infrastructure to be transformed into greener and safer pedestrian trails to encourage exercise. More importantly, digital sensor systems in emergency vehicles and ambulances in the neighborhood can help prevent the possibility of getting stuck in traffic jams by transmitting information about their arrival to receptors installed at traffic lights well over 300 meters or so before they reach (Smartnet, n.d.). This helps create a high-speed green corridor for ambulances to reach hospitals without any delay, thanks to infrared sensors, 5G wireless communication, and GPS.

CHICAGO'S SMART HEALTH COMMUNITY ECOSYSTEM

Smart Health Communities (SHCs) can empower proactive health, foster social wellness, encourage digital health tool use, and enable overall healthy ecosystems. Chicago supports establishing a complex and highly interconnected health and wellness community-based ecosystem that is focused on increasing healthcare provisions, easy access, and equitable medical services. As part of Healthy Chicago 2.0 initiatives (Deloitte, 2021) that were flagged off in 2016, the SHCs work towards improving living conditions and enhancing health outcomes through the optimal use of data. The highest impact of the program has been seen in reducing tobacco use, improving health for mothers and babies, and preventing HIV, along with addressing several lifestyle-related NCDs. Highly dependent on technology for innovative tracking, data analytics, and delivery models, the Chicago SHCs also assist with neighborhood health promotion, mental health, knowledge sharing, aspects of clean air, exercise and walkability—all towards a higher quality of life. Healthy Chicago 2.0 leverages surveillance and data to ensure each of the 82 objectives and 30 goals charted are measurable and achievable.

Close on the heels of the SHCs is the Healthy Chicago 2025 program (Chicago.gov, n.d.). A five-year community health improvement plan that was launched in 2020, it is the result of the work of several hundred community members and organizations to reach a stronger, healthier, and vibrant neighborhood.

DENMARK: AN E-HEALTHY AND HAPPY COUNTRY

Named the third happiest country on the planet, Denmark's e-healthcare system is among the most advanced in the world. It is one of the most forward-looking examples of government-supported objectives and efficient collaboration between healthcare stakeholders. One of the most digitized economies in the

European Union (Medical Futurist, 2019), the priority has been on patient health and data security. The country's new digital health strategies are centered around integrated care and prevention.

Interestingly, the Danish National Patient Registry has records dating back 40 years, making it the world's oldest nationwide hospital registry. This rich health databank provides critical information that can be leveraged for preventive care. To supplement this, the newly established Danish National Genome Centre aims to have at least 60,000 whole-genome sequenced by the turn of the decade to be able to create better health for the future. The Copenhagen Healthtech Cluster has been set up with the goal to network with data registers for real-time information for doctors in Operation Rooms.

Besides these lofty projects, the push for e-health in Denmark since well over two decades has encouraged the adoption of several e-health technologies (Kierkegaard, 2013). The Danish healthcare system allows for free and equal access to digitized medical practitioner services. The patient portal allows citizens to access their data and book appointments. The government is considering bringing together public registries and specialized data sets for specific purposes to establish close collaboration between companies and the healthcare sector towards greater innovation.

FUTURE RESEARCH DIRECTION

On the road to global progress, sustainable Smart healthcare overtakes the drive. The digitalization of healthcare is long-winding, with innovative and immensely beneficial potential. However, there is scope for further research to evaluate the right path for each country, based on several regional and demographic factors. Countries like Estonia, Spain, Denmark, and Singapore, among others, have explored ways in which lives of citizens and healthcare providers can be improved and eased. Smart cities and countries that are open to innovation can use these as benchmarks to further the cause of long-term, sustainable preventive health.

If the global healthcare system has to be made sustainable, it has to be intelligent and adaptable to the constant evolution of the healthcare scenario. Future research needs to focus on the genomics revolution that can make the best use of emerging technologies and stop serious NCDs among entire generations. Along with this, the global demographics dynamics needs to be studied to arrive at newer and optimal models of healthcare.

THE TRANSITION FROM CURATIVE TO PREVENTIVE CARE

There needs to be a transition from curative or responsive care to preventive care. Personalized medicine based on patient experience and medical history can be achieved through deep genomic studies, changing the course of healthcare. The ability for more precise diagnostics can pave the way towards personalization and newer ways of healing. The answer should not be to always turn to medical care for help, but rather to minimize the need for healthcare through a good measure of prevention, extra care, and overall wellbeing. While this can reduce the burden on healthcare providers, it can also boost the health of a nation, its citizens, and its economic stability. Preventive healthcare would require that research focus on more trusting data sharing methods, enhanced interoperability between key contributing factors, greater provision for equitable access, ways to maneuver change in consumer behavior, and more astute scientific breakthroughs.

POLICY IMPLEMENTATION

Preventive care is the answer to sustainable health in Smart cities, where governments can bring in policy changes through regulation and political leadership. Collaborating with technology companies and encouraging enriching partnerships between public and private healthcare players can be a first step. Building security and a trustworthy privacy infrastructure can establish the required faith in the system.

The World Health Organization has been canvassing for better healthcare across the globe. In a 2005 resolution (WHO, 2021), the World Health Assembly urged its member states to draw up long-term strategies to implement e-health services. Following up on this, in 2013 and 2018, member states were issued directives to adopt e-health interoperability and a global strategy on digital health. The purpose of such a strategy is mainly to reinforce healthcare systems by using Smart health technologies towards empowering patients, supporting care providers, and "achieving the vision of health for all."

Governments would do well to restructure timeworn healthcare policies and consider incentivizing development plans by both public sector and private healthcare organizations. Such policies can also help institute community-driven hubs that can strengthen healthcare reach, support medical capabilities, and increase operational efficiency of care. National digital health roadmaps should be chalked out with sustainability and interoperability as the driving force. Smart technologies for shared, reusable, and secure systems could be used to meet the objectives. The target should be an increasing adoption of sustainable e-health systems that are cost-effective and efficient. Fast-tracking these efforts can ensure healthier populations across the world.

CONCLUSION

As we head towards a more interconnected and intelligent world, the future of health is set to be data-intensive and digitally-driven. Health and healthcare will no longer be about treating illnesses but about sustaining wellbeing and preventing disease. And with the pervasiveness of data and digitization, healthcare is likely to be more equitable, affordable, accessible, and accurate. In an idyllic tomorrow, the citizen will be at the center of healthcare.

Smart City innovations show immense potential in reinventing and enhancing healthcare services. According to some earlier analyses, establishing Smart healthcare services and hospitals using AI is expected to drive the growth of the global Smart hospitals market to cross USD 103 billion before the turn of this decade (Business Wire, 2020). Cutting-edge technologies such as AI, IoT, and Blockchain are set to turn health challenges into opportunities to make healthcare management more efficient and effective. Additionally, the demand for home care and elderly care as well as remote patient monitoring will benefit from the technology-based healthcare revolution.

With cities of the future speckled by data-capturing technologies, citizens will form a greater part of the interoperability function, acting as sensors, alerting city services to wellbeing challenges and environmental issues (Fonseca et al., 2021). The aim of Smart Cities is to primarily increase the quality of life by providing better services and finding solutions to environmental sustainability, through greater citizen participation (Berntzen et al., 2018). A well-rounded approach to health and sustainability will hold the answer to the future of a resilient Smart City ecosystem, where the synergy between citizens and all other stakeholders is at its best possible. Such a holistic solution can be realized through improvement in the existing models of healthcare delivery, including a "more innovative, preventive, proactive, evidence-based, person-centered, and wellness-driven" strategy (Knickman & Snell, 2002; Johnson et al., 2007; Swan 2009).

As the world transitions with the lessons learnt from health emergencies and pandemics, adopting new responses and solutions to deal with emerging challenges, it is yet another step towards driving sustainable growth at scale. Each Smart City, though governed by similar technology principles, has its own personality and its unique vision for a better future. Citizen participation, commitment to effective collaboration, and greater levels of coordination between contributing agencies can help secure sustainable benefits for a healthier world. After all, our health is in our hands. Much like the question posed by WHO (2020), "If the purpose for (urban) planning is not human and planetary health, then what is it for?"

REFERENCES

Altmann-Richer, L. (2018). *Using Predictive Analytics to Improve Health Care Demand Forecasting*. Institute and Faculty of Actuaries.

Austin, J., Klein, K., Mattek, N., & Kaye, J. (2017). Variability in medication taking is associated with cognitive performance in nondemented older adults. *Alzheimer's & Dementia: Diagnosis, Assessment & Disease Monitoring*, *6*(1), 210–213. doi:10.1016/j.dadm.2017.02.003 PMID:28349120

Berntzen, L., Johannessen, M. R., Böhm, S., Weber, C., & Morales, R. (2018). Citizens as sensors: Human sensors as a smart city data source. *Proceedings of the SMART 2018 International Conference on Smart Cities, Systems, Devices and Technologies*, 11–18.

Bronswijk, J. E. M. H. V., Bouma, H., Fozard, J. L., Kearns, W. D., Davison, G. C., & Tuan, P.-C. (2009). Defining gerontechnology for R&D purposes. *Gerontechnology (Valkenswaard)*, *8*(1). Advance online publication. doi:10.4017/gt.2009.08.01.002.00

Brunkard, P. (2019, July 16). *In the heart of the [smart] city, home is where the heart is*. Smart Cities World. Retrieved December 20, 2021, from https://www.smartcitiesworld.net/opinions/opinions/in-the-heart-of-the-smart-city-home-is-where-the-heart-is

Business Wire. (2020, October 7). Global Smart Hospitals Market to surpass US$ 103.6 billion by 2027, says Coherent Market Insights (CMI). *Business Wire*. Retrieved December 28, 2021, from https://www.businesswire.com/news/home/20201007005443/en/Global-Smart-Hospitals-Market-to-Surpass-US-103.6-Billion-by-2027-Says-Coherent-Market-Insights-CMI

CBRE. (2017). *The snowball effect of healthy offices. CBRE healthy offices*. Retrieved December 30, 2021, from https://cbrenl.uberflip.com/i/823602-the-snowball-effect-of-healthy-offices/11?utm_source=infographic-EN&utm_medium=Landingspagina&utm_content=E-paper-EN&utm_campaign=healthy%2Boffices&utm_term=Prospects

CDC. (n.d.). *Centers for Disease Control and Prevention*. Retrieved March 11, 2022, from https://www.cdc.gov/

Cheek, P., Nikpour, L., & Nowlin, H. D. (2005). Aging well with Smart Technology. *Nursing Administration Quarterly*, *29*(4), 329–338. doi:10.1097/00006216-200510000-00007 PMID:16260997

Chicago.gov. (n.d.). *Healthy Chicago 2025*. City of Chicago. Retrieved March 16, 2022, from https://www.chicago.gov/city/en/depts/cdph/provdrs/healthy_communities/svcs/healthy-chicago-2025.html

Cook, D. J., Duncan, G., Sprint, G., & Fritz, R. L. (2018). Using Smart City technology to make healthcare smarter. *Proceedings of the IEEE*, *106*(4), 708–722. doi:10.1109/JPROC.2017.2787688 PMID:29628528

Costa, D. G., & Peixoto, J. P. J. (2020). COVID-19 pandemic: A review of smart cities initiatives to face new outbreaks. *IET Smart Cities*, *2*(2), 64–73. doi:10.1049/iet-smc.2020.0044

Deloitte. (2019). *Command center approach to drive efficiencies at Cascais*. Cascais - Deloitte Client Stories. Retrieved March 18, 2022, from https://www2.deloitte.com/content/dam/Deloitte/us/Documents/process-and-operations/us-cons-smart-cities-client-stories-cascais.pdf

Deloitte. (2021). *Smart Health Communities*. Deloitte. Retrieved March 16, 2022, from https://www2.deloitte.com/global/en/pages/public-sector/articles/urban-future-with-a-purpose/smart-health-communities.html

Dhawan, V. (2021, January 18). *Technology and innovation in Healthcare Support Singapore's vision of a smart nation*. OpenGov Asia. Retrieved March 10, 2022, from https://opengovasia.com/technology-and-innovation-in-healthcare-support-singapores-vision-of-a-smart-nation/

El-Gazzar, R., & Stendal, K. (2020). Blockchain in health care: Hope or hype? *Journal of Medical Internet Research*, *22*(7), e17199. Advance online publication. doi:10.2196/17199 PMID:32673219

Fleming, N. S., Becker, E. R., Culler, S. D., Cheng, D., McCorkle, R., da Graca, B., & Ballard, D. J. (2013). The impact of Electronic Health Records on workflow and financial measures in Primary Care Practices. *Health Services Research*, *49*(1pt2), 405–420. doi:10.1111/1475-6773.12133 PMID:24359533

Fonseca, D., Sanchez-Sepulveda, M., Necchi, S., & Peña, E. (2021). Towards smart city governance. case study: Improving the interpretation of quantitative traffic measurement data through citizen participation. *Sensors (Basel)*, *21*(16), 5321. doi:10.339021165321 PMID:34450772

Gao, Z., Wang, S., & Gu, J. (2020). Public participation in smart-city governance: A qualitative content analysis of public comments in urban China. *Sustainability*, *12*(20), 8605. doi:10.3390u12208605

Gibbons, C. (2020, August 8). *How the smart city movement and it are Transforming Health*. HIMSS. Retrieved December 27, 2021, from https://www.himss.org/resources/how-smart-city-movement-and-it-are-transforming-health

Gilkeson, S., & Tilton, J. (2019, June 14). *Blockchain and healthcare innovation: A city, academia, and technology firms join forces to find blockchain solutions to address public health challenges*. JD Supra. Retrieved March 11, 2022, from https://www.jdsupra.com/legalnews/blockchain-and-healthcare-innovation-a-34497/

Gupta, N. (2017, June). *How IoT Can Significantly Improve Healthcare in the Context of Smart City*. Retrieved March 12, 2022, from https://www.iotworldtoday.com/files/2018/09/June_2017_JoI_Improving_Smart_City_Healthcare.pdf

Hansen, M. (2008). Smart card technology and healthcare information: A dynamic duo. *Computers, informatics, nursing. CIN*, *26*(5), 254–257. doi:10.1097/01.NCN.0000304842.44584.42 PMID:18769179

Health Europa. (2019, January 11). *Learning from the Estonian e-health System*. Health Europa. Retrieved March 16, 2022, from https://www.healtheuropa.eu/estonian-e-health-system/89750/

Huisinga, J. M., Mancini, M., St. George, R. J., & Horak, F. B. (2012). Accelerometry reveals differences in gait variability between patients with multiple sclerosis and healthy controls. *Annals of Biomedical Engineering*, *41*(8), 1670–1679. doi:10.100710439-012-0697-y PMID:23161166

Hwang, E., Cummings, L., Sixsmith, A., & Sixsmith, J. (2011). Impacts of home modifications on aging-in-place. *Journal of Housing for the Elderly*, *25*(3), 246–257. doi:10.1080/02763893.2011.595611

Jin, J., Gubbi, J., Marusic, S., & Palaniswami, M. (2014). An information framework for creating a smart city through internet of things. *IEEE Internet of Things Journal*, *1*(2), 112–121. doi:10.1109/JIOT.2013.2296516

Johnson, C. (2016). *The future of connected home health*. https://www.plextek.com/wp-content/uploads/2016/07/The-future-of-connected-home-health.pdf

Johnson, R. W., Toohey, D., & Wiener, J. M. (2007, May 1). *Meeting the long-term care needs of the baby boomers*. Urban Institute. Retrieved December 31, 2021, from https://www.urban.org/research/publication/meeting-long-term-care-needs-baby-boomers

Kanowitz, S. (2021, January 21). *How smart cities can mitigate the impact of health crises*. GCN. Retrieved December 21, 2021, from https://gcn.com/emerging-tech/2021/01/how-smart-cities-can-mitigate-the-impact-of-health-crises/315958/

Kenney, S. (2021, October 21). Blockchain has potential for the future of Personal Health Records, Electronic Health Records. *Pharmacy Times*. Retrieved December 29, 2021, from https://www.pharmacytimes.com/view/blockchain-has-potential-for-the-future-of-personal-health-records-electronic-health-records

Kharas, H., & Remes, J. (2018, June 27). *As cities go smart, vulnerable populations are pushed further to the margins*. McKinsey & Company. Retrieved December 20, 2021, from https://www.mckinsey.com/mgi/overview/in-the-news/as-cities-go-smart-vulnerable-populations-are-pushed

Kierkegaard, P. (2013). EHealth in Denmark: A case study. *Journal of Medical Systems*, *37*(6), 9991. Advance online publication. doi:10.100710916-013-9991-y PMID:24166019

Knickman, J. R., & Snell, E. K. (2002). The 2030 problem: Caring for aging baby boomers. *Health Services Research*, *37*(4), 849–884. doi:10.1034/j.1600-0560.2002.56.x PMID:12236388

Kristoffersson, A., & Lindén, M. (2020). Wearable sensors for monitoring and preventing Noncommunicable Diseases: A systematic review. *Information | Ubiquitous Sensing for Smart Health Monitoring, 11*(11), 521. doi:10.3390/info11110521

Lee, H.-A., Kung, H.-H., Udayasankaran, J. G., Kijsanayotin, B., & Marcelo, B. (2020). An architecture and management platform for blockchain-based Personal Health Record Exchange: Development and Usability Study. *Journal of Medical Internet Research*, *22*(6), e16748. Advance online publication. doi:10.2196/16748 PMID:32515743

Lewin, A. (2020, July 8). *Inside Estonia's pioneering Digital Health Service*. Sifted. Retrieved March 16, 2022, from https://sifted.eu/articles/estonia-digital-health/#:~:text=Estonia's%20health%20service%20has%20been,to%20look%20at%20those%20records

Mancini, M., Schlueter, H., El-Gohary, M., Mattek, N., Duncan, C., Kaye, J., & Horak, F. B. (2016). Continuous monitoring of turning mobility and its association to Falls and Cognitive Function: A pilot study. *The Journals of Gerontology. Series A, Biological Sciences and Medical Sciences*, *71*(8), 1102–1108. doi:10.1093/gerona/glw019 PMID:26916339

Marr, B. (2016, December 13). Big Data in Healthcare: Paris hospitals predict admission rates using machine learning. *Forbes*. Retrieved March 12, 2022, from https://www.forbes.com/sites/bernardmarr/2016/12/13/big-data-in-healthcare-paris-hospitals-predict-admission-rates-using-machine-learning/?sh=1520501479a2

McCormick, K. (2018). *The business case for Healthy Buildings: Insights from Early Adopters*. Urban Land Institute. Retrieved December 30, 2021, from https://www.buildup.eu/en/practices/publications/business-case-healthy-buildings-insights-early-adopters

Medical Futurist. (2019, February 7). *Where is digital health heading in Denmark?* The Medical Futurist. Retrieved March 16, 2022, from https://medicalfuturist.com/danish-digital-health-strategy/

Millard, N. (2020, March 24). *How South Korea's smart city startups curbed the spread of Covid-19*. Retrieved March 10, 2022, from https://e27.co/how-south-koreas-smart-city-startups-curbed-the-spread-of-covid-19-20200323/

Minor, B. D., Doppa, J. R., & Cook, D. J. (2017). Learning activity predictors from sensor data: Algorithms, evaluation, and applications. *IEEE Transactions on Knowledge and Data Engineering*, *29*(12), 2744–2757. doi:10.1109/TKDE.2017.2750669 PMID:29456436

Noury, N., Herve, T., Rialle, V., Virone, G., Mercier, E., Morey, G., Moro, A., & Porcheron, T. (2000). Monitoring behavior in home using a smart fall sensor and position sensors. *1st Annual International IEEE-EMBS Special Topic Conference on Microtechnologies in Medicine and Biology Proceedings*. 10.1109/MMB.2000.893857

OECD. (2019). Case Study 8: Estonia e-government and the creation of a comprehensive data infrastructure for public services and agriculture policies implementation. In *Digital Opportunities for better agricultural policies* (pp. 207–213). Organisation for Economic Co-Operation and Development. doi:10.1787/510a82b5-en

Petersen, J., Larimer, N., Kaye, J. A., Pavel, M., & Hayes, T. L. (2012). SVM to detect the presence of visitors in a smart home environment. *2012 Annual International Conference of the IEEE Engineering in Medicine and Biology Society*. 10.1109/EMBC.2012.6347324

Petersen, J., Thielke, S., Austin, D., & Kaye, J. (2015). Phone behaviour and its relationship to loneliness in older adults. *Aging & Mental Health*, *20*(10), 1084–1091. doi:10.1080/13607863.2015.1060947 PMID:26133745

PWC. (2019). *Estonia – the Digital Republic secured by Blockchain*. PWC. Retrieved March 16, 2022, from https://www.pwc.com/gx/en/services/legal/tech/assets/estonia-the-digital-republic-secured-by-blockchain.pdf

Roehrs, A., da Costa, C. A., da Rosa Righi, R., da Silva, V. F., Goldim, J. R., & Schmidt, D. C. (2019). Analyzing the performance of a blockchain-based personal health record implementation. *Journal of Biomedical Informatics*, *92*, 103140. doi:10.1016/j.jbi.2019.103140 PMID:30844481

Smartnet. (n.d.). *Aiming Life: Automated Green Corridor for quickest passage of ambulances on roadways, saving crucial time and lives*. Smartnet. Retrieved December 30, 2021, from https://smartnet.niua.org/content/c96337fd-d153-4a65-8fdb-be1e485187a3

Stone, A. (2020, August 12). *How smart technology can create healthy office buildings*. Technology Solutions That Drive Government. Retrieved December 30, 2021, from https://fedtechmagazine.com/article/2020/08/how-smart-technology-can-create-healthy-office-buildings-perfcon

Swan, M. (2009). Emerging patient-driven health care models: An Examination of Health Social Networks, consumer personalized medicine and quantified self-tracking. *International Journal of Environmental Research and Public Health*, *6*(2), 492–525. doi:10.3390/ijerph6020492 PMID:19440396

Tanner, B., Tilse, C., & de Jonge, D. (2008). Restoring and sustaining home: The impact of home modifications on the meaning of home for older people. *Journal of Housing for the Elderly*, *22*(3), 195–215. doi:10.1080/02763890802232048

TechNative. (2018, September 11). *Smart cities: How they'll make us healthier*. TechNative. Retrieved December 22, 2021, from https://technative.io/smart-cities-how-theyll-make-us-healthier/

The Economist. (n.d.). *Wellbeing in the smart office - interact - empowering spaces*. The Economist Intelligence Unit. Retrieved December 30, 2021, from https://empoweringspaces.economist.com/wellbeing-in-the-smart-office/

Totty, M. (2017, April 17). The rise of the smart city. *The Wall Street Journal*. Retrieved December 29, 2021, from https://www.wsj.com/articles/the-rise-of-the-smart-city-1492395120

Uludag, C. (2016, October 23). *Why do we need being smart in healthcare?* SMART Cities. Retrieved December 20, 2021, from https://ilab.city/why-do-we-need-being-smart-in-healthcare/

United Nations, Department of Economic and Social Affairs, Population Division. (2019). *World population ageing 2019 - Highlights* (ST/ESA/SER.A/430). Retrieved December 30, 2021, from https://www.un.org/en/development/desa/population/publications/pdf/ageing/WorldPopulationAgeing2019-Report.pdf

USE. (2016). *27 Delvalle: An urban health living lab*. Urban Sustainability Exchange. Retrieved March 16, 2022, from https://use.metropolis.org/case-study/become-expert/2176

Wachowicz, M. (2013, October 28). *New Frontiers for Geomatics*. GIM International. Retrieved December 19, 2021, from https://www.gim-international.com/content/article/new-frontiers-for-geomatics

WHO. (2018). *Social Determinants of Health*. World Health Organization. Retrieved December 18, 2021, from https://www.who.int/health-topics/social-determinants-of-health#tab=tab_1

WHO. (2020a). *Urban planning crucial for Better Public Health in cities.* World Health Organization. Retrieved December 27, 2021, from https://www.who.int/news-room/feature-stories/detail/urban-planning-crucial-for-better-public-health-in-cities

WHO. (2020b). *Integrating health in urban and Territorial Planning: A sourcebook.* UN-HABITAT and World Health Organization. Retrieved December 31, 2021, from https://apps.who.int/iris/handle/10665/331678?locale-attribute=en&

WHO. (2021). *Global Strategy on Digital Health 2020-2025.* Geneva: World Health Organization. Retrieved March 18, 2022, from https://apps.who.int/iris/bitstream/handle/10665/344249/9789240020924-eng.pdf

Wicklund, E. (2020, March 23). *MD City integrates telehealth with Smart City Hub to monitor coronavirus.* mHealthIntelligence. Retrieved March 11, 2022, from https://mhealthintelligence.com/news/md-city-integrates-telehealth-with-smart-city-hub-to-monitor-coronavirus

Zaobao, L. (2021, May 3). *Using technology to allow seniors to age-in-place.* Retrieved March 12, 2022, from https://scis.smu.edu.sg/news/2021/may/03/using-technology-allow-seniors-age-place

ADDITIONAL READING

Corburn, J. (2009). *Toward the healthy city: People, places, and the politics of Urban Planning.* MIT Press. doi:10.7551/mitpress/7583.001.0001

Dannenberg, A. L., Dannenberg, A. L., Frumkin, H., Jackson, R. J., Abrams, R. F., Malizia, E., & Wendel, A. (2012). *Making Healthy Places.* Island Press.

Lockwood, A. H. (2017). *Heat advisory: Protecting health on a warming planet.* MIT Press.

Schifman, M. R., & Fisher, T. (2018). *Building a sustainable home: Practical green design choices for your health, wealth, and soul.* Skyhorse Publishing.

Singha, S. (2020). *Future healthcare design.* RIBA Publishing. doi:10.4324/9780429347955

KEY TERMS AND DEFINITIONS

Electronic Health Records: Electronic health records or EHRs are a collection of systematized health and medical records of a patient or the population at large, stored in a compact, digital format.

Personal Health Records: Personal health records or PHRs are a type of EHR. PHRs are a collection of medical records or health documents that are electronically maintained by the individuals themselves or their caregivers. They contain the individuals' complete medical histories.

Population Health: Population health is the collective health of a group of individuals within a given region.

Sustainable Health: Sustainable health is the commitment to taking care of health and maintaining it either at a personal level, within a community or neighborhood, or for an entire city.

Chapter 12
Urban Health in Vulnerable Environments:
Water, Sanitation, and Hygiene Improvement Actions in School Areas in Makeni, Sierra Leone

Juan Arana
https://orcid.org/0000-0002-5031-8020
Universidad CEU San Pablo, Spain

Luis Perea
Universidad CEU San Pablo, Spain

Adela Salas
Independent Researcher, Spain

ABSTRACT

Adequate access to sanitation and safe water is a main challenge to improve urban health in low-income countries. Diseases derived of precarious hygiene conditions are a major burden in African rapidly growing urban areas. These challenges are embedded in complex stakeholder networks and need to be addressed through a holistic approach. It is argued that schools are a key objective for sanitation, water, and hygiene actions in the city. Concerns for improvement of hygiene conditions among school children have risen in the context of contemporary sanitary crisis. Furthermore, schools play a pivotal role between the public and the private realms and a potential to foster change. The study focuses on WASH actions implemented in schools of the city of Makeni, Sierra Leone. It seeks to define a set of context specific recommendations to improve the health-related conditions in the school environment.

DOI: 10.4018/978-1-6684-4030-8.ch012

INTRODUCTION

Eleventh Sustainable Development Goal (SDG11) of the UN 2030 Agenda calls for "inclusive, safe, resilient and sustainable" cities, but the very notion of sustainable cities is, as we know, problematic. And this is specially so in vulnerable urban environments. When confronted with the rapid and chaotic growth of some African cities, these concepts take a very different meaning than that we may find in first world contexts. In Herbert Girardet´s seminal definition, the sustainable city is "organised so as to enable all of its citizens to meet their own needs" in a sustainable way (Girardet, 1999).

Sustainable environments in this sense, necessarily address necessities and equity among all citizens. In urban contexts where there is a demand of basic services such as water or sanitation, the lack of access to these infrastructures is cause to inequity and vulnerability. This takes us to World Health Organization (WHO) definition of *healthy cities*, that points out how the Healthy City tackles the needs of its inhabitants by enhancing community capacities:

A Healthy City is continually creating and improving those physical and social environments and expanding those community resources which enable people to mutually support each other in performing all the functions of life and in developing to their maximum potential. (Goldstein & Kickbusch, 1996)

It is therefore at the community scale where the response to necessities is organised through enabling environments that foster mutual support. Improvement of common resources in the neighbourhood environment is, thus, a key aspect to generate healthier and more sustainable cities.

The 2030 Agenda´s SDG6, on the other hand, emphasizes the need of clean water and sanitation to "ensure availability and sustainable management of water and sanitation for all". Access to adequate sanitation is a main challenge to improve urban health worldwide, significantly so in developing countries. According to the U.N., inadequate water, sanitation, and hygiene are related to large disease burdens concerning malnutrition, diarrhoea, or soil-transmitted parasites; deficits in this area led to 870,000 deaths worldwide in 2016 (United Nations, 2019). The U.N considers the efforts on this aspect must be doubled in order to achieve the 2030 Agenda´s goal of basic sanitation services for all. The space where these aspects meet is no other but the city´s public realm. It is in the public space, in the common resources and shared communal infrastructures, where the organization of the city as an enabling environment can be most directly addressed.

This is especially important in the informal urban contexts. Here, the public realm is more relevant than ever, setting the standards in terms of hygiene and sanitation. But also, because the urban infrastructure in informal contexts forms a significant part of the public space. Most of the access points to safe drinking water in places as Makeni, Sierra Leone, are public. These infrastructures generate gathering spaces around them that eventually become everyday landmarks in the urban fabric. In a landscape of limited formal resources, it is the infrastructure networks such as water, sewage, sanitation, public lighting, or green spaces, that project a greater visibility of the public realm by structuring everyday life and the urban landscape around them. It has been pointed out that place centred communities achieve greater relevance for those with reduced social mobility (Nash & Christie, 2003).

In the studied Sierra Leonian context this would include infrastructures such as community meeting spaces or commons as the cereals drying facilities around which the neighbourhood gathers at certain times. In most neighbourhoods and villages community centres are key infrastructures for social relations, together with circular small, covered structures called *baffa*. Following this idea, the school appears as one of the main spaces of the community public realm. Primary and secondary schools are a critical scenario for health and sanitation infrastructures, community identity and placemaking. SDG4 stresses the importance of education spaces as centres of the community to "ensure inclusive and quality education for all and promote lifelong learning" (United Nations, 2019). U.N points out that many homes, health centres and schools lack water and soap, significantly increasing risk of diseases: "Access to safe drinking water, sanitation and hygiene can save millions of lives per year and improve school attendance" (United Nations, 2019). 2020 COVID19 pandemics has raised awareness on the importance of proper hygiene in public areas to prevent the spread of contagious diseases. Sanitation requirements have a significant impact in public areas and facilities, crucially those concerning education. One third of public schools worldwide lacked proper water supply and sanitation in 2016 (United Nations, 2019). On top of that, lack of safe and adequate sanitation installations in school centres has a special impact on female schoolers attendance during their menstruation, increasing gender inequality. Therefore, tackling water and sanitation in the public realm is key to improve sustainability and equity in vulnerable urban environments, and the implementation of such actions in school facilities is a critical entrance point to the betterment of urban communities in the neighbourhood scale.

Some key precedents can be found in the nationwide program to implement water and sanitation provision in Schools in Uganda from mid-1990´s to mid-2000´s managed to reduce the ratio pupils/cubicle from 700:1 to 100:1. For that purpose, the UNICEF School and Community Hygiene and Water Programme developed a design of improved latrine cubicle with prefabricated components and fostered hygiene education to change behaviours (Rugumayo, 2004). In this line, the implementation of School Health Clubs in cases as the program funded by the Dutch NGO SIMAVI in Kimusu District, Kenya, complement hardware provision of water supply and sanitary facilities with hygiene education and promotion of management (Rop & Ufanisi, 2004). These clubs include training of groups of students who form in turn fellow students. The members foster hygiene behaviour and supervise the maintenance of the facilities. In Bangladesh, the program School Health Nutrition, promoted by Save the Children in 2009, identified schools that lacked WASH facilities and provided them with latrines and hand washing facilities. This initiative included the organization of management committees and student brigades for the maintenance of the facilities. The responsibility on daily care fostered an enhanced sense of pride and ownership (Save the Children, 2009). Other approaches have emphasized a strategy of leveraging resources, such as WASH SDG programme in Negelle Arsi, Ethiopia promoted by BBBBC. These programmes leverage resources for the construction of improved latrines through microfinance entities that provide loans to households or schools (Hailegiorgis, Ebrahim, Lemecha, & IRC, 2022).

HEALTH, WASH AND THE CITY

The links between water, sanitation, and health (WASH) in the city have been widely explored since the hygienist urbanism of the 19th century. At the time of the initial period of industrialization in Europe and the United States, the rapid deterioration of environmental urban conditions led to the development of public health policies. The increase in illnesses affecting the working-class neighbourhoods in the 1830´s came together with social unrest. The rise of mortality rates and health problems were seen as a threat for labour availability. As a result, first social surveys were organized to gather information on the population´s health conditions and hygiene regulations were developed. The urban discipline was born as a specialized technique for physical environment control and new health legislation was put in place such as the Public Health Act in the UK in 1848 or the Melun Act of unhealthy urban habitation in France in 1850-51, applied by Haussmann in Paris (Ragon, 1978; Sica, 1976). This situation led both to the implementation of centralized water and sanitation systems and the rise of working-class housing projects.

The challenges confronted nowadays by cities in developing countries do not necessarily have much in common with the times of hygienist urbanism except for the extremely rapid growth of urban population and the burgeoning difficulties concerning health and habitation in the most vulnerable areas. The complexity of the situation linked to urban health includes huge socio-economic challenges, deficient legislation frameworks, property and tenancy precariousness, environmental threats, and complex governance networks of stakeholders. It is in this context that water and sanitation problems are of utter significance to tackle urban health. But these issues cannot be isolated from the urban, social, political, and environmental context. Solutions to water and sanitation problems such as centralized sewage systems may not be directly translated to the context of cities in developing countries (Strande, 2014) just as top-down, patronizing models can hardly be applied today. Some contemporary visions on health and the city tend point towards a systems approach. In contrast with a linear concept of health as the result of a given situation and the application of a set of solutions, the systems approach considers it integrated with urban life and the urban system. In this view, the city is seen as a complex system of networks and flows. The different processes and stakeholders interact in nonlinear ways. Within this vision, health must be studied within a complex field of actions and interactions of urban systems. The system approach has been characterized by three conditions: the construction of conceptual models that consider feedbacks from one system to another, the use of system tools such as simulation models and the integration of different kinds of data (Gatzweiler et al., 2017).

This perspective traditionally relays heavily on data availability and simulation models which may not be always appropriate for the achievement of certain objectives. In the context of water improvement related projects in developing countries, it has been argued that a hard systems method is excessively rigid to foster the necessary diversity of disciplines (Cable, Seok, Alemayehu, & Foster, 2009). The systems approach advocates for the use of both analytical and qualitative, context-based information and it specifically calls for a "*goal-centred* approach rather than a technologically-centred approach" (Gibson, Scherer, & Gibson, 2007, p. 35). Nevertheless, the focus on analytical modelling of the hard systems approach may prevent the required engagement of a variety of stakeholders (Cable et al., 2009) and thus, lack the bottom-up component. On the other hand, a more holistic definition of systems approaches calls for "efforts to resolve complex problems by understanding the systems involved and applying appropriate corrective actions" (Rietveld et al., 2016). Such perspective widens the approach to the health-related challenges in the urban context by proposing the study of the problem and its possible solutions as embedded in nonlinear dynamic processes that require transdisciplinary thinking. It also comprehends the co-production of knowledge through the participation of different stakeholders, using both expert analysis and community knowledge around a specific goal or necessity.

Addressing water and sanitation in vulnerable contexts, the juxtaposition of different networks and processes, make it necessary to have a transversal perspective of the different layers of the problem. There are a variety of technologies operating at the same time, projects, and infrastructure for centralized management together with decentralized informal domestic waste and faecal sludge management (FSM), as well as formal and informal stakeholders. A system-level approach has been proposed in order to analyse the processes throughout its different aspects. This perspective would consider the technology but also the management systems and the planning and governance process (Strande, 2014).

Systems analysis has been used (dos Muchangos, Tokai, & Hanashima, 2016) to map out the material flow analysis of Maputo City, from the use of natural resources to the disposal and management of the waste. The study analyses the interactions between different stakeholders, the governance and quantifies the waste flow. It establishes the relation of waste generated and the outcome result in terms of formal and informal dumping, material recovery and uncollected waste. In such approach, flows of wastewater, rainwater or formal and informal sludge treatment can be visualized together with governance aspects of the process (Rietveld et al., 2016). Therefore, through holistic analysis, conclusions and recommendations are drawn regarding the main barriers detected and a connection with policies and decision making is established. Such a concept approach is also applied at a smaller scale, drawing conclusions on urban space FSM-related problems. The study of one informal block in Maputo City shows how urban space interacts with waste management. Difficult access through narrow and unpaved streets to domestic sanitary installations prevents the primary waste disposal systems causing waste accumulation that, combined with lack of drainage, results in water contamination (Subtil, Matos, & Santos, 2018).

Governance, Management, and Technology

In the first place, from the governance point of view, urban health must be understood as socially determined factor as is proposed by the Social Determinants of Urban Health perspective. Within this vision, research in urban health must unpack the relation between governance structures and health. This critically includes local and national governments. The determinants of urban health in the different layers of governance need to be identified together with the interconnections and systems that have an impact in health. In order to achieve these goals, coalitions must be established, and a participatory planning procedure needs to be put in place (Vearey, Luginaah, Magitta, Shilla, & Oni, 2019). Micro level action has been pointed out as the necessary basis for urban services improvement efforts in Sub Saharan Africa (Hoelzel, 2021), especially when the presence of government is scarce or hesitant. Nevertheless, government support is also considered critical by UN Habitat, who identify six different areas that constitute an enabling environment for the planning process of sanitation projects. Government support is considered critical for the success of the planning process. It is an aspect often assumed but not necessarily defined. Assessing government support involves checking if sanitation issues are clearly stated as a priority by local and national administrations and checking if there is a will to decentralize the interventions in order to achieve these goals. Legal framework, including environmental law, land tenancy, health codes and technical standards, is key to understand the margins of action in the implementation of sanitation planning. To implement a participatory approach, institutional arrangements must be created that constitute coalitions between different layers of stakeholders, including households, schools, authorities, private providers, and community organizations. Development of skills and capacity is also crucial as an enabling factor. Financial assessment must be assessed early in the process. Lastly, there must be socio-cultural acceptance from the community to participate in such process and take decisions to change environment, service, and habits (Lüthi, Morel, Tilley, & Ulrich, 2011).

In second place, the management of water and sanitation systems involves a complexity of networks and interrelations that exceed the technological solution. Considerations such as centralized versus decentralized management or public versus private operation have enormous implications in the resolution of necessities and inequities in urban health. In the case of FSM, centralized sewage systems have been considered in the past as a priority to solve sanitation problems in developing countries. Centralized treatment and disposal through sewage infrastructures has traditionally been considered a safer technology in opposition to on-site management. Nevertheless, such systems are often far from affordable for cities in low income developing countries, its infrastructure is costly, and its legal implementation finds problems related to informal urban growth and land tenure. The limited viability of such solutions can generate the opposite effect, widening the inequity gaps (Strande, 2014).

Thirdly, on the technology aspect, on-site technology is increasingly viewed as a more viable and affordable approach, but the entire service chain must be considered. In this sense, solutions cannot be limited to the household scale, but they must be consistent with the complete FSM structure. This comprehends all its phases: collection, transportation, treatment and use or disposal. Specially the last phase conditions the process from the household infrastructure. The type of end product used in the treatment and whether there is an ulterior use may require a specific process. It is in this sense that any intervention must be done in the framework of an FSM plan.

The threats of a deficient FSM include contamination of the environment due to inadequate or informal disposal or treatment and the subsequent human health and wellbeing problems. Some possible hindrances may come from excessive costs of the system or lack of equitable access, difficulty of collection and transport due to cramped spaces, narrow streets or unpaved roads, large distances to treatment facilities or lack of them, lack of financial viability for private operators, lack of coordination or even design mistakes resulting in useless infrastructures (Strande, 2014). These problems can cause informal direct dumping and prevent a correct maintenance of household infrastructure, thus increasing the risk of environmental contamination and associated diseases.

WASH ACTIONS IN MAKENI

Sanitation and Urban Health in Makeni

Sierra Leone ranks among the countries with a lowest human development index; 52% of its population living under the poverty line ("The World Bank," 2021). According to the 2016 PNUD report on Human Development, Sierra Leone is the country with the lowest life expectancy, 51,3 years. Access to safe water and improved sanitation is one of the main habitability challenges for the country. In Sierra Leone, only 6,1% of the population in urban areas out of Freetown, have access to drinkable piped water and 0,2% live in household with connection to a sewerage system (*World Cities Report 2016: Urbanization and Development - Emerging Futures*, 2016) and 16% of Sierra Leonese population have access to improved sanitation, shared among various households in most cases. In urban areas the most common improved system is the simple latrine with slab (8,4%), followed by the toilet with cistern and septic tank (7,6%). In rural areas the dominant improved systems are the latrine with slab and the ventilated improved latrine (VIP). Shared systems are used by 54% of population in urban areas and 31% in rural areas.

This situation is critical in the city of Makeni. In the interior of Sierra Leone, the Northern Province and the Bombali District present a more pressing precariousness than the rest of the country (*Sierra Leone 2015 Population and Housing Census*, 2016). Whilst in the urban areas of Freetown and other significant cities, tap water is the most common water supply system, 89,1% of households in Makeni used water from wells according to the most recent disaggregated figures, while 62,7% used shared toilets, consisting in most of cases (93%) of simple latrine systems (Figure 1) (Perea Moreno, Salas Ruiz, & Arana Giralt, 2018).

Figure 1. Poor state of sanitation facilities, Makeni Sierra Leone
Source: HD-LAB

The precarious situation of sanitation, the scarcity of water in the dry season and the deficiencies in hygiene, generate health problems throughout the city. Despite efforts by the City Council and NGO´s, FSM presents serious deficiencies. The sludge treatment plant is not fully operative due to unrepaired constructional faults and while there is a project by Makeni City Council to implement a public private partnership for the sludge transportation system to the plant, it is not yet operational. Maintenance of septic tanks and latrines is manual and there is not a reliable system to empty, transport and dispose the waste.

Background Participation Process

A strategic planning process in the city of Makeni is being developed through a collaboration of education institutions (CEU San Pablo University and University of Makeni) with Makeni City Council. Within this process a series of participation workshops were organized between 2013 and 2019 with the presence of authorities, education institutions and civil society (Perea Moreno et al., 2018).

The workshops brought together different stakeholders to discuss the situation of the city and sought, among other issues, to detect priorities. They counted with the presence of both administrative and consuetudinary authorities of the *chiefdom* system. Representatives from SALWACO, the agency for water and sanitation were also present as well as technicians from local institutions. The necessity to emphasize training was one of the outcomes of the 2019 workshop together with the need for transparent processes of information and decision making.

Neighbourhood Upgrading Programme

Together with the strategic planning process, a Neighbourhood Upgrading Programme was established in 2013 by University CEU San Pablo in cooperation with University of Makeni and Makeni City Council. The programme has conducted a detailed habitability study in different neighbourhoods and villages of Makeni. Its main goals include field analysis of current conditions through data collection, capacity building and detection of priorities and funding options. The neighbourhood analysis comprehends collaborative cadastral mapping of the area, transect walks, interviews, and questionnaires. The methodology includes an organized set of topics to understand the main characteristics and vulnerability assessment of the neighbourhood, including access to water and sanitation. This approach focuses on the gathering of a quantitative data sequence to be used in future monitoring and evaluation (Perea Moreno, 2015). The results of the survey show that in the analysed neighbourhoods only 10% of the population have access to improved sanitation with half of the households using shared latrines. The dominant system is a basic not ventilated pit latrine. Some problems presented by this type of construction are flies, mosquitoes, odours, and water contamination due to a high groundwater level, floods, short distances to water sources and poor maintenance. Sanitation in public facilities in the community primary school of Masuba community has only two latrines for 500 students while the school of Ropolon with 500 students has no sanitation facilities.

Different actions were implemented in the context of a systems approach considring the different intervention scales in the city. In first place, a strategic planning approach to the city, rising awareness and involving the stakeholders in a participation process to define actions and priorities.

In second place, the analysis at the community scale through the neighbourhood upgrading programme has been crucial for the assessment of the situation of sanitation in relation to habitability, health, and the urban realm. The neighbourhood mapping process brought to light the precariousness of the access to water and sanitation in the most vulnerable areas of the city as well as the peri urban communities. Lastly, the focus on school facilities as a privileged public realm within the community brings the opportunity to make small significant actions that have an impact in their context. The school is thus, an important node of everyday life, with direct connections to health, education, and wellbeing within vulnerable urban environments. These actions are framed by two projects: The first one, funded by Manos Unidas, seeks to implement sanitation projects in three schools of Caritas Makeni, a project in which University CEU San Pablo has collaborated with the design and technical development. The second project consists of a research on habitability funded by Madrid Council for a new community in the corridor Panlap-Kunshu including an action on sanitation, water and hygiene.

Water, Sanitation and Hygiene Actions in Schools

Necessities Addressed by the Project

The general situation of water and sanitation (WASH) in schools is also deficient. Not only the quality of the facilities is poor, but the provision of them in relation to the school population is below standards. The rate of students per latrine is above admissible standards in similar contexts and there is usually no access to safe drinking water. This results in open defecation and a higher risk of disease among children.

Health related outcomes of the deficient access to safe water and sanitation are the water contamination, attraction of mosquitoes, cholera, diarrhoea and malaria. The extension of these deficiencies to the urban scale constitutes a mayor challenge, as pointed out by WHO.

"Many homes, healthcare facilities and schools also still lack soap and water for handwashing. This puts the health of all people – but especially young children – at risk for diseases, such as diarrhoea. As a result, every year, 361 000 children under 5 years of age die due to diarrhoea. Poor sanitation and contaminated water are also linked to transmission of diseases such as cholera, dysentery, hepatitis A, and typhoid" (*Drinking-water*, 2019). At the same time, solid waste management is a persistent problem. A large proportion of plastics and common debris is dumped in areas around the households. As documented in the field studies by research group by research group HD-LAB (Habitability and Development Laboratory) of University CEU San Pablo, solid waste accumulates in the back yards or in specific

Table 1. Rate of students per latrine. Based on field research conducted by CEU University and University of Makeni

School	Students per latrine
Bombali Bana School	125
Every Nation Academy	54
Al Harrken Islamic Primary School	100
Alhadi Islamic Primary School	90
Gish International Elementary Primary and Secondary School	65
St Francis School	150
Masuba School	150
Ropolon School	500 students. No sanitation infrastructure
Yalisandra School	600 students. No sanitation infrastructure

locations of the neighbourhoods. Despite significant financial efforts to implement a centralized solid waste management plan, the precariousness of the situation remains. The garbage accumulation often blocks street drainage, causing floods and worsening environmental contamination in combination with poorly maintained sanitation infrastructures.

A field study of school facilities conducted by HD-LAB shows that the average amount of students per latrine in Makeni is higher than 100 (Table 1), while desirable standards in similar contexts are between 20 and 30 students per latrine (Chatterley & Thomas, 2013). This overall situation implies frequent open defecation. Lack of access to a steady supply of clean, safe water worsens the problems of health, together with equity, gender inequality and school absenteeism.

Absenteeism also represents an important hindrance to equity and progress in this context. According to 2015 census, school attendance in Sierra Leona among children 6 to 14 years old is 37,8% and 35,6% for girls, dropping to 23% above 15 years old children -19,9% for girls-. In the Northern Province, attendance figures are the lowest in the country with a 28,8% and a 41,7% of the population who have never attended school. The field study showed it takes some students close to two hours to walk from their dwellings to the school in Makeni. Walkability throughout the city presents important deficiencies and there are no safe routes for children who often must walk in narrow roadsides among traffic with little if any safety measures. Household work claims the children presence during school hours and boys and girls walk the streets after school, selling fruits, vegetables, and other products. But absenteeism is also related to water, sanitation, and hygiene installations. The lack of hygienic conditions for washing during menstruation period is a recurrent cause for girls not attending class and this can be greatly improved through the betterment of the facilities. In this sense, lack of access to safe water can also condition the assistance to school, requiring children to make long detours to gather water.

Use of latrines by children in school is limited. In most cases the rate of latrines per student is bellow standards and part of the latrine provision of the schools is reserved to teaching staff. But not only the scarce provision of facilities makes its use difficult, precarious maintenance and lack of hygiene discourage the use of the sanitation infrastructure by children provoking open defecation in the back of the latrines or in the vicinity of the school.

In most of the schools visited during the field work in different locations of the city the situation presents grave deficiencies in water and sanitation facilities. The sanitation system consists usually of simple well latrines, hardly ever emptied. They are often unused by children due to odours and poor state of the installation, being open defecation in the surroundings a common practice. It is therefore critical to revert this situation through actions that can be used as replicable models. Although hygiene training among children, teachers and families is a key aspect to reverse these dynamics, the improvement of water and sanitation systems is necessary to turn the schools into better and safer environments from a sanitary perspective.

Technical Solution

The proposed solution represents a qualitative leap in access to water, sanitation and hygiene for the schools in the area. Instead of the precarious situation that results from the use of simple latrine solutions, the user interface of the model consists of a pour flush toilet with a septic tank storage system and a rainwater collection tank. The second phase development of the initial prototype within the project funded by Manos Unidas, a 60 m borehole and a tower water tank are added to provide access to safe water.

Such solution for improved water, sanitation and hygiene facilities in schools is based on a tested and integrated system (Tilley et al., 2014). The goal is to establish a model that allows to move towards safer flush water sanitation facilities with drinking water and sinks for handwashing, using rainwater through collection tanks. The integration of a water-based technology allows for a qualitative improvement in hygiene. It allows for handwashing and safe water access as well as providing a water seal to prevent odours and a safer user interface.

The system requires steady provision of water and periodic emptying of the septic tanks. This implies the project needs to address key challenges in management, maintenance, funding, and sensitization. Water recycling is included in the facilities through a central tank connected directly to the toilets. This water circuit is intended for toilet flushing of blackwaters to the septic tank. A different circuit of drinking water is established from a borehole to an elevated water tank connected to the sinks. During the dry season the borehole can supply water to the central tank, although the flush circuit should preferably operate with non-drinking water from other sources. Within its context, the action seeks to dignify the sanitation facilities. It integrates the installations in the school environment, connecting them to the classrooms through paved areas, adding vegetation, benches, and relation spaces. The sole improvement of the facilities is not enough to make a durable improvement. The action in rural and urban schools completes a wider strategy together with training, capacity building, water committees and coordination and maintenance units, integrating the action with a holistic system approach and enhancing its sustainability.

Governance Approach

In terms of governance, the program considers two aspects: up-down and bottom-up. Regarding the first aspect, the sanitation projects are aligned with government´s programs for the promotion of health and human development such as the sanitation and water project in Freetown, promoted by the African Development Bank Group. Furthermore, the COVID pandemic situation has resulted in an increased attention from the administrations towards hygiene practices. At the local level, authorities are involved in the development of the project in the framework of the ongoing neighbourhood upgrading programme. Makeni City Council is pushing an FSM plan to foster a private public collaboration in the collection and transport of sludge from septic tanks to the treatment plan. Regarding facilities, the council identified the provision of public toilets amongst the 10 top priorities for its inhabitants. Nevertheless, the main operators in charge of the FSM processes are central government institutions as SALWACO and coordination between different levels cannot always be taken for granted.

The involvement of the local administration in the project includes the selection of critical locations for water and sanitation actions. Besides administration authorities, any action related to FSM involves various actors as NGO´s and religious institutions who frequently fund and oversee the schools. It needs to be mentioned the importance of consuetudinary authorities. The Paramount Chieftaincy is of the utmost importance in any decision-making process. It is especially so if considered that Paramount Chiefs and noble local families own most of the land and are seen at the same time as leading community figures.

In second place, the project of WaSH for schools proposes to involve citizens through the creation of water committees in each school, joining parents' associations, the school community, and the project coordination units.

Involvement of stakeholders as students, parents and teachers from the beginning is considered crucial. The school community acts as a multiplier of behaviours and is therefore a key player in the betterment of both public and private standards in sanitation (Lüthi et al., 2011).

Bottom-up governance is also a key factor for the sustainability of the facilities. Many of the sanitation installations suffer decay, poor maintenance, and vandalism. School Management Committees, formed by teachers, parents and students as well as members of the community and authorities oversee the maintenance and management. Sensitization and awareness rising in water, sanitation and hygiene is another aspect to be implemented within the community scale. Students of UNIMAK (University of Makeni) will collaborate with the WASH committees both as a form of training and sensitization of the university community on pressing challenges related to water, sanitation, and hygiene.

Phase 1. Prototypes of Sanitation Facilities

Although there is a small percentage of population using an improved ventilated latrine system, the area's high water table and the weather conditions during the rainy season contraindicate the use of permeable collection systems. Together with a foreseeable increase of water use as the City Council tries to improve access to pipe water in the city, it seems possible to move towards watertight systems that avoid environmental contamination. A first latrine prototype is designed by CEU San Pablo University and installed in the schools of St. Francis and Masuba, in Makeni (Figure 2). The model consists of four pour flush toilets with collection tank, sink and an access ramp. The prototype will be a reference for future actions and will be improved in following models. On site monitoring of the performance will be key to improve the prototype and add other solutions such as the installation of safe drinking water elements.

Regarding FSM, this user interface is supported by the existence of a sludge treatment plant 8,5 km from the city. The project is implemented in coordination with main responsible institutions at state and local levels: SALWACO, City Council and NGO´s. But the action in schools cannot be limited to the implementation of facilities. The schools are community nodes and any improvement in WASH must necessarily involve a wider approach: In an initial phase the candidate schools for the intervention and the actions to be implemented are identified together with the community, the administration, and the school authorities. The water, sanitation, and hygiene actions in the school and its context include the following aspects: Construction of new sanitation block including accessible features in one of the latrines. Reform of existing latrines improving the ventilation conditions, maintenance work and installation of septic

Figure 2. Prototype of WASH facilities in ST. Francis Primary School, Makeni 2019
Source: HD-LAB

tanks or emptying of the existing tanks. Accessibility improvement actions in the surroundings of latrine blocks. Reform and maintenance of existing water wells attending to possible structural deficiencies, access, and underground water availability. Construction of new wells to provide non-drinking water to the school and the latrines. Construction of water towers for the wells. Construction of boreholes or deeper wells to access safe drinking water all year through.

Phase 2. Technical Aspects for Improved WASH Facilities in Schools

Once the first two prototypes are installed, the second phase of the project involves implementation of improved WASH facilities in other schools of Makeni. After construction and monitoring of the first units and in accordance with detected necessities and technical options (Tilley et al., 2014), a set of technical requirements is laid out for the actions to be implemented in the next phase.

Site location: The latrine blocks need to be allocated as far as possible from the classrooms and kitchen. Minimum distance between the water sources and the septic tanks or the infiltration system is stablished in 15-30 meters. The connection of the new facilities to the school areas must be clear and accessible. The facilities need to provide privacy but there must be at the same time a visual contact of the latrines from the school. The surroundings of the latrine block need to allow access area for the vacuum trucks and the streets around the access point need to have adequate width and pavement. The intervention is

Figure 3. Prototype project for WASH facilities
Source: Díaz, Muñoz and Navarrete (2021)

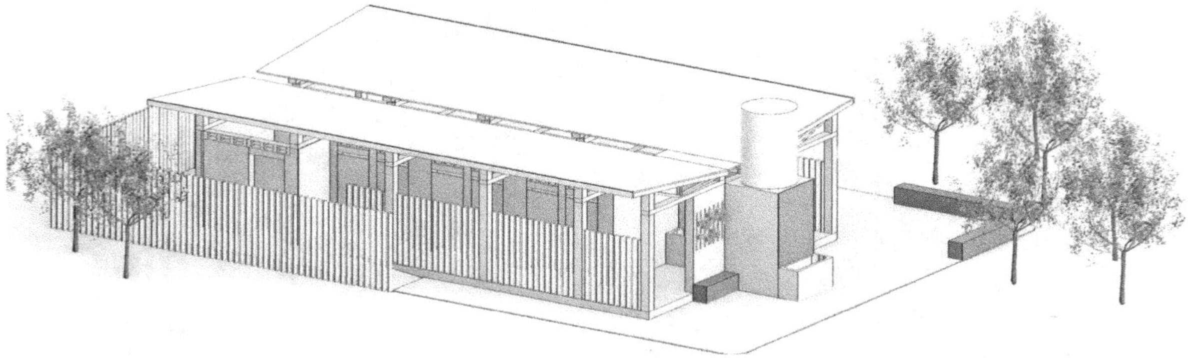

also an opportunity to organize the spatial configuration of the school premises, including paved paths, sitting space, covered areas or vegetation.

Construction and spatial requirements: The studied building solution consists of sand-cement block walls. It also includes bamboo framework perimeter walls to provide privacy in the perimeter. Given the raised floor slab, ramps must be provided to improve accessibility. One latrine unit must be accessible for physically disabled users. This unit will be also used for teachers. A specific private space for changing and washing has been detected as a necessity in order to facilitate access to school to girls during menstruating period. Responding to the detected necessities, the team formed by Díaz, Muñoz and Navarrete (2021) have developed a technical solution that will be used as model during this phase of the project. The scheme proposes a block of eight latrine units surrounded by a bamboo perimeter wall to ensure users privacy (Figure 3).

Sanitation and hygiene equipment**:** Based on the field research and technical assessment of the first prototypes and other different solutions, the CEU San Pablo university led research team has adopted an approach based in the implementation of pour flush toilets as user interface. In this system, water is poured usually using a bucket. Each unit has a water tap connected to the water collection tank to provide flush water. Through the curved pipe, the water seal prevents odours as well as insects coming from the tank, but this system needs a water input to drag the excreta. If water is poured from a height, a volume of 2-3 litres is considered enough (Tilley et al., 2014). Handwashing basins outside the cabins are provided with water taps connected to the drinking water circuit from the borehole well. The surfaces must be easy to clean. Industrial ceramic solutions for sanitation equipment are still hardly affordable even for public facilities, although there are also local hand-made solutions made of recycled ceramics. Frequent cleaning of the facilities must be organized by the responsible stakeholders' committees. The water collection tank will be attached or integrated to the block. It will be connected to a circuit of non-drinking water to be used in the latrines.

Collection and Storage: The studied solution for the project is a septic tank with a drainage field. Some requirements for a septic tank solution are availability of water all through the year and a functioning FSM end treatment system. As opposite to pit latrines, septic tanks are watertight and must be regularly desludged. Although septic tanks can be seen in the city, especially in public buildings, the conveyance and end treatment system has not been reliable in the past, resulting in the tanks been seldom emptied.

As referred by local technicians, sometimes tanks have been built with holes in the bottom to avoid them from filling up. The liquid effluent of the septic tank can be dispersed or transported. If the available space allows it, the dispersion can be done through a soak pit or a leach field. The field distributes the effluent from the septic tank to the subsurface soil. This system can be applied in the studied schools of Makeni since it requires open and not very dense areas. The infiltration area must be located 30 m away from water sources and can be planted with vegetation if trees or other deep-rooted plants are avoided. Further treatment systems can be studied in order to reduce pathogenic load (Tilley et al., 2014). An objective of the project is to test the viability of the system within contemporary FSM improvements in the city. This will greatly depend on maintenance, availability of water and FSM efficiency.

Conveyance: Access of vacuum trucks to the septic tanks needs to be available. This is possible in most of the schools studied but may be a problem in the urban context of private households. For mechanical emptying, trucks need to be within 30 meters from the tank. The sludge treatment plant is located 8,5 km from the city. Transportation is one of the issues addressed by the FSM improvement program undertaken by Makeni City Council together with SALWACO and some NGO´s operating in the country. The goal is to have a reliable float of vehicles for sludge emptying and transportation to the treatment plant. To this purpose, the aim is to foster private stakeholders' collaboration that would promote business creation and employment. As referred by local authorities, after an initial phase this initiative has found some obstacles. Until solutions to this aspect do not become affordable and operational, human-powered emptying and transport will continue to be a standard practice. A semi centralized system as transfer stations may be an intermediate solution to make affordable the transportation to the treatment plant.

CONCLUSION

Revision of systems approach applications to health-related WASH challenges in low-income cities of developing countries reveals the potential of a holistic vision that encompasses the complex network of stakeholders and the various processes of management, governance and technology involved. In this sense, it seems sensible to approach the problem combining both top-down and a bottom-up approaches. Therefore, a WASH strategy will necessarily seek alignment with priorities explicitly set by both local and state authorities as well as other institutions including consuetudinary authorities, NGO´s and religious institutions. On the other hand, the development of a governance system will require the involvement of the community from the beginning, defining the vision and detecting necessities. Local high education institutions can play here a pivotal role through the engagement of its teachers and students.

At the neighbourhood scale, the focus of WASH efforts in schools has proven to be a valuable approach to foster change through small actions. In the first place, they are priority environments for hygiene actions, especially after sanitary emergency situations as the Ebola outbreak or the COVID pandemic situation. In the second place, schools as part of the public realm and everyday life of the community, have the potential to set an example and become an asset in the dissemination of cultural and technical knowledge on health and WASH, promoting the translation of solutions from the institution to the private realm. Finally, the school is a community node, a privileged forum to involve the families and the institutions in a process of betterment of the community health-related challenges.

From the technical aspect, the project intends to be part of a transition from the predominant simple pit latrine towards safer and more adequate technologies in the context of the city of Makeni. Nevertheless, threats detected during this phase of the project include FSM efficiency, sustainability of the actions and governance. An efficient and affordable FSM system is necessary to foster any change in sanitation and hygiene practices. Maintenance and sustainability through time is a challenge for any implemented action centred in a technologic leap, especially if there is an excessive centralization or it relays on discontinuous funding programs. At the present rate of urban growth, problems can scale to an even greater dimension.

Therefore, it is necessary to achieve in these actions a balanced governance throughout the process, to co create a collective knowledge on health-related everyday life aspects, engaging community and key local stakeholders in sustainable solutions to avoid preventable illnesses, to improve wellbeing and to reduce inequity.

REFERENCES

Cable, D., Seok, J., Alemayehu, R., & Foster, A. T. (2009). Using a systems approach in managing and evaluating water and sanitation development projects. *2009 IEEE Systems and Information Engineering Design Symposium, SIEDS '09*. 10.1109/SIEDS.2009.5166170

Chatterley, C., & Thomas, A. (2013). *Snapshot of WASH in Schools in Eastern and Southern Africa: A Review of Data*. Evidence and Inequities in the Region.

Diaz Gallardo, B., Muñoz de Solano Sánchez, S., & Navarrete Suárez, F. (2021). *Prototipo WASH Makeni. ICHaB*. Universidad Politécnica de Madrid.

dos Muchangos, L. S., Tokai, A., & Hanashima, A. (2016). Application of material flow analysis to municipal solid waste in Maputo City, Mozambique. *Waste Management & Research*, *35*(3), 253–266. doi:10.1177/0734242X16678067 PMID:27909211

Drinking-water. (2019). Retrieved from https://www.who.int/news-room/fact-sheets/detail/drinking-water

Gatzweiler, F. W., Zhu, Y.-G., Diez Roux, A. V., Capon, A., Donnelly, C., Salem, G., … Zhang, Y. (2017). *Advancing Health and Wellbeing in the Changing Urban Environment*. doi:10.1007/978-981-10-3364-3_1

Gibson, J. E., Scherer, W. T., & Gibson, W. F. (2007). *How to Do Systems Analysis*. Wiley. doi:10.1002/9780470130599

Girardet, H. (1999). *Creating Sustainable Cities*. Green Books.

Goldstein, G., & Kickbusch, I. (1996). A healthy city is a better city. *World Health*, *49*(1).

Hailegiorgis, B., Ebrahim, M., Lemecha, G., & IRC. (2022). *Levering resources for WASH: lessons from WASH SDG programme in Shashamane and Negelle Arsi woredas in Ethiopia*. Retrieved from https://www.ircwash.org/

Hoelzel, F. (2021). *Local Governance Strategies of Otumara Community in Lagos to Access Water*. Academic Press.

Lüthi, C., Morel, A., Tilley, E., & Ulrich, L. (2011). *Community-led urban environmental sanitation planning: Clues complete guidelines for decision-makers with 30 tools. Swiss Federal Institute of Aquatic Science and Technology*. Eawag.

Nash, V., & Christie, I. (2003). *Making Sense of Community*. IPPR. Retrieved from https://www.ippr.org/files/images/media/files/publication/2011/05/making_sense_of_community_1272.pdf

Perea Moreno, L. (2015). *Hacia un análisis cuantitativo de la Ciudad Informal: una aproximación desde la Habitabilidad Básica y la experiencia en Makeni (Sierra Leona)*. Arquitectura. Retrieved from https://oa.upm.es/39873/

Perea Moreno, L., Salas Ruiz, A., & Arana Giralt, J. (2018). Grandes retos y aplicación de herramientas participativas en la planificación de la ciudad informal. El caso de Makeni, Sierra Leona. *Revista Urbano*.

Ragon, M. (1978). *Histoire mondiale de l'architecture et de l'urbanisme modernes*. Casterman.

Rietveld, L. C., Siri, J. G., Chakravarty, I., Arsénio, A. M., Biswas, R., & Chatterjee, A. (2016). Improving health in cities through systems approaches for urban water management. *Environmental Health: A Global Access Science Source, 15*. doi:10.1186/s12940-016-0107-2

Rop, R., & Ufanisi, M. (2004). Kenya - School Health Clubs: Can they change hygiene behaviours? In M. Snel & IRC (Eds.), *The Worth of School Sanitation and Hygiene Education (SSHE)*. Retrieved from https://www.ircwash.org/

Rugumayo, A. (2004). Scaling up School Sanitation Programmes. In M. Snel & IRC (Eds.), The Worth of School Sanitation and Hygiene Education (SSHE). Academic Press.

Save the Children. (2009). *Improving water and sanitation in schools and communities. Successes and lessons learned from Nasirnagar, Bangladesh*. Retrieved from https://www.ircwash.org/

Sica, P. (1976). *Storia dell'urbanistica. L' Ottocento*. Editori Laterza.

Sierra Leone 2015 Population and Housing Census. (2016). *Statistics Sierra Leone*. Retrieved from https://www.statistics.sl/images/StatisticsSL/Documents/final-results_-2015_population_and_housing_census.pdf

Strande, L. (2014). The Global Situation. In L. Strande & D. Brdjanovic (Eds.), *Faecal sludge management: Systems approach for implementation and operation* (pp. 1–18). IWA Publishing. doi:10.2166/9781780404738

Subtil, M., Matos, J. S., & Santos, R. F. (2018). Conceptual Approach for the Modernization of Sanitation Systems in Peri-Urban Slums and Its Application in Maputo City, Mozambique. INCREaSE. doi:10.1007/978-3-319-70272-8_48

The World Bank. (2021). Retrieved January 8, 2022, from https://data.worldbank.org/country/sierra-leone

Tilley, E., Ulrich, L., Lüthi, C., Reymond, P., Schertenleib, R., & Zurbrügg, C. (2014). *Compendium of sanitation systems and technologies*. Eawag. Retrieved from https://www.eawag.ch/en/department/sandec/publications/compendium/

United Nations. (2019). *The Sustainable Development Goals Report 2019*. UN. Retrieved from https://unstats.un.org/sdgs/report/2019/The-Sustainable-Development-Goals-Report-2019.pdf

Vearey, J., Luginaah, I., Magitta, N. F., Shilla, D. J., & Oni, T. (2019). Urban health in Africa: A critical global public health priority. *BMC Public Health, 19*(1), 340. Advance online publication. doi:10.118612889-019-6674-8 PMID:30909893

World Cities Report 2016: Urbanization and Development - Emerging Futures. (2016). *United Nations Fund for Population Activities*. Retrieved from https://unhabitat.org/world-cities-report

Compilation of References

Ababneh, A., Matalkah, F., & Aqel, R. (2020). Synthesis of kaolin-based alkali-activated cement: Carbon footprint, cost and energy assessment. *Journal of Materials Research and Technology*, *9*(4), 8367–8378. doi:10.1016/j.jmrt.2020.05.116

Abdel-Gawwad, H. A., García, S. R. V., & Hassan, H. S. (2018). Thermal activation of air cooled slag to create one-part alkali activated cement. *Ceramics International*, *44*(12), 14935–14939. doi:10.1016/j.ceramint.2018.05.089

Abdel-Gawwad, H. A., & Khalil, K. A. (2018). Application of thermal treatment on cement kiln dust and feldspar to create one-part geopolymer cement. *Construction & Building Materials*, *187*, 231–237. doi:10.1016/j.conbuildmat.2018.07.161

Abimbola, T., Lim, M., Hillestad, T., Xie, C., & Haugland, S. A. (2010). Innovative corporate social responsibility: The founder's role in creating a trustworthy corporate brand through "green innovation". *Journal of Product and Brand Management*, *19*(6), 440–451. doi:10.1108/10610421011085758

Adesanya, E., Ohenoja, K., Di Maria, A., Kinnunen, P., & Illikainen, M. (2020). Alternative alkali-activator from steel-making waste for one-part alkali-activated slag. *Journal of Cleaner Production*, *274*, 123020. doi:10.1016/j.jclepro.2020.123020

Adhikari, U., Chabrelie, A., Weir, M., Boehnke, K., McKenzie, E., Ikner, L., Wang, M., Wang, Q., Young, K., Haas, C. N., Rose, J., & Mitchell, J. (2019). A case study evaluating the risk of infection from Middle Eastern Respiratory Syndrome Coronavirus (MERS-CoV) in a Hospital Setting Through Bioaerosols. *Risk Analysis*, *39*(12), 2608–2624. doi:10.1111/risa.13389 PMID:31524301

AENOR Confía. (2022). *ISO 14006 Certification ecodesign management.* https://www.aenor.com/certification/environment/ecodesign

Ahrens, M. (2019). *Smoke Alarms in U.S. Home Fire.* https://www.nfpa.org/-/media/Files/News-and-Research/Fire-statistics-and-reports/Detection-and-signaling/ossmokealarms.pdf

AIA. (2021). *Market one.* The American Institute of Architects. Retrieved January 5, 2022, from https://www.aia.org/showcases/6388129-market-one

Aithal & Aithal. (2016). Nanotechnology Innovations and Commercialization – Opportunities, Challenges & Reasons for Delay. *Proceedings of National Conference on Changing Perspectives of Management, IT, and Social Sciences in Contemporary Environment*, 14, 1-12. 10.5815/ijem.2016.06.02

Aithal & Aithal. (2018). Nanotechnology based Innovations and Human Life Comfortability –Are we Marching towards Immortality? *International Journal of Applied.*

Akbari, H., Pomerantz, M., & Taha, H. (2001). Cool surfaces and shade trees to reduce energy use and improve air quality in urban areas. *Urban Environment*, *70*(3), 295–310. doi:10.1016/S0038-092X(00)00089-X

Aktas, B., & Ozorhon, B. (2015). Green building certification process of existing buildings in developing countries: Cases from Turkey. *Journal of Management Engineering*, *31*(6), 05015002. doi:10.1061/(ASCE)ME.1943-5479.0000358

Alaloul, W. S., Musarat, M. A., Tayeh, A. B., Sivalingam, S., Rosli, M. F. Bin, Haruna, S., & Khan, M. I. (2020). Mechanical and deformation properties of rubberized engineered cementitious composite (ECC). *Case Studies in Construction Materials, 13*.

Alanazi, H., Yang, M., Zhang, D., & Gao, Z. (2016). Bond strength of PCC pavement repairs using metakaolin-based geopolymer mortar. *Cement and Concrete Composites, 65*, 75–82.

Alawattage, C., & Fernando, S. (2017). Postcoloniality in corporate social and environmental accountability. *Accounting, Organizations and Society*, *60*, 1–20. doi:10.1016/j.aos.2017.07.002

Albitar, M., Mohamed Ali, M. S., Visintin, P., & Drechsler, M. (2017). Durability evaluation of geopolymer and conventional concretes. *Construction & Building Materials*, *136*, 374–385. doi:10.1016/j.conbuildmat.2017.01.056

Ali, M., & Armstrong, P. (2008). Overview of sustainable design factors in high-rise buildings. *Proc. of the CTBUH 8th World Congress*, 282–291.

Aliabdo, A. A., Abd Elmoaty, A. E. M., & Mohamed, M. F. (2018). Permeability indices and corrosion resistance of geopolymer and Portland cement concretes. *Magazine of Concrete Research*, *70*(12), 595–609. doi:10.1680/jmacr.17.00151

Ali-Tourdert, F., & Helmut, M. (2006). Numerical study on the effects of aspect ratio and orientation of an urban street canyon on outdoor thermal comfort and dry climate. Building and Environment, 41(2), 94-108.

Allard, F., & Santamouris, M. (2003). *Natural Ventilation in Buildings, a design handbook*. James and James.

Allen, W. L. III. (2012). Environmental Reviews and Case Studies: Advancing Green Infrastructure at All Scales: From Landscape to Site. *Environmental Practice*, *14*(1), 17–25. doi:10.1017/S1466046611000469

Al-Majidi, M. H., Lampropoulos, A., Cundy, A., & Meikle, S. (2016). Development of geopolymer mortar under ambient temperature for in situ applications. *Construction & Building Materials*, *120*, 198–211. doi:10.1016/j.conbuildmat.2016.05.085

Almalkawi, A. T., Balchandra, A., & Soroushian, P. (2019). Potential of Using Industrial Wastes for Production of Geopolymer Binder as Green Construction Materials. *Construction & Building Materials*, *220*, 516–524. doi:10.1016/j.conbuildmat.2019.06.054

Almalkawi, A. T., Hamadna, S., & Soroushian, P. (2017). One-part alkali activated cement based volcanic pumice. *Construction & Building Materials*, *152*, 367–374. doi:10.1016/j.conbuildmat.2017.06.139

Al-mashhadani, M. M., Canpolat, O., Aygörmez, Y., Uysal, M., & Erdem, S. (2018). Mechanical and microstructural characterization of fiber reinforced fly ash based geopolymer composites. *Construction & Building Materials*, *167*, 505–513. doi:10.1016/j.conbuildmat.2018.02.061

Almutairi, A. L., Tayeh, B. A., Adesina, A., Isleem, H. F., & Zeyad, A. M. (2021). Potential applications of geopolymer concrete in construction: A review. *Case Studies in Construction Materials, 15*, e00733.

Alonso, M. M., Gascó, C., Morales, M. M., Suárez-Navarro, J. A., Zamorano, M., & Puertas, F. (2019). Olive biomass ash as an alternative activator in geopolymer formation: A study of strength, radiology and leaching behaviour. *Cement and Concrete Composites*, *104*, 103384. doi:10.1016/j.cemconcomp.2019.103384

Alsarraf, A. A., & Alobaidi, M. M. (2020). Advanced techniques in design openings of smart facades in buildings. *IOP Conference Series. Materials Science and Engineering*, *757*, 012003. doi:10.1088/1757-899X/757/1/012003

Altmann-Richer, L. (2018). *Using Predictive Analytics to Improve Health Care Demand Forecasting*. Institute and Faculty of Actuaries.

Ancajima, R. (n.d.). *Tecnologías Ancestrales - Sistemas Hidráulicos Pre Incas e Incas* [conference session]. Obtained from: https://www.minam.gob.pe/diadiversidad/wp-content/uploads/sites/63/2015/01/resumen1.pdf

Anderson, R., Amodeo, M., & Harzfeld, J. (2010). Changing business cultures from within. The World watch Institute.

Aram, F., Solgi, E., Higueras García, E., Mosavi, A. R., & Várkonyi-Kóczy, A. (2019). The Cooling Effect of Large-Scale Urban Parks on Surrounding Area Thermal Comfort. *Energies*, *12*, 3904. doi:10.3390/en12203904

Arch2O.com. (n.d.). *10 most sustainable office buildings worldwide*. Arch20 - In Depth. Retrieved January 4, 2022, from https://www.arch2o.com/sustainable-office-buildings-worldwide/

Architecture 2030. (n.d.). *Why the building sector?* Architecture 2030. Retrieved January 5, 2022, from https://architecture2030.org/why-the-building-sector/

Ariffin, M. A. M., Bhutta, M. A. R., Hussin, M. W., Mohd Tahir, M., & Aziah, N. (2013). Sulfuric acid resistance of blended ash geopolymer concrete. *Construction & Building Materials*, *43*, 80–86. doi:10.1016/j.conbuildmat.2013.01.018

Askarian, M., Tao, Z., Adam, G., & Samali, B. (2018). Mechanical properties of ambient cured one-part hybrid OPC-geopolymer concrete. *Construction & Building Materials*, *186*, 330–337. doi:10.1016/j.conbuildmat.2018.07.160

Askarian, M., Tao, Z., Samali, B., Adam, G., & Shuaibu, R. (2019). Mix composition and characterisation of one-part geopolymers with different activators. *Construction & Building Materials*, *225*, 526–537. doi:10.1016/j.conbuildmat.2019.07.083

Austin, J., Klein, K., Mattek, N., & Kaye, J. (2017). Variability in medication taking is associated with cognitive performance in nondemented older adults. *Alzheimer's & Dementia: Diagnosis, Assessment & Disease Monitoring*, *6*(1), 210–213. doi:10.1016/j.dadm.2017.02.003 PMID:28349120

Autodesk, Inc. (2008). *Improving Building Industry Results through Integrated Project Delivery and Building Information Modeling*. White Paper. Available online at www.autodesk.com

Aviv, D., Chen, K. W., Teitelbaum, E., Sheppard, D., Pantelic, J., Rysanek, A., & Meggers, F. (2021). A fresh (air) look at ventilation for COVID-19: Estimating the global energy saving potential of coupling natural ventilation with new radiant cooling strategies. *Applied Energy*, *292*, 116848. doi:10.1016/j.apenergy.2021.116848 PMID:33776191

Axley, J. W., & Emmerich, S. J. (2002). A method for evaluating the suitabilityof climate for the natural ventilation of commercial buildings. Proceedings: Indoor Air.

Bannova, O. K., & Kristiansen, T. H. (2014). Technology transfer: Current trends in incorporating new technologies into the housing industry. *Proceedings of the International Astronautical Congress, IAC*, *13*(November), 9751–9756.

Bansal, P. (2005). Evolving sustainably: A longitudinal study of corporate sustainable development. *Strategic Management Journal*, *26*(3), 197–218. doi:10.1002mj.441

Barbosa, V. F. F., MacKenzie, K. J. D., & Thaumaturgo, C. (2000). Synthesis and characterisation of materials based on inorganic polymers of alumina and silica: Sodium polysialate polymers. *International Journal of Inorganic Materials*, *2*(4), 309–317. doi:10.1016/S1466-6049(00)00041-6

Barnet, M. L. (2007). Stakeholder influence capacity and the variability of financial returns to corporate social responsibility. *Academy of Management Review*, *33*(3), 794–816. doi:10.5465/amr.2007.25275520

Barney, J. (1991). Firm Resources and Sustained Competitive Advantage. *Journal of Management, 17*(1), 99–120. doi:10.1177/014920639101700108

Barredo Cano, J. I. (1996). *Sistemas de información geográfica y evaluación multicriterio en la ordenación del territorio*. Editorial RA-MA.

Bazazzadeh, H., Świt-Jankowska, B., Fazeli, N., Nadolny, A., Safar ali najar, B., Hashemi safaei, S., & Mahdavinejad, M. (2021). Efficient shading device as an important part of Daylightophil architecture; a designerly framework of high-performance architecture for an office building in Tehran. *Energies, 14*(24), 8272. doi:10.3390/en14248272

Beavis, L. (2014, May 15). PWC creates the most sustainable building in the world. *The Guardian*. Retrieved January 5, 2022, from https://www.theguardian.com/sustainable-business/sustainability-case-studies-pwc-one-embankment-place

Becqué, R., Mackres, E., Layke, J., Aden, N., Liu, S., Managan, K., Nesler, C., Mazur-Stommen, S., Petrichenko, K., & Graham, P. (2016). *Accelerating Building Efficiency: Eight actions for urban leaders*. Accelerating Building Efficiency. Retrieved December 13, 2021, from https://publications.wri.org/buildingefficiency/

Bellum, R. R., Muniraj, K., & Madduru, S. R. C. (2020). Influence of slag on mechanical and durability properties of fly ash-based geopolymer concrete. *Journal of the Korean Ceramic Society, 57*(5), 530–545. doi:10.100743207-020-00056-7

Benedict, M. A. (2002). ET Green Infrastructure: Smart Conservation for the 21st Century. *The Renewable Resources Journal, 20*(3), 12-17. Obtained from: https://www.merseyforest.org.uk/files/documents/1365/2002+Green+Infrastructure+Smart+Conservation+for+the+21st+Century.pdf

Beng, T. H. (2019, January 24). *A greener Singapore: Proptech, tax incentives crucial*. Retrieved January 7, 2022, from https://www.businesstimes.com.sg/opinion/a-greener-singapore-proptech-tax-incentives-crucial

Berardi, U., GhaffarianHoseini, A. H., & GhaffarianHoseini, A. (2014). State-of-the-art analysis of the environmental benefits of green roofs. *Applied Energy, 115*, 411–428. doi:10.1016/j.apenergy.2013.10.047

Berlin, G. (2020). *Capital of Green Trends: How Berlin leads the Way in Urban Sustainability*. Available online: https://about. visitberlin.de/en/greenberlin

Berlin, N. (2020). *Stadt der grünen Trends-Wie Berlin den Weg der urbanen Nachhaltigkeit geht*. Available online: https://about.visitberlin.de/nachhaltiges-berlin

Bernardini & Irvine. (2007). The 'nature' of urban sustainability: private or public greenspaces? *Transactions on Ecology and the Environment,* 661-673.

Berntzen, L., Johannessen, M. R., Böhm, S., Weber, C., & Morales, R. (2018). Citizens as sensors: Human sensors as a smart city data source. *Proceedings of the SMART 2018 International Conference on Smart Cities, Systems, Devices and Technologies*, 11–18.

Biörck, J., Blanco, J. L., Mischke, J., Ribeirinho, M. J., Rockhill, D., Sjödin, E., & Strube, G. (2020). *How construction can emerge stronger after coronavirus*. McKinsey & Company. https://www.mckinsey.com/business-functions/operations/our-insights/how-construction-can-emerge-stronger-after-coronavirus

Birch, R. L. (1980). Radburn and the American Planning Movement. Departmental Papers (City and Regional Planning).

Bissing-Olson, M. J., Iyer, A., Fielding, K. S., & Zacher, H. (2013). Relationships between daily affect and pro-environmental behavior at work: The moderating role of pro-environmental attitude. *Journal of Organizational Behavior, 175*(2), 156–175. doi:10.1002/job.1788

Blutstein, H., & Rodger, A. (2001). The sustainable tall building of the third millennium. *Proc. of the CTBUH 6th World Congress*, 131–142. 10.4324/NOE0415232418.ch6

Bock, T. (2007). Hybrid construction automation and robotics. *Automation and Robotics in Construction - Proceedings of the 24th International Symposium on Automation and Robotics in Construction, Isarc*. 10.22260/ISARC2007/0002

Bock, T. (2014). Ki-NEW-Matics. *The 31st International Symposium on Automation and Robotics in Construction and Mining (ISARC 2014) Ki-NEW-Matics*, 1–8.

Bock, T., & Linner, T. (2015). *Robotics Industrialization - Automation and robotic technologies for customized componet, module, and building prefabrication.* Cambridge University Press.

Bock, G. W., Zmud, R. W., Kim, Y. G., & Lee, J. N. (2005). Behavioral intention formation in knowledge sharing: Examining the roles of extrinsic motivators, social-psychological forces, and organizational climate. *Management Information Systems Quarterly*, *2005*(29), 87–112. doi:10.2307/25148669

Bock, T. (2015). Construction Robotics enabling innovative disruption and social supportability. *32nd International Symposium on Automation and Robotics in Construction and Mining: Connected to the Future, Proceedings.*

Bock, T.-A. (2017). Robot-Oriented Design. *Proceedings of the 5th International Symposium on Automation and Robotics in Construction (ISARC)*. 10.22260/ISARC1988/0019

Boiral, O., & Paillé, P. (2012). Organizational citizenship behaviour for the environment: Measurement and validation. *Journal of Business Ethics*, *109*(4), 431–445. doi:10.100710551-011-1138-9

Bondar, D. (2009). *Alkali activation of Iranian natural pozzolans for producing geopolymer cement and concrete*. University of Sheffield.

Bourbia, F., & Boucheriba, F. (2010). Impact of street design on urban microclimate for semiarid climate. *Renewable Energy*, *35*(2), 343–347. doi:10.1016/j.renene.2009.07.017

Bowen, M. (2021, June 2). *Four ways smart building technology creates sustainable environments*. Intelligent CIO North America. Retrieved December 10, 2021, from https://www.intelligentcio.com/north-america/2021/05/28/four-ways-smart-building-technology-creates-sustainable-environments/

Brands, R. A., & Fernandez-Mateo, I. (2017). Leaning out: How negative recruitment experiences shape women's decisions to compete for executive roles. *Administrative Science Quarterly*, *62*(3), 405–442. doi:10.1177/0001839216682728

BRINNER. 2021). *What is an electrostatic filter*. Retrieved from https://brinner.es/filtros-electrostaticos/

Bröde, P., Krüger, E., Rossi, A., & Fiala, D. (2012). Predicting urban outdoor thermal comfort by the Universal Thermal Climate Index UTCI—A case study in Southern Brazil. *International Journal of Biometeorology*, *56*(3), 471–480. doi:10.100700484-011-0452-3 PMID:21604151

Bronswijk, J. E. M. H. V., Bouma, H., Fozard, J. L., Kearns, W. D., Davison, G. C., & Tuan, P.-C. (2009). Defining gerontechnology for R&D purposes. *Gerontechnology (Valkenswaard)*, *8*(1). Advance online publication. doi:10.4017/gt.2009.08.01.002.00

Browning, W. D., & Ryan, C. O. (2020). *Nature Inside: A Biophilic Design Guide* (1st ed.). RIBA Publishing., doi:10.4324/9781003033011

Bruck, D. (2001). The who, what, where and why of waking to fire alarms: A review. *Fire Safety Journal*, *36*(7), 623–639. doi:10.1016/S0379-7112(01)00025-X

Brudermann, T., & Sangkakool, T. (2017). Green roofs in temperate climate cities in Europe - An analysis of key decision factors. *Urban Forestry & Urban Greening, 21*, 224–234. doi:10.1016/j.ufug.2016.12.008

Brundtland, G. (1987). *Our Common Future: The World Commission on Environment and Development.* Oxford University Press.

Brunkard, P. (2019, July 16). *In the heart of the [smart] city, home is where the heart is.* Smart Cities World. Retrieved December 20, 2021, from https://www.smartcitiesworld.net/opinions/opinions/in-the-heart-of-the-smart-city-home-is-where-the-heart-is

Burszta-Adamiak, E., & Fiałkiewicza, W. (2019). Review of green roof incentives as motivators for the expansion of green infrastructure in European cities. *Scientific Review – Engineering and Environmental Sciences, 28*(4), 641–652. doi:10.22630/PNIKS.2019.28.4.5

Burszta-Adamiak, E., Stańczyk, J., & Łomotowski, J. (2019). Hydrological performance of green roofs in the context of the meteorological factors during the 5-year monitoring period. *Water and Environment Journal: the Journal / the Chartered Institution of Water and Environmental Management, 33*(1), 144–154. doi:10.1111/wej.12385

Burton, J. (2021, November 29). *3D Digital Twins, IOT, and the future of decarbonization.* IIoT World. Retrieved January 5, 2022, from https://www.iiot-world.com/industrial-iot/connected-industry/3d-digital-twins-iot-and-the-future-of-decarbonization/

Business Wire. (2020, October 7). Global Smart Hospitals Market to surpass US$ 103.6 billion by 2027, says Coherent Market Insights (CMI). *Business Wire.* Retrieved December 28, 2021, from https://www.businesswire.com/news/home/20201007005443/en/Global-Smart-Hospitals-Market-to-Surpass-US-103.6-Billion-by-2027-Says-Coherent-Market-Insights-CMI

Cable, D., Seok, J., Alemayehu, R., & Foster, A. T. (2009). Using a systems approach in managing and evaluating water and sanitation development projects. *2009 IEEE Systems and Information Engineering Design Symposium, SIEDS '09.* 10.1109/SIEDS.2009.5166170

Cantó López, M. T. (2014). *La ordenación de la infraestructura verde en el sudeste Ibérico (Comunidad Valenciana, España).* Universidad de Alicante. doi:10.14198/cdbio.2014.45.03

Caparrós-Martínez, J. L., Milán-García, J., Martínez-Vázquez, R. M., & de Pablo Valenciano, J. (2021). Green Infrastructures and Grand Environmental Challenges: A Review of Research Trends by Keyword. *Agronomy, 11*(4), 782. . doi:10.3390/agronomy11040782

Castagnino, S., Rothballer, C., Abreu, J., & Zupancic, T. (2018). *6 ways the construction industry can build for the future.* World Economic Forum (WEF). https://www.weforum.org/agenda/2018/03/how-construction-industry-can-build-its-future/

CBRE. (2017). *The snowball effect of healthy offices. CBRE healthy offices.* Retrieved December 30, 2021, from https://cbrenl.uberflip.com/i/823602-the-snowball-effect-of-healthy-offices/11?utm_source=infographic-EN&utm_medium=Landingspagina&utm_content=E-paper-EN&utm_campaign=healthy%2Boffices&utm_term=Prospects

CDC. (n.d.). *Centers for Disease Control and Prevention.* Retrieved March 11, 2022, from https://www.cdc.gov/

Celerier, H., Jouin, J., Gharzouni, A., Mathivet, V., Sobrados, I., Tessier-Doyen, N., & Rossignol, S. (2019). Relation between working properties and structural properties from 27Al, 29Si and 31P NMR and XRD of acid-based geopolymers from 25 to 1000°C. *Materials Chemistry and Physics, 228*, 293–302.

Celerier, H., Jouin, J., Mathivet, V., Tessier-Doyen, N., & Rossignol, S. (2018). Composition and properties of phosphoric acid-based geopolymers. *Journal of Non-Crystalline Solids*, *493*, 94–98. doi:10.1016/j.jnoncrysol.2018.04.044

Centro de Estudios Ambientales (CEA). (2014). *The urban green Infrastructure of Vitoria- Gasteiz*. Obtained from: https://www.vitoria-gasteiz.org/wb021/http/contenidosEstaticos/adjuntos/eu/34/21/53421.pdf

CfUT. (2019, September 19). *Executive Summary | Climate Emergency, Urban Opportunity*. Coalition for Urban Transitions. Retrieved January 10, 2022, from https://urbantransitions.global/en/publication/climate-emergency-urban-opportunity/executive-summary

Chang, N. J., & Fong, C. M. (2010). Green product quality, green corporate image, green customer satisfaction, and green customer loyalty. *African Journal of Business Management*, *4*(13), 2336–2344.

Chang, T. W., Chen, Y. S., Yeh, Y. L., & Li, H. X. (2020). Sustainable consumption models for customers: Investigating the significant antecedents of green purchase behavior from the perspective of information asymmetry. *Journal of Environmental Planning and Management*, *2020*, 1–21.

Chatterley, C., & Thomas, A. (2013). *Snapshot of WASH in Schools in Eastern and Southern Africa: A Review of Data*. Evidence and Inequities in the Region.

Cheek, P., Nikpour, L., & Nowlin, H. D. (2005). Aging well with Smart Technology. *Nursing Administration Quarterly*, *29*(4), 329–338. doi:10.1097/00006216-200510000-00007 PMID:16260997

Cheng, T. W., Lee, M. L., Ko, M. S., Ueng, T. H., & Yang, S. F. (2012). The heavy metal adsorption characteristics on metakaolin-based geopolymer. *Applied Clay Science*, *56*, 90–96. doi:10.1016/j.clay.2011.11.027

Chen, L., & Ng, E. (2012). Outdoor thermal comfort and outdoor activities: A review of research in the past decade. *Cities (London, England)*, *29*(2), 118–125. doi:10.1016/j.cities.2011.08.006

Chen, Y. S. (2008). The driver of green innovation and green image–green core competence. *Journal of Business Ethics*, *81*(3), 531–543. doi:10.100710551-007-9522-1

Chen, Y. S., & Chang, C. H. (2013). The determinants of green product development performance: Green dynamic capabilities, green transformational leadership, and green creativity. *Journal of Business Ethics*, *2013*(116), 107–119. doi:10.100710551-012-1452-x

Chen, Y. S., Lai, S. B., & Wen, C. T. (2006). The influence of green innovation performance on corporate advantage in Taiwan. *Journal of Business Ethics*, *2006*(67), 331–339. doi:10.100710551-006-9025-5

Chen, Y. S., Lin, S. H., Lin, C. Y., Hung, S. T., Chang, C. W., & Huang, C. W. (2020). Improving green product development performance from green vision and organizational culture perspectives. *Corporate Social Responsibility and Environmental Management*, *2020*(27), 222–231. doi:10.1002/csr.1794

Chen, Z., Xiao, F., Shi, J., & Li, A. (2021). Dynamic Model Development for Vehicle Air Conditioners Based on Physics-guided Deep Learning. *International Journal of Refrigeration*.

Chicago.gov. (n.d.). *Healthy Chicago 2025*. City of Chicago. Retrieved March 16, 2022, from https://www.chicago.gov/city/en/depts/cdph/provdrs/healthy_communities/svcs/healthy-chicago-2025.html

Chindaprasirt, P., Chareerat, T., & Sirivivatnanon, V. (2007). Workability and strength of coarse high calcium fly ash geopolymer. *Cement and Concrete Composites*, *29*(3), 224–229. doi:10.1016/j.cemconcomp.2006.11.002

Chiou, T. Y., Chan, H. K., Lettice, F., & Chung, S. H. (2011). The influence of greening the suppliers and green innovation on environmental performance and competitive advantage in Taiwan. *Transp. Res. Logist. Transp. Rev*, *2011*(47), 822–836. doi:10.1016/j.tre.2011.05.016

Chowdhury, A. A., Rasul, M. G., & Khan, M. M. K. (2008). Thermal-comfort analysis and simulation for various low-energy cooling-technologies applied to an office building in a subtropical climate. *Applied Energy*, *85*(6), 449–462. doi:10.1016/j.apenergy.2007.10.001

Cities Alliance. (2020). *Affordable Housing and Land in Latin America*. Cities Alliance / Cities without Slums. https://www.citiesalliance.org/newsroom/events/affordable-housing-and-land-latin-america

Coca, J., & Fernández, F. (2011). La Renovación del Manzanares: Transformaciones y Reciclajes Urbanos. In Proyecto, Progreso, Arquitectura. Universidad de Sevilla.

Cohen, A. J., Brauer, M., Burnett, R., Anderson, H. R., Frostad, J., Estep, K., Balakrishnan, K., Brunekreef, B., Dandona, L., Dandona, R., Feigin, V., Freedman, G., Hubbell, B., Jobling, A., Kan, H., Knibbs, L., Liu, Y., Martin, R., Morawska, L., ... Forouzanfar, M. H. (2017). Estimates and 25-year trends of the global burden of disease attributable to ambient air pollution: An analysis of data from the Global Burden of Diseases Study 2015. *Lancet*, *389*(10082), 1907–1918. doi:10.1016/S0140-6736(17)30505-6 PMID:28408086

Collins, A., Watts, S., & McAlister, M. (2008). The economics of sustainable tall buildings. *Proc. of the CTBUH 8th World Congress*, 175–185.

Confía, A. E. N. O. R. (2022). ISO 14006 Certification ecodesign management. https://www.aenor.com/certification/environment/ecodesign

Convención Relativa a los Humedales de Importancia Internacional Especialmente como Hábitats de Aves Acuáticas (RAMSAR). (1971). *¿Qué son los humedales?* Convención de Ramsar.

Cook, D. J., Duncan, G., Sprint, G., & Fritz, R. L. (2018). Using Smart City technology to make healthcare smarter. *Proceedings of the IEEE*, *106*(4), 708–722. doi:10.1109/JPROC.2017.2787688 PMID:29628528

Coolen, H., & Meesters, J. (2012). Private and public green spaces: Meaningful but different settings. *Journal of Housing and the Built Environment*, *27*(1), 49–67. doi:10.100710901-011-9246-5

Cornelius, B., Cornelius, A., Crisafi, L., Collins, C., McCarthy, S., Foster, C., Shannon, H., Bennett, R., Brown, S., Rodriguez, K., & Bachini, S. (2020). Mass air medical repatriation of coronavirus disease 2019 patients. *Air Medical Journal*, *39*(4), 251–256. doi:10.1016/j.amj.2020.04.005 PMID:32690299

Costa, D. G., & Peixoto, J. P. J. (2020). COVID-19 pandemic: A review of smart cities initiatives to face new outbreaks. *IET Smart Cities*, *2*(2), 64–73. doi:10.1049/iet-smc.2020.0044

Cox, L. (2020, February 17). Researchers claim solar efficiency breakthrough for flexible 'skin'. *The Guardian*. Retrieved January 4, 2022, from https://www.theguardian.com/environment/2020/feb/18/researchers-claim-solar-efficiency-breakthrough-for-flexible-skin

Cubillos-González, R. A., & Cardoso, G. T. (2020). Clean technology transfer and innovation in social housing production in Brazil and Colombia. A framework from a systematic review. *Sustainability (Switzerland)*, *12*(4), 1–12. doi:10.3390u12041335

Cubillos-González, R.-A. (2021). Barreras y oportunidades de la Construcción 4.0 frente al cambio climático. In M. E. Molar Orozco, M. G. Vázquez Jiménez, & M. J. Velázquez Lozano (Eds.), *Un mundo sostenible en el siglo xxi: Alternativas para el rescate del planeta* (pp. 125–153). Universidad Autónoma de CoaHuila. http://www.investigacionyposgrado.uadec.mx/publicacion/un-mundo-sostenible-en-el-siglo-xxi-alternativas-para-el-rescate-del-planeta/

Dai, F. C., Lee, C. F., & Zhang, X. H. (2001). GIS-based geo-environmental evaluation for urban land-use planning: A case study. *Engineering Geology, 61*(4), 257–271. doi:10.1016/S0013-7952(01)00028-X

Darko, A., Chan, A. P. C., Ameyaw, E. E., He, B.-J., & Olanipekun, A. O. (2017). Examining issues influencing green building technologies adoption: The United States green building experts' perspectives. *Energy and Building, 144*, 320–332. doi:10.1016/j.enbuild.2017.03.060

Das, S. K., Mishra, J., Singh, S. K., Mustakim, S. M., Patel, A., Das, S. K., & Behera, U. (2020). Characterization and utilization of rice husk ash (RHA) in fly ash – Blast furnace slag based geopolymer concrete for sustainable future. *Materials Today: Proceedings, 33*, 5162–5167. doi:10.1016/j.matpr.2020.02.870

Davidovits, J. (1991). Geopolymers: Inorganic polymeric new materials. *Journal of Thermal Analysis and Calorimetry, 37*(8), 1633–1656. doi:10.1007/BF01912193

Davydova, B. V., & Frías, U. H. G. (2016). Detección de señales de cambio 22. climático en Jalisco, México. Memorías en Extenso de AIDIS, 2.

Davydova, B.V., & Frías, U.H.G. (2016). Detección de señales de cambio 22. climático en Jalisco, México. *Memorías en Extenso de AIDIS, 2.*

de la República, C. (1997). *Ley de Areas Naturales Protegidas*. Gobierno del Perú.

De Silva, P., Sagoe-Crenstil, K., & Sirivivatnanon, V. (2007). Kinetics of geopolymerization: Role of Al2O3 and SiO2. *Cement and Concrete Research, 37*(4), 512–518. doi:10.1016/j.cemconres.2007.01.003

de Soto, B. G., & Skibniewski, M. J. (2020). Future of robotics and automation in construction. In A. Sawhney, M. Riley, & J. Irizarry (Eds.), *Construction 4.0: An Innovation Platform for the Built Environment*. Routledge. doi:10.1201/9780429398100-15

Deloitte. (2019). *Command center approach to drive efficiencies at Cascais*. Cascais - Deloitte Client Stories. Retrieved March 18, 2022, from https://www2.deloitte.com/content/dam/Deloitte/us/Documents/process-and-operations/us-cons-smart-cities-client-stories-cascais.pdf

Deloitte. (2021). *Smart and sustainable buildings and infrastructure*. Deloitte. Retrieved January 5, 2022, from https://www2.deloitte.com/global/en/pages/public-sector/articles/urban-future-with-a-purpose/smart-and-sustainable-buildings-and-infrastructure.html

Deloitte. (2021). *Smart Health Communities*. Deloitte. Retrieved March 16, 2022, from https://www2.deloitte.com/global/en/pages/public-sector/articles/urban-future-with-a-purpose/smart-health-communities.html

Dhawan, V. (2021, January 18). *Technology and innovation in Healthcare Support Singapore's vision of a smart nation*. OpenGov Asia. Retrieved March 10, 2022, from https://opengovasia.com/technology-and-innovation-in-healthcare-support-singapores-vision-of-a-smart-nation/

Diaz Gallardo, B., Muñoz de Solano Sánchez, S., & Navarrete Suárez, F. (2021). *Prototipo WASH Makeni. ICHaB*. Universidad Politécnica de Madrid.

Digital Europe. (2020, December 4). *Digital Building Transformations for Europe's green renovations: Sustainable Living and working*. DIGITALEUROPE. Retrieved January 5, 2022, from https://www.digitaleurope.org/resources/digital-building-transformations-for-europes-green-renovations-sustainable-living-and-working/

Dijst, M., Worrell, E., Böcker, L., Lyons, G., Mokhtarian, P. L., Newman, P., & Perrels, A. (2018). Exploring urban metabolism—Towards an interdisciplinary perspective. *Resources, Conservation and Recycling, 1322*, 190–203. doi:10.1016/j.resconrec.2017.09.014

Dodge Data and Analytics. (2016). *World Green Building Trends 2016 | Developing Markets Accelerate Global Green Growth*. Smart Market Report. Retrieved January 7, 2022, from https://www.infrastructureusa.org/world-green-building-trends-developing-markets-accelerate-global-green-growth/

Domínguez Arribas, L. E., Serrano Lanzarote, B., García-Prieto Ruiz, A., Ortega Madrigal, L., & Soto Francés, L. (2015). *Estudio del Potencial de ahorro energético y reducción de emisiones de CO2 en la Comunitat Valenciana*. www.five.es

Dong, J., Zhang, H., Klapare, S., Wang, J., & Luo, Y. (2020). 16: Socioeconomic benefit of contaminated site remediation. In *Sustainable Remediation of Contaminated Soil and Groundwater* (pp. 433–450). Butterworth-Heinemann. doi:10.1016/B978-0-12-817982-6.00016-1

dos Muchangos, L. S., Tokai, A., & Hanashima, A. (2016). Application of material flow analysis to municipal solid waste in Maputo City, Mozambique. *Waste Management & Research, 35*(3), 253–266. doi:10.1177/0734242X16678067 PMID:27909211

Drinking-water. (2019). Retrieved from https://www.who.int/news-room/fact-sheets/detail/drinking-water

Duan, P., Yan, C., Zhou, W., & Ren, D. (2016). Development of fly ash and iron ore tailing based porous geopolymer for removal of Cu(II) from wastewater. *Ceramics International, 42*(12), 13507–13518. doi:10.1016/j.ceramint.2016.05.143

Du, K., Xie, C., & Ouyang, X. (2017). A comparison of carbon dioxide (CO2) emission trends among provinces in China. *Renewable & Sustainable Energy Reviews, 73*, 19–25. doi:10.1016/j.rser.2017.01.102

Duxson, P., Fernández-Jiménez, A., Provis, J. L., Lukey, G. C., Palomo, A., & van Deventer, J. S. J. (2007). Geopolymer technology: The current state of the art. *Journal of Materials Science, 42*(9), 2917–2933. doi:10.100710853-006-0637-z

Duxson, P., Provis, J. L., Lukey, G. C., & van Deventer, J. S. J. (2007). The role of inorganic polymer technology in the development of 'green concrete.'. *Cement and Concrete Research, 37*(12), 1590–1597. doi:10.1016/j.cemconres.2007.08.018

Echeverría Trueba, J. B., Rubio Poves, N., Rodríguez Oyarbide, I., Jönsson, J., Vigne, G., & Sánchez Peña, J. (2020). Retos de la seguridad en caso de incendio de levantes en edificios de vivienda existentes. *Informes de la Construcción, 559.*

Edenhofer, O. (2015). *Climate change 2014: Mitigation of climate change: Working Group III Contribution to the fifth assessment report of the intergovernmental panel on climate change*. Cambridge University Press.

EIA. (2016, May). *International Energy Outlook 2016*. International Energy Outlook 2016 - With Projections to 2040. Retrieved February 28, 2022, from https://www.eia.gov/outlooks/ieo/pdf/0484(2016).pdf

Elefante, C. (2012). The Greenest Building Is... One That Is Already Built. *Forum Journal, 27*(1), 62-72. https://www.muse.jhu.edu/article/494514

El-Gazzar, R., & Stendal, K. (2020). Blockchain in health care: Hope or hype? *Journal of Medical Internet Research, 22*(7), e17199. Advance online publication. doi:10.2196/17199 PMID:32673219

Elhakeem, A., & Hegazy, T. (2012). Building Asset Management with deficiency tracking and integrated life cycle optimisation. *Structure and Infrastructure Engineering, 8*(8), 729–738. doi:10.1080/15732471003777071

Elkington, J., 1994. Towards the sustainable corporation: Win-win-win business strategies for sustainable development. *Cal. Manag. Rev., 36*(3), 90-100.

Elsaid, A. M., & Ahmed, M. S. (2021). Indoor Air Quality Strategies for Air-Conditioning and Ventilation Systems with the Spread of the Global Coronavirus (COVID-19) Epidemic: Improvements and Recommendations. *Environmental Research, 199*, 111314. doi:10.1016/j.envres.2021.111314 PMID:34048748

Enerdata. (2017). *Share of new dwellings built according to national nZEB definition or better than nZEB*. ZEBRA2020. www.zebramonitoring.enerdata.eu/overall-building-activities/share-of-new-dwellings-built-according-to-national-nzeb-definition-or-better-than-nzeb.html

EPA. (2006). *Particulate Matter Basics*. Available from. https://www.epa. gov/pm-pollution/particulate-matter-pm-basics#PM

EPA. (2021, June 10). *Climate Smart Brownfields Manual*. Retrieved January 5, 2022, from https://www.epa.gov/sites/default/files/2021-06/documents/final_climate_smart_brownfields_manual_6-10-21_508_complaint.pdf

European Commission. (2019). *Green Infrastructure (GI)—Enhancing Europe's Natural Capital*. Communication from the Commission to the European Parliament, the Council, The European Economic and Social Committee and the Committee of the Regions.

European Environment Agency. (2005). *Communication from the Commission to the Council and the European Parliament—Thematic Strategy on Air Pollution* (COM (2005) 0446 Final).

Eurostat. (n.d.). *Housing statistics*. Retrieved December 18, 2019, from https://ec.europa.eu/eurostat/statistics-explained/index.php?title=Housing_statistics#Type_of_dwelling

EY.com. (2017, September 1). *Smart buildings: how sustainability strategies can impact investor ROI*. EY UK. Retrieved December 13, 2021, from https://www.ey.com/en_uk/real-estate-hospitality-construction/smart-buildings-how-sustainability-strategies-can-impact-investor-roi

Fariña Tojo, J. (2002). La ciudad y el medio natural. Madrid: Akal.

Fariña, J. (2000). *La ciudad y el medio natural*. AKAL.

Fazli, W., Hakim, I. L., Rozaida, G., & Muhammad, A. (2019). *An Efficient Artificial Intelligence Hybrid Approach for Energy Management in Intelligent Buildings. KSII Transactions on Internet and Information Systems*. doi:10.3837/tiis.2019.12.007

Feng, D., Provis, J. L., & Deventer, J. S. J. (2012). Thermal activation of albite for the synthesis of one-part mix geopolymers. *Journal of the American Ceramic Society, 95*(2), 565–572. doi:10.1111/j.1551-2916.2011.04925.x

Feng, D., Tan, H., & van Deventer, J. S. J. (2004). Ultrasound enhanced geopolymerisation. *Journal of Materials Science, 39*(2), 571–580. doi:10.1023/B:JMSC.0000011513.87316.5c

Fernandes, L. (2021, July 8). *Smart buildings and IOT: Supporting a safe return to the Office*. Quocirca. Retrieved March 4, 2022, from https://quocirca.com/content/smart-buildings-and-iot-supporting-a-safe-return-to-the-office/

Fernandes, A. A. R., & Solimun. (2017). The mediating effect of strategic orientation and innovations on the effect of environmental uncertainties on the performance of business in the Indonesian aviation industry. *International Journal of Law and Management, 59*(6), 1269–1278. doi:10.1108/IJLMA-10-2016-0087

Fernández-Vigil, M., & Echeverría, J. B. (2019). Elderly at Home: A Case for the Systematic Collection and Analysis of Fire Statistics in Spain. *Fire Technology, 55*(6), 2215–2244. doi:10.100710694-019-00852-6

Ferrão, P., & Fernández, J. E. (2013). *Sustainable Urban Metabolism*. MIT Press. doi:10.7551/mitpress/8617.001.0001

Fiala, D., Havenith, G., Bröde, P., Kampmann, B., & Jendritzky, G. (2012). UTCI-Fiala multi-node model of human heat transfer and temperature regulation. *International Journal of Biometeorology*, *56*(3), 1–13. doi:10.100700484-011-0424-7 PMID:21503622

Fialkiewicz, W., Burszta-Adamiak, E., Kolonko-Wiercik, A., Manzardo, A., Loss, A., Mikovits, C., & Scipioni, A. (2018). Simplified direct water footprint model to support urban water management. *Water (Basel)*, *10*(5), 630. doi:10.3390/w10050630

Fichte, J., & Hollfelder, M. (2020, April 13). *Smart buildings: Making buildings smarter, greener, and more energy-efficient - page 2 of 3*. EE Times Europe. Retrieved January 3, 2022, from https://www.eetimes.eu/smart-buildings-making-buildings-smarter-greener-and-more-energy-efficient/2/

Fix Development. (2013). *Clock shadow building*. Fix Development. Retrieved January 5, 2022, from https://www.fixdevelopment.com/clock-shadow-building/

Fleming, N. S., Becker, E. R., Culler, S. D., Cheng, D., McCorkle, R., da Graca, B., & Ballard, D. J. (2013). The impact of Electronic Health Records on workflow and financial measures in Primary Care Practices. *Health Services Research*, *49*(1pt2), 405–420. doi:10.1111/1475-6773.12133 PMID:24359533

Fong, C. M., & Chang, N. J. (2012). The impact of green learning orientation on proactive environmental innovation capability and firm performance. *African Journal of Business Management*, *6*(32), 727–735.

Fonseca, D., Sanchez-Sepulveda, M., Necchi, S., & Peña, E. (2021). Towards smart city governance. case study: Improving the interpretation of quantitative traffic measurement data through citizen participation. *Sensors (Basel)*, *21*(16), 5321. doi:10.339021165321 PMID:34450772

Forschungsanstalt Landschaftsentwicklung Landschaftsbau (FLL). (2002). *Dachbegrünungsrichtlinie. Richtlinien für die Planung, Ausführung und Pfl ege von Dachbegrünungen* [Green roof policy. Guidelines for the planning, execution, and maintenance of green roofs]. Bonn: Forschungsanstalt Landschaftsentwicklung Landschaftsbau.

Framil, L., & Ovanessoff, A. (2018). *Here's why Latin Americans will thrive in the age of the robots*. World Economic Forum (WEF). https://www.weforum.org/agenda/2018/03/here-s-why-latin-americans-will-thrive-in-the-age-of-the-robots/

Franchini, T. (Dir.) & Raventós, T. (Coord.). (2018). Temas de planeamiento urbano. Madrid, Spain: CEU Ediciones.

Franz, M., Pahlen, G., Nathanail, P., Okuniek, N., & Koj, A. (2006). Sustainable development and brownfield regeneration. What defines the quality of derelict land recycling? *Environmental Sciences*, *3*(2), 135–151. doi:10.1080/15693430600800873

Fundación Mapfre & APTB. (2010). *Víctimas de incendios en España 2010*. https://www.fundacionmapfre.org/fundacion/es_es/images/informe-victimas-incendios-en-espana-2010-actualizado_tcm1069-211526.pdf

Fundación Mapfre & APTB. (2011). *Víctimas de Incendios en España 2011*. https://www.fundacionmapfre.org/fundacion/es_es/images/informe-victimas-incendios-en-espana-2011_tcm1069-211527.pdf

Fundación Mapfre & APTB. (2014). *Víctimas de incendios en España 2012 y 2013*. https://www.fundacionmapfre.org/fundacion/es_es/images/informe-victimas-incendios-en-espana-2012-2013_tcm1069-211528.pdf

Fundación Mapfre & APTB. (2015). *Víctimas de incendios en España en 2014*. https://www.fundacionmapfre.org/fundacion/es_es/images/informe-victimas-incendios-en-espana-2014_tcm1069-211529.pdf

Fundación Mapfre & APTB. (2016). *Víctimas de incendios en españa en 2015*. https://www.fundacionmapfre.org/fundacion/es_es/images/informe-victimas-incendios-espana-2015_tcm1069-388743.pdf

Fundación Mapfre & APTB. (2017). *Víctimas de Incendios en España en 2016.* doi:10.2865/5721

Fundación Mapfre & APTB. (2018). *Estudio de víctimas de incendios en España en 2017.* https://www.fundacionmapfre.org/fundacion/es_es/images/informe-victimas-incendios-espana-2017_tcm1069-522308.pdf

Fundación Peruana para la Conservación de la Naturaleza. (2010). *Documento base para la elaboración de una estrategia de conservación de los humedales de la costa peruana.* ProNaturaleza.

Fung, K. L. (2018). *Expanding the green network on rooftops: A study of integrating green roofs as a part of urban green innovation space planning* (Order No. 10932095). Available from ProQuest One Academic. (2124999351). Retrieved from http://wdg.biblio.udg.mx:2048/login?url=https://www.proquest.com/dissertations-theses/expanding-green-network-on-rooftops-study/docview/2124999351/se-2?accountid=28915 https://www.proquest.com/docview/2124999351?pqorigsite=gscholar&fromopenview=true

Galdeano-Gómez, E., Céspedes-Lorente, J., & Martínez-del-Río, J. (2008). Environmental performance and spillover effects on productivity: Evidence from horticultural firms. *Journal of Environmental Management, 88*(4), 1552–1561. doi:10.1016/j.jenvman.2007.07.028 PMID:17825476

Gao, Z., Wang, S., & Gu, J. (2020). Public participation in smart-city governance: A qualitative content analysis of public comments in urban China. *Sustainability, 12*(20), 8605. doi:10.3390u12208605

García Chávez, J. R., & Fuentes, V. (2005). *Viento y Arquitectura: El viento como factor de diseño arquitectónico* (3rd ed.). Trillas.

García de Soto, B., Agustí-Juan, I., Joss, S., & Hunhevicz, J. (2019). Implications of Construction 4.0 to the workforce and organizational structures. *International Journal of Construction Management, 0*(0), 1–13. doi:10.1080/15623599.2019.1616414

Gartner Inc. (2019). *Hype cycle for Smart City Technologies and Solutions, 2019.* Gartner. Retrieved January 3, 2022, from https://www.gartner.com/en/documents/3955915/hype-cycle-for-smart-city-technologies-and-solutions-201

Gatzweiler, F. W., Zhu, Y.-G., Diez Roux, A. V., Capon, A., Donnelly, C., Salem, G., … Zhang, Y. (2017). *Advancing Health and Wellbeing in the Changing Urban Environment.* doi:10.1007/978-981-10-3364-3_1

Geraie, M. S., & Rad, F. M. (2015). Mediator role of the organizational identity green in relationship between total quality management and perceived innovation with sustainable competitive advantage. *International Journal of Biology, Pharmacy and Allied Sciences, 2015*(4), 266–276.

Geros, V., Santamouris, M., Karatasou, S., Tsangrassoulis, A., & Papanikolaou, N. (2005). On the cooling potential of night ventilation techniques in the urban environment. *Energy and Building, 37*(3), 243–257. doi:10.1016/j.enbuild.2004.06.024

Ge, Y., Yuan, Y., Wang, K., He, Y., & Cui, X. (2015). Preparation of geopolymer-based inorganic membrane for removing Ni2+ from wastewater. *Journal of Hazardous Materials, 299*, 711–718. doi:10.1016/j.jhazmat.2015.08.006 PMID:26295686

Giannotti, E., Vásquez, A., Galdámez, E., Velásquez, P., & Devoto, C. (2021). Green infrastructure planning for climate emergency: Learning from the "stgo+" project, Santiago de Chile. *Cuadernos de Geografia: Revista Colombiana de Geografia, 30*(2), 359–375. doi:10.15446/rcdg.v30n2.88749

Gibbons, C. (2020, August 8). *How the smart city movement and it are Transforming Health.* HIMSS. Retrieved December 27, 2021, from https://www.himss.org/resources/how-smart-city-movement-and-it-are-transforming-health

Gibbs, A., & Peiser, R. (2022, January 25). ResilientHub: A study in getting office buildings to net zero. *Urban Land Magazine.* Retrieved March 3, 2022, from https://urbanland.uli.org/sustainability/resilienthub-a-study-in-getting-office-buildings-to-net-zero/?Site=ULI2015

Gibson, J. E., Scherer, W. T., & Gibson, W. F. (2007). *How to Do Systems Analysis*. Wiley. doi:10.1002/9780470130599

Gilkeson, S., & Tilton, J. (2019, June 14). *Blockchain and healthcare innovation: A city, academia, and technology firms join forces to find blockchain solutions to address public health challenges*. JD Supra. Retrieved March 11, 2022, from https://www.jdsupra.com/legalnews/blockchain-and-healthcare-innovation-a-34497/

Girardet, H. (1999). *Creating Sustainable Cities*. Green Books.

Giridharan, R. S., Lau, S., Ganesan, C., & Givoni, B. (2007). Urban design factors influencing heat island intensity in high-rise high-density environments of Hong Kong. *Building and Environment*, *42*(10), 3669–3684. doi:10.1016/j.buildenv.2006.09.011

Gobierno del Perú. (1993). *Constitución Política del Perú*. Gobierno del Perú.

Goldstein, G., & Kickbusch, I. (1996). A healthy city is a better city. *World Health*, *49*(1).

González Capitel A., & García-Gutiérrez Mosteiro, J. (2000). *Luis Moya Blanco. Arquitecto 1904-1990*. Ministerio de Fomento.

Goulden, M., Mason, M. A., & Frasch, K. (2011). Keeping women in the science pipeline. *The Annals of the American Academy of Political and Social Science*, *638*(1), 141–162. doi:10.1177/0002716211416925

Gralton, J., Tovey, E., McLaws, M. L., & Rawlinson, W. D. (2011). The role of particle size in aerosolised pathogen transmission: A review. *The Journal of Infection*, *62*(1), 1–13. doi:10.1016/j.jinf.2010.11.010 PMID:21094184

Greene, J. C., Caracelli, V. J., & Graham, W. F. (1989). Toward a conceptual framework for mixed-method evaluation designs. *Educational Evaluation and Policy Analysis*, *11*(3), 255–274. doi:10.3102/01623737011003255

Grevelman, L., & Kluiwstra, M. (2010). Sustainability in project management: A case study on enexis. *PM World Today*, *12*(7).

Grosse, J., Dietrich, B., Garrido, G. I., Habisreuther, P., Zarzalis, N., Martin, H., Kind, M., & Kraushaar-Czarnetzki, B. (2009). Morphological characterization of ceramic sponges for applications in chemical engineering. *Industrial & Engineering Chemistry Research*, *48*(23), 10395–10401. doi:10.1021/ie900651c

Gunn, A., & Mintrom, M. (2016). Higher Education Policy Change in Europe: Academic Research Funding and the Impact Agenda. *European Education*, *48*(4), 241–257. doi:10.1080/10564934.2016.1237703

Gupta, N. (2017, June). *How IoT Can Significantly Improve Healthcare in the Context of Smart City*. Retrieved March 12, 2022, from https://www.iotworldtoday.com/files/2018/09/June_2017_JoI_Improving_Smart_City_Healthcare.pdf

Gupta, J., & Vegelin, C. (2016). Sustainable development goals and inclusive development. *International Environmental Agreement: Politics, Law and Economics*, *16*(3), 433–448. doi:10.100710784-016-9323-z

Hadavi, S. (2017). Direct and indirect effects of the physical aspects of the environment on mental well-being. *Environment and Behavior*, *2017*(49), 1071–1104. doi:10.1177/0013916516679876

Hailegiorgis, B., Ebrahim, M., Lemecha, G., & IRC. (2022). *Levering resources for WASH: lessons from WASH SDG programme in Shashamane and Negelle Arsi woredas in Ethiopia*. Retrieved from https://www.ircwash.org/

Hajimohammadi, A., Ngo, T., Mendis, P., Nguyen, T., Kashani, A., & van Deventer, J. S. J. (2017). Pore characteristics in one-part mix geopolymers foamed by H2O2: The impact of mix design. *Materials & Design*, *130*, 381–391. doi:10.1016/j.matdes.2017.05.084

Hajimohammadi, A., & van Deventer, J. S. J. (2017). Solid reactant-based geopolymers from rice hull ash and sodium aluminate. *Waste and Biomass Valorization*, *8*(6), 2131–2140. doi:10.100712649-016-9735-6

Halls, J. N., & Magolan, J. L. (2019). A methodology to assess land use development, flooding, and wetland change as indicators of coastal vulnerability. *Remote Sensing*, *11*(19), 2260. Advance online publication. doi:10.3390/rs11192260

Hamada, H. M., Al-attar, A. A., Yahaya, F. M., Muthusamy, K., Tayeh, B. A., & Humada, A. M. (2020). Effect of high-volume ultrafine palm oil fuel ash on the engineering and transport properties of concrete. *Case Studies in Construction Materials, 12*.

Hansen, M. (2008). Smart card technology and healthcare information: A dynamic duo. *Computers, informatics, nursing. CIN*, *26*(5), 254–257. doi:10.1097/01.NCN.0000304842.44584.42 PMID:18769179

Hardcastle, J. L. (2016, April 14). *Why water management will take smart buildings to the next level.* Environment + Energy Leader. Retrieved March 6, 2022, from https://www.environmentalleader.com/2016/04/why-water-management-will-take-smart-buildings-to-the-next-level/

Harg, I. M. (2000). *Proyectar con la naturaleza.* Gustavo Gili Barcelona.

Hashimoto, S., Machino, T., Takeda, H., Daiko, Y., Honda, S., & Iwamoto, Y. (2015). Antimicrobial activity of geopolymers ion-exchanged with copper ions. *Ceramics International*, *41*(10, 10, Part A), 13788–13792. doi:10.1016/j.ceramint.2015.08.061

Hassan, H. S., Abdel-Gawwad, H. A., Vásquez-García, S. R., Israde-Alcántara, I., Flores-Ramirez, N., Rico, J. L., & Mohammed, M. S. (2019). Cleaner production of one-part white geopolymer cement using pre-treated wood biomass ash and diatomite. *Journal of Cleaner Production*, *209*, 1420–1428. doi:10.1016/j.jclepro.2018.11.137

Hatcher, J. (2021, June 30). The making of a smart building: Revamping current structures. *Smart Buildings Magazine.* Retrieved January 4, 2022, from https://smartbuildingsmagazine.com/features/the-making-of-a-smart-building-revamping-current-structures

Hazem, N., Fatehi, A., & Bakr, A. (2021). reviewed paper Construction Automation: Post-Pandemic Integrated Robotized Construction Sites. *Journal of Construction Engineering and Management*, *134*(8), 201–211.

Health Europa. (2019, January 11). *Learning from the Estonian e-health System.* Health Europa. Retrieved March 16, 2022, from https://www.healtheuropa.eu/estonian-e-health-system/89750/

Hedayat, K. (2020, June 23). *The return to workplace is happening. Is your building ready?* Schneider Electric Blog. Retrieved January 5, 2022, from https://blog.se.com/building-management/2020/06/23/the-return-to-workplace-is-happening-is-your-building-ready/

He, J., Jie, Y., Zhang, J., Yu, Y., & Zhang, G. (2013). Synthesis and characterization of red mud and rice husk ash-based geopolymer composites. *Cement and Concrete Composites*, *37*, 108–118. doi:10.1016/j.cemconcomp.2012.11.010

Helmboldt, O., Keith Hudson, L., Misra, C., Wefers, K., Heck, W., Stark, H., Danner, M., & Rösch, N. (2000). Aluminum compounds, inorganic. Ullmann's Encyclopedia of Industrial Chemistry, 1–17.

Helmy, A. I. I. (2016). Intermittent curing of fly ash geopolymer mortar. *Construction & Building Materials*, *110*, 54–64. doi:10.1016/j.conbuildmat.2016.02.007

Hendrickson, V. L. (2018, April 26). *Where are the best property tax breaks for Green Homes Around the world?* Mansion Global. Retrieved January 7, 2022, from https://www.mansionglobal.com/articles/where-are-the-best-property-tax-breaks-for-green-homes-around-the-world-95175

HEPA Filter. (2021). *HEPA filter applications*. Retrieved from https://filtroshepa.com.mx/default.html

Higueras, E., Roman, E., & Fariña, J. (2021). Guidelines for Healthier Public Spaces for the Elderly Population: Recommendations in the Spanish Context. Handbook of Quality of Life and Sustainability.

Higueras, E. (2006). *Urbanismo Bioclimático*. Gustavo Gili.

Hilkevics, S., & Hilkevics, A. (2017). The comparative analysis of technology transfer models. *Entrepreneurship and Sustainability Issues*, *4*(4), 540–558. doi:10.9770/jesi.2017.4.4(11)

Hodge, T. (1997). Toward a conceptual framework for assessing progress toward sustainability. *Social Indicators Research*, *1997*(40), 5–98. doi:10.1023/A:1006847209030

Hoelzel, F. (2021). *Local Governance Strategies of Otumara Community in Lagos to Access Water*. Academic Press.

Horvath, A. (2004). Construction materials and the environment. *Annual Review of Environment and Resources*, *29*(1), 181–204. doi:10.1146/annurev.energy.29.062403.102215

Hossain, F. (2019). Chapter Seven - Best Management Practices. In Sustainable design and build: Building, energy, roads, bridges, water and Sewer Systems (pp. 419–431). Butterworth-Heinemann.

Hough, M. (1998). *Naturaleza y ciudad. Planificación urbana y procesos ecológicos*. Editorial Gustavo Gili. Barcelona Junta de Andalucía. *SIMA, Sistema de información Multiterritorial de Andalucía*. https://www.juntadeandalucia.es/institutodeestadisticaycartografia/sima/ficha.htm?mun=29067

Howard, E. (1898). *To-morrow: A peaceful path to real reform*. Swan Sonnenschein & CO.

Hua, X., Provis, J., Van Deventer, J., & Krivenko, P. (2008). Characterization of Aged Slag Concretes. *ACI Materials Journal*, *105*, 131–139.

Huisinga, J. M., Mancini, M., St. George, R. J., & Horak, F. B. (2012). Accelerometry reveals differences in gait variability between patients with multiple sclerosis and healthy controls. *Annals of Biomedical Engineering*, *41*(8), 1670–1679. doi:10.100710439-012-0697-y PMID:23161166

Hu, R., Pan, W., & Bock, T. (2020). Towards Dynamic Vertical Urbanism. *International Journal of Industrialized Construction*, *1*(1), 34–47. doi:10.29173/ijic208

Hu, S., Wang, H., Zhang, G., & Ding, Q. (2008). Bonding and abrasion resistance of geopolymeric repair material made with steel slag. *Cement and Concrete Composites*, *30*(3), 239–244. doi:10.1016/j.cemconcomp.2007.04.004

Huseien, G. F., Mirza, J., Ismail, M., & Hussin, M. W. (2016). Influence of different curing temperatures and alkali activators on properties of GBFS geopolymer mortars containing fly ash and palm-oil fuel ash. *Construction and Building Materials, 125*, 1229–1240.

Huseien, G. F., Mirza, J., Ismail, M., Ghoshal, S. K., & Ariffin, M. A. M. (2018). Effect of metakaolin replaced granulated blast furnace slag on fresh and early strength properties of geopolymer mortar. *Ain Shams Engineering Journal*, *9*(4), 1557–1566. doi:10.1016/j.asej.2016.11.011

Hwang, E., Cummings, L., Sixsmith, A., & Sixsmith, J. (2011). Impacts of home modifications on aging-in-place. *Journal of Housing for the Elderly*, *25*(3), 246–257. doi:10.1080/02763893.2011.595611

IEA & UNEP. (2018). *2018 Global Status Report: towards a zero-emission, efficient and resilient buildings and construction sector.* Author.

IEA. (2017). *Energy Technology Perspectives 2017.* IEA/OECD. www.iea.org/et

IEA. (2019). *Multiple Benefits of Energy Efficiency.* IEA. https://www.iea.org/reports/multiple-benefits-of-energy-efficiency

Ihnatenko, M., & Novak, N. (2018). Development of regional programs for the development of agrarian enterprises with organic production based on the European and international experience. *Baltic Journal of Economic Studies, 4*(4), 126–133. doi:10.30525/2256-0742/2018-4-4-126-133

IIoT-World. (2021, October 27). *The role of 3D Digital Twins and IIOT in decarbonization of buildings.* IIoT World. Retrieved January 5, 2022, from https://www.iiot-world.com/industrial-iot/connected-industry/the-role-of-3d-digital-twins-and-iiot-on-the-road-to-decarbonization-of-buildings/

IISD. (2017, June 7). *Mitigation update: Building Sector Must Reach Net Zero Carbon by 2050.* IISD - SDG. Retrieved January 3, 2022, from https://sdg.iisd.org/news/mitigation-update-building-sector-must-reach-net-zero-carbon-by-2050/

ILO. (2015). *News.* https://www.ilo.org/global/topics/green-jobs/news/WCMS_422575/lang--en/index.htm

INAIGEM. (2016). *Para la Formulación, seguimiento y evaluación del plan Operativo Institucional-Poi del Instituto Nacional de Investigación en Glaciares y Ecosistemas de Montaña- INAIGEM.* Obtained from: https://cdn.www.gob.pe/uploads/document/file/568900/23970810750073625520200318-16266-o1mrp9.pdf

Instituto Nacional de Estadística. (2011). *Censos de Población y Viviendas 2011.* Cifras de Población y Censos Demográficos. https://www.ine.es/censos2011_datos/cen11_datos_inicio.htm

Instituto Nacional de Estadística. (2014). *Proyección de la Población de España 2014–2064.* Author.

Ioannidou, D., Meylan, G., Sonnemann, G., & Habert, G. (2017). Is gravel becoming scarce? Evaluating the local criticality of construction aggregates. *Resources, Conservation and Recycling, 126*, 25–33. doi:10.1016/j.resconrec.2017.07.016

Jabbour, C. J. C., Santos, F. C. A., Fonseca, S. A., & Nagano, M. S. (2013). Green teams: Understanding their roles in the environmental management of companies located in Brazil. *Journal of Cleaner Production, 2013*(46), 58–66. doi:10.1016/j.jclepro.2012.09.018

Jacomit, A. M., Granja, A. D., & da Silva, V. G. (2009). Can Sustainable Buildings Cost the Same as Conventional Buildings? In *SASBE 2009 proceedings of the 3rd CIB International Conference on Smart and Sustainable Built Environments:* June 15-19, 2009, *Delft, the Netherlands* (pp. 1–2). Delft University of Technology.

Jain, S. K., & Kaur, G. (2004). Green marketing: an Indian perspective. *Decision, 31*(2), 161–209.

Jennifer Graeff, A. (2015). *Supporting Innovation : Case Studies from Latin America on Sustainable Housing and Community Development.* /www.planning.org/publications/document/9149230/

Jin, J., Gubbi, J., Marusic, S., & Palaniswami, M. (2014). An information framework for creating a smart city through internet of things. *IEEE Internet of Things Journal, 1*(2), 112–121. doi:10.1109/JIOT.2013.2296516

Johnson, C. (2016). *The future of connected home health.* https://www.plextek.com/wp-content/uploads/2016/07/The-future-of-connected-home-health.pdf

Johnson, R. W., Toohey, D., & Wiener, J. M. (2007, May 1). *Meeting the long-term care needs of the baby boomers.* Urban Institute. Retrieved December 31, 2021, from https://www.urban.org/research/publication/meeting-long-term-care-needs-baby-boomers

Johnson, S. K. (2017). What 11 CEOs have learned about championing diversity. *Harvard Business Review.*

Johnson, S. K. (2019). *Leaking Talent How People of Color are Pushed Out of Environmental Organizations How People of Color are Pushed Out of Environmental Organizations.* www.diversegreen.org https://diversegreen.org/research/leaking-talent/

Kalandides, A., & Grésillon, B. (2021). The Ambiguities of "Sustainable" Berlin. *Sustainability, 2021*(13), 1666. doi:10.3390u13041666

Kam-Sing Wong, S. (2012). The influence of green product competitiveness on the success of green product innovation: Empirical evidence from the Chinese electrical and electronics industry. *European Journal of Innovation Management, 2012*(15), 468–490. doi:10.1108/14601061211272385

Kang, M., Choi, Y., Kim, J., Lee, K. O., Lee, S., Park, K., Park, J., & Seo, I. (2020). COVID-19 impact on city and region : What ' s next after lockdown? *International Journal of Urban Sciences, 24*(3), 297–315. doi:10.1080/12265934.2020.1803107

Kanowitz, S. (2021, January 21). *How smart cities can mitigate the impact of health crises.* GCN. Retrieved December 21, 2021, from https://gcn.com/emerging-tech/2021/01/how-smart-cities-can-mitigate-the-impact-of-health-crises/315958/

Kelly, R. (n.d.). Implementing sustainable property tax reform in developing countries. *Taxation and Development: The Weakest Link?*, 326–363. doi:10.4337/9781783474332.00018

Kemp, J. (2019). *Urbanisation - Selected Indicators.* Thomson Reuters. Retrieved February 28, 2022, from https://fingfx.thomsonreuters.com/gfx/ce/7/7352/7334/URBANISATION.pdf

Kenney, S. (2021, October 21). Blockchain has potential for the future of Personal Health Records, Electronic Health Records. *Pharmacy Times.* Retrieved December 29, 2021, from https://www.pharmacytimes.com/view/blockchain-has-potential-for-the-future-of-personal-health-records-electronic-health-records

Kharas, H., & Remes, J. (2018, June 27). *As cities go smart, vulnerable populations are pushed further to the margins.* McKinsey & Company. Retrieved December 20, 2021, from https://www.mckinsey.com/mgi/overview/in-the-news/as-cities-go-smart-vulnerable-populations-are-pushed

Kierkegaard, P. (2013). EHealth in Denmark: A case study. *Journal of Medical Systems, 37*(6), 9991. Advance online publication. doi:10.100710916-013-9991-y PMID:24166019

King, J., & Perry, C. (2017, February). *Smart buildings: Using smart technology to save energy in ...* American Council for an Energy-Efficient Economy. Retrieved January 7, 2022, from https://www.aceee.org/sites/default/files/publications/researchreports/a1701.pdf

Klöckner, M., Haage, M., Nilsson, K., Robertsson, A., & Andersson, R. (2020). Parallel kinematic construction robot for aec industry. *Proceedings of the 37th International Symposium on Automation and Robotics in Construction, ISARC 2020: From Demonstration to Practical Use - To New Stage of Construction Robot, Isarc*, 1488–1495. 10.22260/ISARC2020/0206

Knibbs, L. D., Cole-Hunter, T., & Morawska, L. (2011). A review of commuter exposure to ultrafine particles and its health effects. *Atmospheric Environment, 45*(16), 2611–2622. doi:10.1016/j.atmosenv.2011.02.065

Knickman, J. R., & Snell, E. K. (2002). The 2030 problem: Caring for aging baby boomers. *Health Services Research, 37*(4), 849–884. doi:10.1034/j.1600-0560.2002.56.x PMID:12236388

Knight Frank Research Reports. (2021, September). *The Sustainability Series.* Knight Frank. Retrieved December 13, 2021, from https://www.knightfrank.com.hk/research/the-sustainability-series-september-2021-8395.aspx

Koerner, D. (2021, October 22). *How are digital twins used in building automation?* Smart Building Blog. Retrieved January 4, 2022, from https://www.75f.io/blog/how-are-digital-twins-used-in-building-automation

Kolomiets, T.V., & Tomashuk, I.V. (2021). Entrepreneurship and development of rural areas in Ukraine. *Colloquium-Journal*, *9*(96), 29-42.

Kovtun, M., Kearsley, E. P., & Shekhovtsova, J. (2015). Dry powder alkali-activated slag cements. *Advances in Cement Research*, *27*(8), 447–456. doi:10.1680/jadcr.14.00078

Kristoffersson, A., & Lindén, M. (2020). Wearable sensors for monitoring and preventing Noncommunicable Diseases: A systematic review. *Information | Ubiquitous Sensing for Smart Health Monitoring*, *11*(11), 521. doi:10.3390/info11110521

Kucher, A. (2019). *Sustainable soil management in the formation of competitiveness of agricultural enterprises*. Academic Publishing House «Talent». . doi:10.13140/RG.2.2.19554.07366

Kucher, L., Heldak, M., & Orlenko, A. (2018). Project management in organic agricultural production. *Agricultural and Resource Economics*, *4*(3), 104–128. doi:10.22004/ag.econ.281753

Kumar, V., & Christodoulopoulou, A. (2014). Sustainability and branding: An integrated perspective. *Industrial Marketing Management*, *43*(1), 6–15. doi:10.1016/j.indmarman.2013.06.008

Kupwade-Patil, K., & Allouche, E. N. (2013). Examination of chloride-induced corrosion in reinforced geopolymer concretes. *Journal of Materials in Civil Engineering*, *25*(10), 1465–1476. doi:10.1061/(ASCE)MT.1943-5533.0000672

Landsberg, H. E. (1947). Microclimatology. *The Architectural Forum*, *86*(3), 114–119.

Larwood, L., Falbe, C. M., Kriger, M. P., & Miesing, P. (1995). Structure and meaning of organizational vision. *Academy of Management Journal*, *1995*(38), 740–769.

Law, D. W., Adam, A. A., Molyneaux, T. K., Patnaikuni, I., & Wardhono, A. (2015). Long term durability properties of class F fly ash geopolymer concrete. *Materials and Structures*, *48*(3), 721–731. doi:10.161711527-014-0268-9

Lee, S. (2019). *Robots en América Latina: ¿cuántos son, dónde están y cuánto tributan?* BID Mejorando Vidas. https://blogs.iadb.org/gestion-fiscal/es/robots-en-america-latina-cuantos-son-donde-estan-y-cuanto-tributan/

Lee, H.-A., Kung, H.-H., Udayasankaran, J. G., Kijsanayotin, B., & Marcelo, B. (2020). An architecture and management platform for blockchain-based Personal Health Record Exchange: Development and Usability Study. *Journal of Medical Internet Research*, *22*(6), e16748. Advance online publication. doi:10.2196/16748 PMID:32515743

Lee, S. H., Augenbroe, G., Lee, J.-K., & Zhao, F. (2013). A design methodology for energy infrastructures at the campus scale. *Computer-Aided Civil and Infrastructure Engineering*, *28*(10), 753–768. doi:10.1111/mice.12050

Leung, D. Y. C., Caramanna, G., & Maroto-Valer, M. M. (2014). An overview of current status of carbon dioxide capture and storage technologies. *Renewable & Sustainable Energy Reviews*, *39*, 426–443. doi:10.1016/j.rser.2014.07.093

Leung, L., & Weismantle, P. (2008). Sky-sourced sustainability – how super tall buildings can benefit from height. *Proc. of the CTBUH 8th World Congress*, 328–335.

Lewin, A. (2020, July 8). *Inside Estonia's pioneering Digital Health Service*. Sifted. Retrieved March 16, 2022, from https://sifted.eu/articles/estonia-digital-health/#:~:text=Estonia's%20health%20service%20has%20been,to%20look%20at%20those%20records

Liao, S. H., Chang, J. C., Cheng, S. C., & Kuo, C. M. (2004). Employee relationship and knowledge sharing: A case study of a Taiwanese finance and securities firm. *Knowledge Management Research and Practice*, *2004*(2), 24–34. doi:10.1057/palgrave.kmrp.8500016

Li, H., & Kwan, M. (2018). Advancing analytical methods for urban metabolism studies. *Resources, Conservation and Recycling*, *132*, 239–245. doi:10.1016/j.resconrec.2017.07.005

Linner, T., Pan, W., Hu, R., Zhao, C., Iturralde, K., Taghavi, M., Trummer, J., Schlandt, M., & Bock, T. (2020). A technology management system for the development of single-task construction robots. *Construction Innovation*, *20*(1), 96–111. doi:10.1108/CI-06-2019-0053

Lin, Y. H., & Chen, Y. S. (2017). Determinants of green competitive advantage: The roles of green knowledge sharing, green dynamic capabilities, and green service innovation. *Quality & Quantity*, *51*(4), 1663–1685. doi:10.100711135-016-0358-6

Liu, D., Jiang, K., Shalley, C. E., Keem, S., & Zhou, J. (2016). Motivational mechanisms of employee creativity: A meta-analytic examination and theoretical extension of the creativity literature. *Organizational Behavior and Human Decision Processes*, *2016*(137), 236–263. doi:10.1016/j.obhdp.2016.08.001

Liu, J., Chen, J., & Tao, Y. (2015). Innovation Performance in New Product Development Teams in China's Technology Ventures: The Role of Behavioral Integration Dimensions and Collective Efficacy. *Journal of Product Innovation Management*, *2015*(32), 29–44. doi:10.1111/jpim.12177

Llonch, P., Haskell, M. J., Dewhurst, R. J., & Turner, S. P. (2017). Current available strategies to mitigate greenhouse gas emissions in livestock systems: An animal welfare perspective. *Animal*, *2017*(11), 274–284. doi:10.1017/S1751731116001440 PMID:27406001

López-Escamilla, Á., Herrera-limones, R., León-Rodríguez, Á., & Torres-García, M. (2020). Environmental Comfort as a Sustainable Strategy for Housing Integration: The AURA 1.0 Prototype for Social Housing. *Applied Sciences Article*, *10*(21), 7734. doi:10.3390/app10217734

López, F. J., Sugita, S., & Kobayashi, T. (2014). Cesium-adsorbent geopolymer foams based on silica from rice husk and metakaolin. *Chemistry Letters*, *43*(1), 128–130. doi:10.1246/cl.130851

López-Gamero, M. D., Claver-Cortés, E., & Molina-Azorín, J. F. (2008). Complementary resources and capabilities for an ethical and environmental management: A qual/quan study. *Journal of Business Ethics*, *82*(3), 701–732. doi:10.100710551-007-9587-x

Luo, Y., Meng, J., Wang, D., Jiao, L., & Xue, G. (2022). Experimental study on mechanical properties and microstructure of metakaolin based geopolymer stabilized silty clay. *Construction & Building Materials*, *316*, 125662. doi:10.1016/j.conbuildmat.2021.125662

Lüthi, C., Morel, A., Tilley, E., & Ulrich, L. (2011). *Community-led urban environmental sanitation planning: Clues complete guidelines for decision-makers with 30 tools. Swiss Federal Institute of Aquatic Science and Technology*. Eawag.

Lynch, K. (1975). P*lanificación del sitio*. Ed Gustavo Gili.

Maas, C. J., & Hox, J. J. (2005). Sufficient sample sizes for multilevel modeling. *Methodology*, *2005*(1), 86–92. doi:10.1027/1614-2241.1.3.86

Ma, C., Zhao, B., Wang, L., Long, G., & Xie, Y. (2020). Clean and low-alkalinity one-part geopolymeric cement: Effects of sodium sulfate on microstructure and properties. *Journal of Cleaner Production*, *252*, 119279. doi:10.1016/j.jclepro.2019.119279

MAGRAMA. (2014). Manual nacional de recomendaciones para el diseño de tanques de tormenta. Equipo de redacción: Aqualogy.

Ma, H., Ling, F., Yang, J., & Wang, G. (2002). Preparation of mineral polymer from potassium feldspar wastes: An experimental study. *Earth Sciences (Paris)*, *27*(5), 576–583.

Malczewski, J. (2010). Multiple Criteria Decision Analysis and Geographic Information Systems. In M. Ehrgott, J. Figueira, & S. Greco (Eds.), *Trends in Multiple Criteria Decision Analysis. International Series in Operations Research & Management Science* (Vol. 142). Springer. doi:10.1007/978-1-4419-5904-1_13

Maluquer & Salvador. (2014). Manual Nacional de Recomendaciones para el diseño de Tanques de Tormenta. Ministerio de Agricultura, Alimentación y Medioambiente.

Manchester City Counsil. (2015). *Manchester's great outdoors – a green and blue infrastructure strategy for Manchester*. Obtained from: https://www.manchester.gov.uk/downloads/download/6314/manchester_green_and_blue_strategy

Mancini, M., Schlueter, H., El-Gohary, M., Mattek, N., Duncan, C., Kaye, J., & Horak, F. B. (2016). Continuous monitoring of turning mobility and its association to Falls and Cognitive Function: A pilot study. *The Journals of Gerontology. Series A, Biological Sciences and Medical Sciences*, *71*(8), 1102–1108. doi:10.1093/gerona/glw019 PMID:26916339

Manjunath, R., Narasimhan, M. C., & Umesha, K. M. (2019). Studies on high performance alkali activated slag concrete mixes subjected to aggressive environments and sustained elevated temperatures. *Construction & Building Materials*, *229*, 116887. doi:10.1016/j.conbuildmat.2019.116887

Marchi, V. D. (2012). Environmental innovation and R&D cooperation: Empirical evidence from Spanish manufacturing firms. *Research Policy*, *41*(3), 614–623. doi:10.1016/j.respol.2011.10.002

Marín Perellón, F. J., Ortega Vidal, J., & Aragoneses López, D. (2009). *El Canal Real de Manzanares*. Ayuntamiento de Madrid.

Markets & Markets. (2020, February). *Smart buildings market*. Markets and Markets. Retrieved March 4, 2022, from https://www.marketsandmarkets.com/Market-Reports/smart-building-market-1169.html

Marpakwar, C. (2018, December 1). *Home buyers to get property tax rebates in Green Buildings*. Mumbai Mirror. Retrieved January 7, 2022, from https://mumbaimirror.indiatimes.com/mumbai/civic/home-buyers-to-get-property-tax-rebates-in-green-buildings/articleshow/66893701.cms

Marr, B. (2016, December 13). Big Data in Healthcare: Paris hospitals predict admission rates using machine learning. *Forbes*. Retrieved March 12, 2022, from https://www.forbes.com/sites/bernardmarr/2016/12/13/big-data-in-healthcare-paris-hospitals-predict-admission-rates-using-machine-learning/?sh=1520501479a2

Marshall, S. W., Runyan, C. W., Bangdiwala, S. I., Linzer, M. A., Sacks, J. J., & Butts, J. D. (1998). Fatal Residential Fires: Who dies and who survives? *Journal of the American Medical Association*, *279*(20), 1633–1637. doi:10.1001/jama.279.20.1633 PMID:9613913

Martínez & Muñoz de Pablo. (2018). El palimpsesto del Ensanche de Madrid = The Madrid´s expansion palimpses. *EGA. Revista de expresión gráfica arquitectónica, 23*(33), 118-129.

Martínez, C., López, P., Rojas, C., Qüense, J., Hidalgo, R., & Arenas, F. (2020). A sustainability index for anthropized and urbanized coasts: The case of Concón Bay, central Chile. *Applied Geography (Sevenoaks, England)*, *116*, 102166. doi:10.1016/j.apgeog.2020.102166

Maruani, T., & Amit-Cohen, I. (2007). Open Space planning models: A review of approaches and methods. *Landscape and Urban Planning*, *81*(1-2), 1–13. doi:10.1016/j.landurbplan.2007.01.003

Matava, C. T., Yu, J., & Denning, S. (2020). Clear plastic drapes may be effective at limiting aerosolization and droplet spray during extubation: implications for COVID-19. *Canadian Journal of Anesthesia/Journal canadien d'anesthésie, 67*(7), 902-904.

Matzarakis, A. (2001). Assessing climate for tourism purposes: Existing methods and tools for the thermal complex. *Proceedings of the first international workshop on climate, tourism and recreation*, 101–112.

Matzarakis, A., Rutz, F., & Mayer, H. (2007). Modelling radiation fluxes in simple and complex environments—Application of the RayMan model. *International Journal of Biometeorology*, *51*(4), 323–334. doi:10.100700484-006-0061-8 PMID:17093907

Mazur, K. V., & Tomashuk, I. V. (2019). Governance, and regulation as an indispensable condition for developing the potential of rural areas. *Baltic Journal of Economic Studies*, *5*(5), 67–78. doi:10.30525/2256-0742/2019-5-5-67-78

McCormick, K. (2018). *The business case for Healthy Buildings: Insights from Early Adopters*. Urban Land Institute. Retrieved December 30, 2021, from https://www.buildup.eu/en/practices/publications/business-case-healthy-buildings-insights-early-adopters

McHarg, I. L. (1995). *Desing With Nature*. Wiley Blackwell.

McKinsey Global Institute. (2017). *Reinventing Construction: A Route To Higher Productivity*. McKinsey & Company. doi:10.1080/19320248.2010.527275

McLellan, B. C., Williams, R. P., Lay, J., van Riessen, A., & Corder, G. D. (2011). Costs and carbon emissions for geopolymer pastes in comparison to ordinary portland cement. *Journal of Cleaner Production*, *19*(9), 1080–1090. doi:10.1016/j.jclepro.2011.02.010

Meadows, C. (2021, November 20). What is Smart Building Technology Worth to the market? *LEDs Magazine*. Retrieved March 4, 2022, from https://www.ledsmagazine.com/blogs/article/14214312/what-is-smart-building-technology-worth-to-the-market

Medical Futurist. (2019, February 7). *Where is digital health heading in Denmark?* The Medical Futurist. Retrieved March 16, 2022, from https://medicalfuturist.com/danish-digital-health-strategy/

Meerow, S., & Newell, J. P. (2017). Detroit, Spatial planning for multifunctional green infrastructure: Growing resilience. *Landscape and Urban Planning*, 62-75.

Melenbrink, N., Kassabian, P., Menges, A., & Werfel, J. (2017). Towards force-aware robot collectives for on-site construction. *Disciplines and Disruption - Proceedings Catalog of the 37th Annual Conference of the Association for Computer Aided Design in Architecture, ACADIA 2017*, 382–391.

Mell, I. C. (2010). *Green infrastructure: concepts, perceptions, and its use in spatial planning* [Doctoral Thesis]. School of Architecture, Planning and Landscape Newcastle University.

Mentens, J., Raes, D., & Hermy, M. (2006). Green roofs as a tool for solving the rainwater runoff problem in the urbanized 21st century? *Landscape and Urban Planning*, *77*(3), 217–226. doi:10.1016/j.landurbplan.2005.02.010

Mesimaki, M., Hauru, K., Kotze, D. J., & Lehvavirta, S. (2017). Neo-spaces for urban livibility? Urbanities' versatile mental image of green roofs in the Helsinki metropolitan area, Finland. *Land Use Policy*, *61*, 587–600. doi:10.1016/j.landusepol.2016.11.021

Millard, N. (2020, March 24). *How South Korea's smart city startups curbed the spread of Covid-19*. Retrieved March 10, 2022, from https://e27.co/how-south-koreas-smart-city-startups-curbed-the-spread-of-covid-19-20200323/

Miller, S. L., Nazaroff, W. W., Jimenez, J. L., Boerstra, A., Buonanno, G., Dancer, S. J., Kurnitski, J., Marr, L. C., Morawska, L., & Noakes, C. (2021). Transmission of SARS-CoV-2 by inhalation of respiratory aerosol in the Skagit Valley Chorale superspreading event. *Indoor Air, 31*(2), 314–323. doi:10.1111/ina.12751 PMID:32979298

Minelli, M., Medri, V., Papa, E., Miccio, F., Landi, E., & Doghieri, F. (2016). Geopolymers as solid adsorbent for CO2 capture. *Chemical Engineering Science, 148*, 267–274. doi:10.1016/j.ces.2016.04.013

Minier, V., & Tsai, O. (2021, December 6). *4 expert views on how to accelerate investment in green buildings and decarbonize cities.* World Economic Forum. Retrieved January 5, 2022, from https://www.weforum.org/agenda/2021/12/4-expert-views-how-to-accelerate-investment-green-buildings/

Ministerio de Fomento. (2019a). *Código Técnico de la Edificación: Documento Básico de Seguridad de utilización y accesibilidad.* https://www.codigotecnico.org/pdf/Documentos/SUA/DBSUA.pdf

Ministerio de Fomento. (2019b). *Código Técnico de la Edificación: Documento Básico de Seguridad en caso de incendio.* https://www.codigotecnico.org/images/stories/pdf/seguridadIncendio/DccSI.pdf

Ministerio de Fomento. (2019c). *Código Técnico de la Edificación. Parte I.* https://www.codigotecnico.org/pdf/Documentos/Parte1/Parte_I_20dic2019.pdf

Ministerio de Obras Públicas. (1939). *Instrucción para el Proyecto y Ejecución de Obras de Hormigón.*

Ministerio de transportes movilidad y agenda urbana. (2020). *Nota sobre el Documento Reconocido CTE-DR/048/14.* https://www.mitma.es/carreteras/trafico-velocidades-y-accidentes-mapa-estimacion-y-evolucion/mapas-de-trafico/2019

Ministerio del Ambiente del Perú (MINAM). (2014). *Estrategia Nacional de Humedales - Document for public consultation.* Obtained from: http://www.minam.gob.pe/consultaspublicas/wp-content/uploads/sites/52/2014/02/RM-N%C2%B0-051-2014-MINAM.pdf

Minor, B. D., Doppa, J. R., & Cook, D. J. (2017). Learning activity predictors from sensor data: Algorithms, evaluation, and applications. *IEEE Transactions on Knowledge and Data Engineering, 29*(12), 2744–2757. doi:10.1109/TKDE.2017.2750669 PMID:29456436

Mirzaei, P. A., & Haghighat, F. (2010). Approaches to study Urban Heat Island – Abilities and limitations. *Building and Environment, 45*(10), 2192–2201. doi:10.1016/j.buildenv.2010.04.001

Mohd Isa, N. K., Abdul Samad, Z., & Alias, A. (2014). A review on sustainability principles of building: Formulation of a theoretical framework. *Journal of Surveying, Construction & Property, 5*(1), 1–16. doi:10.22452/jscp/vol5no1.5

Mohd, H. A., & Norhidayah, S. (2016), Sustainable food production: insights of Malaysian halal small and medium sized enterprises. *International Journal of Production Economics, 181*(B), 303–314. . doi:10.1016/j.ijpe.2016.06.003

Mojtaba, P., Mastani, M., Joybari, A., Parham, A., Mirzaei, J., & Haghighat, F. (2019). *Urban heat island, urban climate maps and urban development policies and action plans.* Environmental Technology & Innovation. doi:10.1016/j.eti.2019.100341

Morawska, L., Tang, J. W., Bahnfleth, W., Bluyssen, P. M., Boerstra, A., Buonanno, G., Cao, J., Dancer, S., Floto, A., Franchimon, F., Haworth, C., Hogeling, J., Isaxon, C., Jimenez, J. L., Kurnitski, J., Li, Y., Loomans, M., Marks, G., Marr, L. C., ... Yao, M. (2020). How can airborne transmission of COVID-19 indoors be minimised? *Environment International, 142*, 105832. doi:10.1016/j.envint.2020.105832 PMID:32521345

Moura, D., Vasconcelos, E., Pacheco-Torgal, F., & Ding, Y. (2011). *Concrete repair with geopolymeric mortars: Influence of mortars composition on their workability and mechanical strength.* Academic Press.

Müller, C., Wyatt, S., Bieller, S., Qu, D., & Song, X. (2019). IFR Press Conference: Shanghai 18th September 2019. *International Federation of Robotics,* 1–32. https://ifr.org/downloads/press2018/IFR World Robotics Presentation - 18 Sept 2019.pdf

Municipalidad Provincial del Santa. (2012). Obtained from: https://eudora.vivienda.gob.pe/OBSERVATORIO/INSTRUM_GESTION/PAT/ESTUDIOS/SANTA/01-PAT%20SANTA-RESUMEN%20EJECUTIVO.pdf

Murray, N. (2017). Urban disaster risk governance. A systemic review. PPI-Centre. UCL Institute of Education.

Namkung, Y., & Jang, S. (2013). Effects of restaurant green practices on brand equity formation: Do green practices really matter? *International Journal of Hospitality Management, 33*(2), 85–95. doi:10.1016/j.ijhm.2012.06.006

Nash, V., & Christie, I. (2003). *Making Sense of Community*. IPPR. Retrieved from https://www.ippr.org/files/images/media/files/publication/2011/05/making_sense_of_community_1272.pdf

Neila, F., Sánchez-Guevara, C., Núñez, M., Román, E., Gómez, G., & De Luxán, M. (2020). *Urban heat island, urban climate and building energy consumption*. MODIFICA Project. García-Maroto Editores S.L.

Neila, J. (1990). *Arquitectura bioclimática*. Munilla Leira.

Ng, E., Tam, I., Ng, A., Givoni, B., Katzschner, L., Kwok, K., Murakami, S., Wong, N. H., Wong, K. S., Cheng, V., Davis, A., Tsou, J. Y., & Chow, B. (2004). *Final Report – Feasibility Study for Establishment of Air Ventilation Assessment System*. Technical Report for Planning Department HKSAR.

Ng, M. K. (2020). Sustainable Development Goals (SDGs) and Pandemic Planning. *Planning Theory & Practice, 21*(4), 507–512. doi:10.1080/14649357.2020.1807130

Nikolopoulou, M., Baker, N., & Steemers, K. (2001). Thermal comfort in outdoor urban spaces: Understanding the human parameter. *Solar Energy, 70*(3), 227–235. doi:10.1016/S0038-092X(00)00093-1

Nikolopoulou, M., & Lykoudis, S. (2007). Use of outdoor spaces and microclimate in a Mediterranean urban area. *Building and Environment, 42*(10), 3691–3707. doi:10.1016/j.buildenv.2006.09.008

Norton, T. A., Parker, S. L., Zacher, H., & Ashkanasy, N. M. (2015). Employee green behavior: A theoretical framework, multilevel review, and future research agenda. *Organization & Environment, 2015*(28), 103–125. doi:10.1177/1086026615575773

Noury, N., Herve, T., Rialle, V., Virone, G., Mercier, E., Morey, G., Moro, A., & Porcheron, T. (2000). Monitoring behavior in home using a smart fall sensor and position sensors. *1st Annual International IEEE-EMBS Special Topic Conference on Microtechnologies in Medicine and Biology Proceedings*. 10.1109/MMB.2000.893857

Nyeko, M. (2012). GIS and Multi-Criteria Decision Analysis for Land Use Resource Planning. *Journal of Geographic Information System, 4*(4), 341–348. Advance online publication. doi:10.4236/jgis.2012.44039

O'Connor, D., & Hou, D. (2020). 1: Green and Sustainable Remediation: concepts, principles, and pertaining research. In *Sustainable remediation of contaminated soil and groundwater: Materials, processes, and assessment* (pp. 43–73). Butterworth-Heinemann. doi:10.1016/B978-0-12-817982-6.00003-3

OECD. (2019). Case Study 8: Estonia e-government and the creation of a comprehensive data infrastructure for public services and agriculture policies implementation. In *Digital Opportunities for better agricultural policies* (pp. 207–213). Organisation for Economic Co-Operation and Development. doi:10.1787/510a82b5-en

OECD. (2020). *Latin American Economic Outlook 2020: Digital Transformation for Building Back Better*. OECD Publishing. doi:10.1787/e6e864fb-en

Oke, T. R. (1987). *Boundary Layer Climates*. Routledge.

Okoye, F. N., Prakash, S., & Singh, N. B. (2017). Durability of fly ash based geopolymer concrete in the presence of silica fume. *Journal of Cleaner Production, 149*, 1062–1067. doi:10.1016/j.jclepro.2017.02.176

Olgyay, V. (1963). *Design with Climate: Bioclimatic Approach to Architectural Regionalism, Manual de Arquitectura Bioclimática para arquitectos y urbanistas*. GG.

Olivia, M., & Nikraz, H. (2012). Properties of fly ash geopolymer concrete designed by Taguchi method. *Materials & Design, 36*, 191–198.

Ones, D. S., & Dilchert, S. (2012a). Employee green behaviors. In D. S. S. E. Jackson (Ed.), *Managing human resource for environmental sustainability* (pp. 85–116). Jossey-Bass.

Ortega Madrigal, L., García-Prieto Ruiz, A., & Serrano Lanzarote, B. (2013). *Catálogo de tipología edificatoria residencial.* Ámbito. https://episcope.eu/fileadmin/tabula/public/docs/brochure/ES_TABULA_TypologyBrochure_IVE.pdf

Oswald, S. L., Mossholder, K. W., & Harris, S. G. (1994). Vision salience and strategic involvement: Implications for psychological attachment to organization and job. *Strategic Management Journal, 1994*(15), 477–489. doi:10.1002mj.4250150605

Özcan, A., & Karakoç, M. B. (2019). The resistance of blast furnace slag-and ferrochrome slag-based geopolymer concrete against acid attack. *International Journal of Civil Engineering, 17*(10), 1571–1583. doi:10.100740999-019-00425-2

Pacheco-Torgal, F., Gomes, J. P., & Jalali, S. (2006). *Bond strength between concrete substance and repair materials: comparisons between tungsten mine waste geopolymeric binder versus current commercial repair products.* Academic Press.

Pacheco-Torgal, F., Moura, D., Ding, Y., & Jalali, S. (2011). Composition, strength and workability of alkali-activated metakaolin based mortars. *Construction & Building Materials, 25*(9), 3732–3745. doi:10.1016/j.conbuildmat.2011.04.017

Palomo, Á., Fernández-Jiménez, A., López-Hombrados, C., & Lleyda, J. L. (2011). Railway sleepers made of alkali activated fly ash concrete. *Revista de Ingeniería de Construcción, 22*(2), 75–80.

Palomo, A., Grutzeck, M. W., & Blanco, M. T. (1999). Alkali-activated fly ashes: A cement for the future. *Cement and Concrete Research, 29*(8), 1323–1329. doi:10.1016/S0008-8846(98)00243-9

Pan, M., Linner, T., Pan, W., Cheng, H., & Bock, T. (2020). Structuring the context for construction robot development through integrated scenario approach. *Automation in Construction, 114*(June). https://doi.org/10.1016/j.autcon.2020.103174

Pasini, M., Brondino, M., Trombin, R., & Filippi, Z. (2021). A participatory interior design approach for a restorative work environment: A research-intervention. *Frontiers in Psychology, 12*, 718446. Advance online publication. doi:10.3389/fpsyg.2021.718446 PMID:34603143

Peduzzi, P. (2014). Sand, rarer than one thinks. *Environmental Development, 11*, 208–218. doi:10.1016/j.envdev.2014.04.001

Pell, A. N. (1996). Fixing the leaky pipeline: Women scientists in academia. *Journal of Animal Science, 74*(11), 2843–2848. doi:10.2527/1996.74112843x PMID:8923199

Peng, M. X., Wang, Z. H., Shen, S. H., & Xiao, Q. G. (2015). Synthesis, characterization and mechanisms of one-part geopolymeric cement by calcining low-quality kaolin with alkali. *Materials and Structures, 48*(3), 699–708. doi:10.161711527-014-0350-3

Peng, M. X., Wang, Z. H., Xiao, Q. G., Song, F., Xie, W., Yu, L. C., Huang, H. W., & Yi, S. J. (2017). Effects of alkali on one-part alkali-activated cement synthesized by calcining bentonite with dolomite and Na2CO3. *Applied Clay Science, 139*, 64–71. doi:10.1016/j.clay.2017.01.020

Perea Moreno, L. (2015). *Hacia un análisis cuantitativo de la Ciudad Informal: una aproximación desde la Habitabilidad Básica y la experiencia en Makeni (Sierra Leona).* Arquitectura. Retrieved from https://oa.upm.es/39873/

Perea Moreno, L., Salas Ruiz, A., & Arana Giralt, J. (2018). Grandes retos y aplicación de herramientas participativas en la planificación de la ciudad informal. El caso de Makeni, Sierra Leona. *Revista Urbano.*

Pereira, Á., & Vence, X. (2012). Key business factors for eco-innovation: An overview of recent firm-level empirical studies. *Cuadernos de Gestión, 12*, 73–103. doi:10.5295/cdg.110308ap

Pérez-Martín, J. C. (2014). *MEREDICTE Método de evaluación de riesgo en caso de incendio en el marco del Código Técnico de la Edificación*. Academic Press.

Pérez-Martín, J. C., Díaz-Díaz, R., & Santos García, R. (2010). Método de evaluación del riesgo de incendio en el marco del Código Técnico de la Edificación. *DYNA Ingeniería e Industria, 85*(4), 303–314. doi:10.6036/3437

Petersen, J., Larimer, N., Kaye, J. A., Pavel, M., & Hayes, T. L. (2012). SVM to detect the presence of visitors in a smart home environment. *2012 Annual International Conference of the IEEE Engineering in Medicine and Biology Society*. 10.1109/EMBC.2012.6347324

Petersen, J., Thielke, S., Austin, D., & Kaye, J. (2015). Phone behaviour and its relationship to loneliness in older adults. *Aging & Mental Health, 20*(10), 1084–1091. doi:10.1080/13607863.2015.1060947 PMID:26133745

Phoo-ngernkham, T., Maegawa, A., Mishima, N., Hatanaka, S., & Chindaprasirt, P. (2015). Effects of sodium hydroxide and sodium silicate solutions on compressive and shear bond strengths of FA–GBFS geopolymer. *Construction & Building Materials, 91*, 1–8. doi:10.1016/j.conbuildmat.2015.05.001

Phoo-ngernkham, T., Sata, V., Hanjitsuwan, S., Ridtirud, C., Hatanaka, S., & Chindaprasirt, P. (2015). High calcium fly ash geopolymer mortar containing Portland cement for use as repair material. *Construction & Building Materials, 98*, 482–488. doi:10.1016/j.conbuildmat.2015.08.139

Phoon, P. H., MacLaren, G., Ti, L. K., Tan, J. S., & Hwang, N. C. (2019). History and current status of cardiac anesthesia in Singapore. *Journal of Cardiothoracic and Vascular Anesthesia, 33*(12), 3394–3401. doi:10.1053/j.jvca.2018.07.018 PMID:30131218

Pinto, T., Praça, I., Vale, Z., Morais, H., & Sousa, T. M. (2013). Strategic bidding in electricity markets: An agent-based simulator with game theory for scenario analysis. *Integrated Computer-Aided Engineering, 20*(4), 335–346. doi:10.3233/ICA-130438

Plattform Produktives Stadtgrün. (2020). Available online: https://www.berlin.de/gemeinschaftsgaertnern/

Pollo, R., Trane, M., & Giovanardi, M. (2020). Urban Metabolism, modelli interdisciplinari e progetto a scala microurbana. Firenze University Press. doi:10.36253/techne-9857

Popescu, G., & Patrascoiu, R. (2012). Brownfield sites - Between abandonment and redevelopment case study: Craiova city. *Human Geographies, 6*(1), 91–97. doi:10.5719/hgeo.2012.61.91

Premier, A. (2012). Dynamic facades and smart technologies for building envelope requalification. *Screencity-International Academic Journal, 1*, 65–69.

Prud'Homme, E., Michaud, P., Joussein, E., Peyratout, C., Smith, A., & Rossignol, S. (2011). In situ inorganic foams prepared from various clays at low temperature. *Applied Clay Science, 51*(1–2), 15–22. doi:10.1016/j.clay.2010.10.016

PwC. (2017). *Building smart - How digital technology can help construction companies achieve more value*. Orange Business Services. Retrieved January 4, 2022, from https://www.orange-business.com/en/library/white-paper/building-smart

PWC. (2019). *Estonia – the Digital Republic secured by Blockchain*. PWC. Retrieved March 16, 2022, from https://www.pwc.com/gx/en/services/legal/tech/assets/estonia-the-digital-republic-secured-by-blockchain.pdf

Quintero Sepúlveda, I. C., Ospina Nieto, Y., Quiroga Parra, D., & Cubillo-González, R.-A. (2021). Relación entre Capacidad de Innovación e Índice de Innovación en América Latina. *Journal of Technology Management & Innovation, 16*(3), 47–56. https://doi.org/https://doi.org/10.4067/S0718-27242021000300047

Ragon, M. (1978). *Histoire mondiale de l'architecture et de l'urbanisme modernes*. Casterman.

Ranjbar, N., Mehrali, M., Maheri, M. R., & Mehrali, M. (2017). Hot-pressed geopolymer. *Cement and Concrete Research, 100*, 14–22. doi:10.1016/j.cemconres.2017.05.010

Raudeliuniene, J., Davidaviˇciene, V., & Jakubaviˇcius, A. (2018). Knowledge management process model. *Entrep. Sustain. Issues, 5*, 542–554.

Real Decreto 1053/1985, de 25 de mayo, sobre ordenación de la estadísticas de las actuaciones de los Servicios contra Incendios y de Salvamento. (1985). In *Boletín Oficial del Estado* (Vol. 158, Issue 12771, pp. 20843–20844).

Real Decreto 2059/1981 por el que se aprueba la Norma Básica de la Edificación. (1981). In *Boletín Oficial del Estado* (Vol. 224, pp. 21707–21727).

Real Decreto 3062/1973, de 19 de octubre, por el que se aprueba la instrucción para el proyecto y la ejecución de obras de hormigón en masa o armado. (1973). In *Boletín Oficial del Estado* (Vol. 2993, pp. 23701–24204). https://www.boe.es/boe/dias/1973/12/07/pdfs/R23701-24204.pdf

Real Decreto 513/2017, de 22 de mayo, por el que se aprueba el Reglamento de instalaciones de protección contra incendios. (2017). In *Boletín Oficial del Estado* (Vol. 139, Issue I, pp. 48349–48386). http://www.boe.es

Real Decreto 842/2013, de 31 de octubre, por el que se aprueba la clasificación de los productos de construcción y de los elementos constructivos en función de sus propiedades de reacción y de resistencia frente al fuego. (2013). In *Boletín Oficial del Estado* (Vol. 281, Issue I, pp. 93493–93527).

Redman, C. (2014). Should sustainability and resilience be combined or remain distinct pursuits? *Ecology and Society, 19*(2), 190–202. doi:10.5751/ES-06390-190237

REHVA. (2020). COVID, R. (19). guidance document, April 3, 2020. REHVA, Federation of European Heating, Ventilation and Air Conditioning Association, Brussels, Belgium.

Reisinger, A., & Clark, H. (2018). How much do direct livestock emissions contribute to global warming? *Global Change Biology, 2018*(24), 1749–1761. doi:10.1111/gcb.13975 PMID:29105912

Research and Markets. (2021, December 13). *Global Smart Water Management Market Forecast to reach USD 22.4 billion by 2026*. Cision PR Newswire. Retrieved March 6, 2022, from https://www.prnewswire.com/news-releases/global-smart-water-management-market-forecast-to-reach-usd-22-4-billion-by-2026--301443307.html

Rieira, P. (2018). *Beneficio social del Pasillo Verde Ferroviario de Madrid*. Instituto Universitario de Estudios Europeos. Universidad Autónoma de Barcelona. Editorial Noesis. (Original work published 1995)

Rietveld, L. C., Siri, J. G., Chakravarty, I., Arsénio, A. M., Biswas, R., & Chatterjee, A. (2016). Improving health in cities through systems approaches for urban water management. *Environmental Health: A Global Access Science Source, 15*. doi:10.1186/s12940-016-0107-2

Rocío Santo-Tomás & Rodríguez Romero. (2020). *City thresholds. The role of urban green infrastructures in Madrid*. ZARCH 14. doi:10.26754/ojs_zarch/zarch.2020144289

Rodríguez Romero, E. J. (Dir.). (2018). Paisajes de aproximación a la ciudad de Madrid. Conarquitectura.

Roehrs, A., da Costa, C. A., da Rosa Righi, R., da Silva, V. F., Goldim, J. R., & Schmidt, D. C. (2019). Analyzing the performance of a blockchain-based personal health record implementation. *Journal of Biomedical Informatics, 92*, 103140. doi:10.1016/j.jbi.2019.103140 PMID:30844481

Rohde, D., Corcoran, J., Sydes, M., & Higginson, A. (2016). The association between smoke alarm presence and injury and death rates: A systematic review and meta-analysis. *Fire Safety Journal, 81*, 58–63. doi:10.1016/j.firesaf.2016.01.008

Rojas, C., Aldana-Domingue, J., Munizaga, J., Moschella, P., Martínez, C., & Stamm, C. (2020). Urban Wetland trends in three Latin American Cities during the latest decades (2002): Concón (Chile), Barranquilla (Colombia) and Lima (Perú). *Wetlands Science & Practice, 37*(4), 283–293.

Rojas, C., Munizaga, J., Rojas, O., Martínez, C., & Pino, J. (2019). Urban development versus wetland loss in a coastal Latin American city: Lessons for sustainable land use planning. *Land Use Policy, 80*, 47–56. doi:10.1016/j.landusepol.2018.09.036

Rop, R., & Ufanisi, M. (2004). Kenya - School Health Clubs: Can they change hygiene behaviours? In M. Snel & IRC (Eds.), *The Worth of School Sanitation and Hygiene Education (SSHE)*. Retrieved from https://www.ircwash.org/

Rosario, D. K., Mutz, Y. S., Bernardes, P. C., & Conte-Junior, C. A. (2020). Relationship between COVID-19 and weather: Case study in a tropical country. *International Journal of Hygiene and Environmental Health, 229*, 113587. doi:10.1016/j.ijheh.2020.113587 PMID:32917371

Rossato, A. (2018). *Is Latin America ready for the Fourth Industrial Revolution?* World Economic Forum (WEF). https://www.weforum.org/agenda/2018/03/industry-4-0-fourth-industrial-revolution-is-latin-america-ready/

Rowe, D. B. (2011). Green Roofs as a means of pollution abatement. *Environmental Pollution, 159*(8-9), 2100–2110. doi:10.1016/j.envpol.2010.10.029 PMID:21074914

Rowles, M., & O'connor, B. (2003). Chemical optimisation of the compressive strength of aluminosilicate geopolymers synthesised by sodium silicate activation of metakaolinite. *Journal of Materials Chemistry, 13*(5), 1161–1165. doi:10.1039/b212629j

Roxburgh, H. (2016, August 23). Inside Shanghai Tower: China's tallest skyscraper claims to be world's Greenest. *The Guardian*. Retrieved January 4, 2022, from https://www.theguardian.com/cities/2016/aug/23/inside-shanghai-tower-china-tallest-building-green-skyscrapers

Rubio del Val, J. (2011). Rehabilitación Urbana en España (1989-2010). Barreras actuales y sugerencias para su eliminación. *Informes de La Construccion, 63*, 5–20. doi:10.3989/ic.11.060

Rugumayo, A. (2004). Scaling up School Sanitation Programmes. In M. Snel & IRC (Eds.), The Worth of School Sanitation and Hygiene Education (SSHE). Academic Press.

Russo, A., & Cirella, G. (2018). Modern compact cities: How much greenery do we need? *International Journal of Environmental Research and Public Health, 15*(10), 2180. doi:10.3390/ijerph15102180 PMID:30301177

Şahin, F., Uysal, M., & Canpolat, O. (2021). Systematic evaluation of the aggregate types and properties on metakaolin based geopolymer composites. *Construction & Building Materials, 278*, 122414. doi:10.1016/j.conbuildmat.2021.122414

Sarni, W. (2010). *Greening brownfields: Remediation through sustainable development*. McGraw-Hill.

Sartori, I., & Hestnes, A. G. (2007). Energy use in the life cycle of conventional and low-energy buildings: A review article. *Energy and Building, 39*(3), 249–257. doi:10.1016/j.enbuild.2006.07.001

Save the Children. (2009). *Improving water and sanitation in schools and communities. Successes and lessons learned from Nasirnagar, Bangladesh.* Retrieved from https://www.ircwash.org/

SAVIArquitectura & Worcester Polythecnic Institute. (2016). *Proyecto prestaRener: Protocolo de Actuación en la Rehabilitación de la Envolvente de Edificios basado en el Diseño Prestacional.* Author.

SAVIArquitectura. (2016). *Proyecto prestaRener. Capítulo 3: Tipologías.* Author.

Savickiene, J., & Miceikiene, A. (2018). *Sustainable economic development assessment model for family farms* (Vol. 64). Agricultural Economics – Czech. doi:10.17221/310/2017-AGRICECON

Sayadi, S., Tsatsaronis, G., & Morozyuk, T. (2016). CLIMA 2016 - 12th REHVA World Congress. In *Exergy-based Control Strategies for HVAC Systems* (Vol. 8). Academic Press.

Schlueter, A., & Thesseling, F. (2009). Building Information Model Based Energy/Exergy Performance Assessment in early design stages. *Automation in Construction, 18*(2), 153–163. doi:10.1016/j.autcon.2008.07.003

Science for Environmental Policy, In Depth Report, The Multifunctionality of Green Infrastructure. (2012). Obtained from: https://ec.europa.eu/environment/nature/ecosystems/docs/Green_Infrastructure.pdf

Shafique, M., Kim, R., & Rafiq, M. (2018). Green roof benefits, opportunities, and challenges– A review. *Renewable and Sustainable Energy Reviews, 90*, 757-773.

Shevchenko, A., & Petrenko, O. (2020). Current state of micro and small agribusiness in Ukraine. *Agricultural and Resource Economics, 6*(1), 146–160. doi:10.51599/are.2020.06.01.10

Shoeb, A., & Nisar, T. (2015). Green Human Resource Management: Policies and practices. *Cogent Business & Management, 2*(1). https://www.tandfonline.com/doi/full/10.1080/23311975.2015.1030817 doi:10.1080/23311975.2015.1030817

Shuai, Q., Xu, Z., Yao, Z., Chen, X., Jiang, Z., Peng, X., An, R., Li, Y., Jiang, X., & Li, H. (2020). Fire resistance of phosphoric acid-based geopolymer foams fabricated from metakaolin and hydrogen peroxide. *Materials Letters, 263*, 127228. doi:10.1016/j.matlet.2019.127228

Sica, P. (1976). *Storia dell'urbanistica. L' Ottocento.* Editori Laterza.

Sierla, S., Azangoo, M., Rainio, K., Papakonstantinou, N., Fay, A., Honkamaa, P., & Vyatkin, V. (2021). Roadmap to semi-automatic generation of Digital Twins for Brownfield Process Plants. *Journal of Industrial Information Integration, 100282*. Advance online publication. doi:10.1016/j.jii.2021.100282

Sierra Leone 2015 Population and Housing Census. (2016). *Statistics Sierra Leone.* Retrieved from https://www.statistics.sl/images/StatisticsSL/Documents/final-results_-2015_population_and_housing_census.pdf

Singh, S. K., Del Giudice, M., Chierici, R., & Graziano, D. (2020). Green innovation and environmental performance: The role of green transformational leadership and green human resource management. *Technological Forecasting and Social Change, 150*, 119762. doi:10.1016/j.techfore.2019.119762

Skydan, O., Nykolyuk, O., Pyvovar, P., & Martynchuk, I. (2020). Methodological approach to the evaluation of agricultural business system flexibility. *Management Theory and Studies for Rural Business and Infrastructure Development, 41*(4), 444–462. doi:10.15544/mts.2019.36

Smart Cities World. (2017, October 2). *The world's most sustainable office.* Smart Cities World. Retrieved January 4, 2022, from https://www.smartcitiesworld.net/news/the-worlds-most-sustainable-office-2167

Smart Energy International. (2021, March 22). *The role of smart buildings in the development of Smart Cities*. Smart Energy International. Retrieved January 4, 2022, from https://www.smart-energy.com/regional-news/asia/the-role-of-smart-buildings-in-the-development-of-smart-cities/

Smartnet. (n.d.). *Aiming Life: Automated Green Corridor for quickest passage of ambulances on roadways, saving crucial time and lives*. Smartnet. Retrieved December 30, 2021, from https://smartnet.niua.org/content/c96337fd-d153-4a65-8fdb-be1e485187a3

Solxenergy. (n.d.). *Solxenergy*. Downloads. Retrieved January 5, 2022, from https://www.solxenergy.com/index.php/downloads

Somna, K., Jaturapitakkul, C., Kajitvichyanukul, P., & Chindaprasirt, P. (2011). NaOH-activated ground fly ash geopolymer cured at ambient temperature. *Fuel*, *90*(6), 2118–2124. doi:10.1016/j.fuel.2011.01.018

Stevenson, M. R., & Lee, A. H. (2003). Smoke alarms and residential fire mortality in the United States: An ecologic study. *Fire Safety Journal*, *38*(1), 43–52. doi:10.1016/S0379-7112(02)00048-6

Stone, A. (2020, August 12). *How smart technology can create healthy office buildings*. Technology Solutions That Drive Government. Retrieved December 30, 2021, from https://fedtechmagazine.com/article/2020/08/how-smart-technology-can-create-healthy-office-buildings-perfcon

Stovin, V., Vesuviano, G., & Kasmin, H. (2012). The hydrological performance of a green roof test bed under UK climatic conditions. *Journal of Hydrology (Amsterdam)*, *414*, 148–161. doi:10.1016/j.jhydrol.2011.10.022

Strande, L. (2014). The Global Situation. In L. Strande & D. Brdjanovic (Eds.), *Faecal sludge management: Systems approach for implementation and operation* (pp. 1–18). IWA Publishing. doi:10.2166/9781780404738

Stubbs, W., & Cocklin, C. (2008). Conceptualizing a sustainability business model. *Org.Env.*, *21*(2), 103-127.

Sturm, P., Gluth, G. J. G., Jäger, C., Brouwers, H. J. H., & Kühne, H.-C. (2018). Sulfuric acid resistance of one-part alkali-activated mortars. *Cement and Concrete Research*, *109*, 54–63. doi:10.1016/j.cemconres.2018.04.009

Suárez, J., & Puertas, J. (2005). Determination of COD, BOD, and Suspended Solid Loads During CSO Events in Some Combined Catchments in Spain. Ecological Engineering, 24, 199-217.

Subtil, M., Matos, J. S., & Santos, R. F. (2018). Conceptual Approach for the Modernization of Sanitation Systems in Peri-Urban Slums and Its Application in Maputo City, Mozambique. INCREaSE. doi:10.1007/978-3-319-70272-8_48

Sukopp, H., & Werner, P. (1989). *Naturaleza en las ciudades. Desarrollo de flora y fauna en áreas urbanas*. Ministerio de Obras Públicas y Urbanismo.

Sutton, R. (2014). Aesthetics for green roofs and green walls. *The Journal of Living Architecture*, *1*(2), 1–20. doi:10.46534/jliv.2014.01.02.001

Swan, M. (2009). Emerging patient-driven health care models: An Examination of Health Social Networks, consumer personalized medicine and quantified self-tracking. *International Journal of Environmental Research and Public Health*, *6*(2), 492–525. doi:10.3390/ijerph6020492 PMID:19440396

Tabilo, E., Burmeister, J., Chávez, C., & Zöckler, C. (2016). *Humedales y aves migratorias en la costa árida del Pacífico Sudamericano. Etapa I: Evaluación Ecológica Rápida*. Centro Neotropical de Entrenamiento en Humedales y Manfred Hermsen Stiftung de Alemania.

Taha, B., & Nounu, G. (2008). Using lithium nitrate and pozzolanic glass powder in concrete as ASR suppressors. *Cement and Concrete Composites*, *30*(6), 497–505. doi:10.1016/j.cemconcomp.2007.08.010

Tammam, Y., Uysal, M., & Canpolat, O. (2021). Effects of alternative ecological fillers on the mechanical, durability, and microstructure of fly ash-based geopolymer mortar. *European Journal of Environmental and Civil Engineering*, 1–24. doi:10.1080/19648189.2021.1925157

Tanner, B., Tilse, C., & de Jonge, D. (2008). Restoring and sustaining home: The impact of home modifications on the meaning of home for older people. *Journal of Housing for the Elderly*, *22*(3), 195–215. doi:10.1080/02763890802232048

Tay, V. (2021, July 9). *DBS to turn Newton office building into net zero development*. Businesstimes.com.sg. Retrieved January 5, 2022, from https://www.businesstimes.com.sg/banking-finance/dbs-to-turn-newton-office-building-into-net-zero-development

Tayeh, B. A., Zeyad, A. M., Agwa, I. S., & Amin, M. (2021). Effect of elevated temperatures on mechanical properties of lightweight geopolymer concrete. *Case Studies in Construction Materials, 15*.

Tchakouté, H. K., & Rüscher, C. H. (2017). Mechanical and microstructural properties of metakaolin-based geopolymer cements from sodium waterglass and phosphoric acid solution as hardeners: A comparative study. *Applied Clay Science*, *140*, 81–87. doi:10.1016/j.clay.2017.02.002

TechNative. (2018, September 11). *Smart cities: How they'll make us healthier*. TechNative. Retrieved December 22, 2021, from https://technative.io/smart-cities-how-theyll-make-us-healthier/

TEP. (2005). *Advancing the delivery of green infrastructure Targeting Issues in England's Northwest*. Issue Paper, England Northwest.

The Economist. (n.d.). *Wellbeing in the smart office - interact - empowering spaces*. The Economist Intelligence Unit. Retrieved December 30, 2021, from https://empoweringspaces.economist.com/wellbeing-in-the-smart-office/

The World Bank. (2021). Retrieved January 8, 2022, from https://data.worldbank.org/country/sierra-leone

Thieme, W. (2019, April 3). *3 practical ways to transform brownfield plants into digital factories*. IIoT World. Retrieved January 4, 2022, from https://iiot-world.com/connected-industry/3-practical-ways-to-transform-brownfield-plants-into-digital-factories/

Thornton, G., Franz, M., Edwards, D., Pahlen, G., & Nathanail, P. (2007). The challenge of sustainability: Incentives for brownfield regeneration in Europe. *Environmental Science & Policy*, *10*(2), 116–134. doi:10.1016/j.envsci.2006.08.008

Thorpe, D. (2015, February 2). *How can cities cut energy use by 50%*. World Economic Forum. Retrieved February 28, 2022, from https://www.weforum.org/agenda/2015/02/how-can-cities-cut-energy-use-by-50/

ThoughtWire. (2019, February 14). *What is OPM and How is it Transforming Buildings?* Thoughts Over the Wire. Retrieved January 4, 2022, from https://blog.thoughtwire.com/what-is-opm-and-how-is-it-transforming-buildings

Tilley, E., Ulrich, L., Lüthi, C., Reymond, P., Schertenleib, R., & Zurbrügg, C. (2014). *Compendium of sanitation systems and technologies*. Eawag. Retrieved from https://www.eawag.ch/en/department/sandec/publications/compendium/

Tindale, L., Coombe, M., Stockdale, J. E., Garlock, E., Lau, W. Y. V., Saraswat, M., . . . Colijn, C. (2020). Transmission interval estimates suggest pre-symptomatic spread of COVID-19. MedRxiv. doi:10.1101/2020.03.03.20029983

Tizouiar, O., & Boussoualim, A. (2018). The approach used for comfortable and high efficient buildings in sustainable cities by different actors. *Renewable Energy for Smart and Sustainable Cities*, 3–13. doi:10.1007/978-3-030-04789-4_1

Toli, A. M., & Murtagh, N. (2020). The concept of sustainability in Smart City definitions. *Frontiers in Built Environment - Sustainable Design and Construction, 6*. doi:10.3389/fbuil.2020.00077

Tomashuk, I.V., & Baldynyuk, V.M. (2021). Identification of problems and prospects of rural infrastructure development of Ukraine. *Economic Sciences*, 13. doi:10.24412/2520-6990-2021-13100-58-70

Totty, M. (2017, April 17). The rise of the smart city. *The Wall Street Journal*. Retrieved December 29, 2021, from https://www.wsj.com/articles/the-rise-of-the-smart-city-1492395120

Triton-Market-Research. (2021). *Latin America Industrial Robots Market 2021-2028*. Research and Markets. https://www.researchandmarkets.com/reports/5459459/latin-america-industrial-robots-market-2021-2028

Tsai, W., & Ghoshal, S. (1998). Social capital and value creation: The role of intrafirm networks. *Academy of Management Journal*, *1998*(41), 464–476.

Tuenge, J., Kelly, K., & Poplawski, M. (2019, January). *Connected Lighting Systems Efficiency Study -- POE cable Energy Losses, Part 1*. US Dept. of Energy. Retrieved January 7, 2022, from https://www.energy.gov/sites/prod/files/2019/01/f58/cls_poe-cable-pt1.pdf

Ugai, T. (2016). Evaluation of sustainable roof from various aspects and benefits of agriculture roofing in urban core. *Procedia: Social and Behavioral Sciences*, *216*, 850–860. doi:10.1016/j.sbspro.2015.12.082

Uludag, C. (2016, October 23). *Why do we need being smart in healthcare?* SMART Cities. Retrieved December 20, 2021, from https://ilab.city/why-do-we-need-being-smart-in-healthcare/

UN Environment Programme and International Energy Agency. (2017). *Towards a zero-emission, efficient, and resilient buildings and construction sector*. Global Status Report 2017.

UN WCED. (1989). *Our common future: Brundtland report*. Oxford University Press.

UN. (2016, May 18). *Habitat launches the World Cities Report 2016, Urbanization and Development: Emerging Futures: UN-Habitat*. UN Habitat For a Better Urban Future. Retrieved January 3, 2022, from https://unhabitat.org/un-habitat-launches-the-world-cities-report-2016

UNDP. (2020). *Environmentally Sustainable Operations*. https://www.undp.org/accountability/social-and-environmental-responsibility/sustainable-operations

UNEP. (2020, December 16). *Building sector emissions hit record high, but low-carbon pandemic recovery can help transform sector – UN report*. Retrieved January 3, 2022, from https://www.unep.org/news-and-stories/press-release/building-sector-emissions-hit-record-high-low-carbon-pandemic

United Nations, Department of Economic and Social Affairs, Population Division. (2019). *World population ageing 2019 - Highlights* (ST/ESA/SER.A/430). Retrieved December 30, 2021, from https://www.un.org/en/development/desa/population/publications/pdf/ageing/WorldPopulationAgeing2019-Report.pdf

United Nations. (2019). *The Sustainable Development Goals Report 2019*. UN. Retrieved from https://unstats.un.org/sdgs/report/2019/The-Sustainable-Development-Goals-Report-2019.pdf

Urban Land Institute. (2021). *State of Green: Greenprint Performance Report* (Vol. 12). Urban Land Institute.

US EIA. (2017, September 14). *International Energy Outlook 2017*. Retrieved January 5, 2022, from https://www.eia.gov/outlooks/ieo/pdf/0484(2017).pdf

US EPA. (n.d.). *Sustainability*. EPA. Retrieved January 3, 2022, from https://www.epa.gov/sustainability/

US GSA. (2011). *Green building performance - general services administration*. GSA Public Buildings Service. Retrieved January 7, 2022, from https://www.gsa.gov/cdnstatic/Green_Building_Performance2.pdf

USE. (2016). *27 Delvalle: An urban health living lab.* Urban Sustainability Exchange. Retrieved March 16, 2022, from https://use.metropolis.org/case-study/become-expert/2176

Valencia-Saavedra, W. G., Mejía de Gutiérrez, R., & Puertas, F. (2020). Performance of FA-based geopolymer concretes exposed to acetic and sulfuric acids. *Construction & Building Materials, 257*, 119503. doi:10.1016/j.conbuildmat.2020.119503

Van Holt, T., Statler, M., Atz, U., Whelan, T., van Loggerenberg, M., & Cebulla, J. (2020). The cultural consensus of sustainability-driven innovation: Strategies for success. *Business Strategy and the Environment, 29*(8), 3399–3409. doi:10.1002/bse.2584

Vásquez, A., Devoto, C., Giannotti, E., & Velásquez, P. (2016). Green Infrastructure Systems Facing Fragmented Cities in Latin America - Case of Santiago, Chile. Procedia Engineering, 161, 1410–1416. doi:10.1016/j.proeng.2016.08.602

Vearey, J., Luginaah, I., Magitta, N. F., Shilla, D. J., & Oni, T. (2019). Urban health in Africa: A critical global public health priority. *BMC Public Health, 19*(1), 340. Advance online publication. doi:10.118612889-019-6674-8 PMID:30909893

Velandia, D. F., Lynsdale, C. J., Provis, J. L., Ramirez, F., & Gomez, A. C. (2016). Evaluation of activated high volume fly ash systems using Na2SO4, lime and quicklime in mortars with high loss on ignition fly ashes. *Construction & Building Materials, 128*, 248–255. doi:10.1016/j.conbuildmat.2016.10.076

Ventura, S., Niemann, B., Miskowiak, D., & Heinzel, M. (1998). *Environmental Corridors: Land Use Planning Guide.* Land Information & Computer Graphics Facility - University of Wisconsin – Madison.

Verma, U. (2018, December 14). *How digital twins unlock value beyond BMS in commercial office buildings.* In-Building Tech. Retrieved January 4, 2022, from https://inbuildingtech.com/smart-buildings/digital-twins-opm-thoughtwire/

Verma, U. (2018, November 21). *PropTech: How digital twins impact OPM in smart buildings.* In-Building Tech. Retrieved January 4, 2022, from https://inbuildingtech.com/uncategorized/digital-twins-proptech/

Vijayaraghavan, K. (2016). Green roofs: A critical review on the role of components, benefits, limitations, and trends. *Renewable & Sustainable Energy Reviews, 57*, 740–752. doi:10.1016/j.rser.2015.12.119

VisitBerlin. (Ed.). (2017). *12 mal Berliner Leben, 12 mal Berlin Erleben. Konzept für einen stadtverträglichen Berlin-Tourismus 2018+.* VisitBerlin. Available online: https://about.visitberlin.de/tourismuskonzept-2018

Wachowicz, M. (2013, October 28). *New Frontiers for Geomatics.* GIM International. Retrieved December 19, 2021, from https://www.gim-international.com/content/article/new-frontiers-for-geomatics

Wang, H., Li, H., & Yan, F. (2005). Synthesis and tribological behavior of metakaolinite-based geopolymer composites. *Materials Letters, 59*(29–30), 3976–3981. doi:10.1016/j.matlet.2004.08.049

Wang, N., & Adeli, H. (2014). Sustainable building design. *Journal of Civil Engineering and Management, 20*(1), 1–10. doi:10.3846/13923730.2013.871330

Wang, Y., Hu, H., Dai, W., & Burns, K. (2021). Evaluation of industrial green development and industrial green competitiveness: Evidence from Chinese urban agglomerations. *Ecological Indicators, 124*, 107371. doi:10.1016/j.ecolind.2021.107371

Warda, L., Tenenbein, M., & Moffatt, M. E. K. (1999). House fire injury prevention update. Part II. A review of the efectiveness of preventive interventions. *Injury Prevention, 5*(3), 217–225. doi:10.1136/ip.5.3.217 PMID:10518271

Washburn, D., Sindhu, U., Balaoura, S., Dines, R. A., Hayes, N. M., & Nelson, L. E. (2010). *Helping CIOs Understand "Smart City" Initiatives.* Forrester Research, Inc.

WBCSD - EEB. (2009). *Energy Efficiency in Buildings*. World Business Council for Sustainable Development. Retrieved February 28, 2022, from https://docs.wbcsd.org/2008/07/EEB-FactsTrends-FullReport.pdf

WEF. (2021, January). *Net Zero Carbon Cities: An Integrated Approach - Insight Report*. World Economic Forum. Retrieved January 5, 2022, from https://www3.weforum.org/docs/WEF_Net_Zero_Carbon_Cities_An_Integrated_Approach_2021.pdf

Wesco. (2020, February 10). *The ROI of sustainable intelligent buildings*. WESCO International, Inc. Retrieved January 11, 2022, from https://blog.wesco.com/the-roi-of-sustainable-intelligent-buildings

WGBC. (2016). *Towards zero-emission efficient and resilient buildings - Global Status Report 2016*. Global Alliance for Buildings and Construction. Retrieved February 28, 2022, from https://www.worldgbc.org/sites/default/files/GABC_Global_Status_Report_V09_november_FINAL.pdf

WGBC. (2018). *World Green Building Trends in 2018: Green keeps growing*. U.S. Green Building Council. Retrieved March 3, 2022, from https://www.usgbc.org/articles/world-green-building-trends-2018-green-keeps-growing

WGBC. (n.d.a). *What is green building?* World Green Building Council. Retrieved January 3, 2022, from https://www.worldgbc.org/what-green-building

WGBC. (n.d.b). *The benefits of Green Buildings*. World Green Building Council. Retrieved February 28, 2022, from https://www.worldgbc.org/benefits-green-buildings#:~:text=The%20world%20over%2C%20evidence%20is,communities%2C%20and%20driving%20economic%20growth

WHO. (2016). *Ambient (outdoor) air quality and health*. https://www.who.int/es/news-room/fact-sheets/detail/ambient-(outdoor)-air-quality-and-health

WHO. (2018). *Social Determinants of Health*. World Health Organization. Retrieved December 18, 2021, from https://www.who.int/health-topics/social-determinants-of-health#tab=tab_1

WHO. (2020a). *Urban planning crucial for Better Public Health in cities*. World Health Organization. Retrieved December 27, 2021, from https://www.who.int/news-room/feature-stories/detail/urban-planning-crucial-for-better-public-health-in-cities

WHO. (2020b). *Integrating health in urban and Territorial Planning: A sourcebook*. UN-HABITAT and World Health Organization. Retrieved December 31, 2021, from https://apps.who.int/iris/handle/10665/331678?locale-attribute=en&

WHO. (2021). *Global Strategy on Digital Health 2020-2025*. Geneva: World Health Organization. Retrieved March 18, 2022, from https://apps.who.int/iris/bitstream/handle/10665/344249/9789240020924-eng.pdf

Wicklund, E. (2020, March 23). *MD City integrates telehealth with Smart City Hub to monitor coronavirus*. mHealth-Intelligence. Retrieved March 11, 2022, from https://mhealthintelligence.com/news/md-city-integrates-telehealth-with-smart-city-hub-to-monitor-coronavirus

Wiering, M., Liefferink, D., Boezeman, D., Kaufmann, M., Crabbé, A., & Kurstjens, N. (2020). The Wicked Problem the Water Framework Directive Cannot Solve. The Governance Approach in Dealing with Pollution of Nutrients in Surface Water in the Netherlands, Flanders, Lower Saxony, Denmark, and Ireland. *Water (Basel)*, *12*(5), 1240. doi:10.3390/w12051240

Wirtenberg, J., Harmon, K. D., Russell, W. G., & Fairfield, K. D. (2007). HR's role in building a sustainable enterprise. *Human Resource Planning*, *30*, 10–20.

Witthoeft, S., & Kosta, I. (2017). *Shaping the Future of Construction. Inspiring innovators redefine the industry.* World Economic Forum (WEF). https://www3.weforum.org/docs/WEF_Shaping_the_Future_of_Construction_Inspiring_Innovators_redefine_the_industry_2017.pdf

Wolfson, A. (2018). *In Sustainability and the City: The Service Approach.* Business Expert Press.

Wong, D. (2019, January 30). NUS launches Singapore's first net-zero energy building to be built from scratch. *The Straits Times*. Retrieved January 4, 2022, from https://www.straitstimes.com/singapore/environment/nus-launches-singapores-first-net-zero-energy-building-to-be-built-from

Wood, A. (2008). Green or grey? The aesthetics of tall building sustainability. *Proc. of the CTBUH 8th World Congress*, 194–202.

World Cities Report 2016: Urbanization and Development - Emerging Futures. (2016). *United Nations Fund for Population Activities.* Retrieved from https://unhabitat.org/world-cities-report

World Economic Forum. (2019). *Making Affordable Housing a Reality in Cities.* World Economic Forum. homeguides.sfgate.com/student-hous

World Health Organization. (2020). *Supporting healthy urban transport and mobility in the context of COVID-19*. Author.

Wysocki, J. (2021). Innovative green initiatives in the manufacturing SME sector in Poland. *Sustainability*, *13*(4), 2386. doi:10.3390u13042386

Xie, T., & Ozbakkaloglu, T. (2015). Behavior of low-calcium fly and bottom ash-based geopolymer concrete cured at ambient temperature. *Ceramics International*, *41*(4), 5945–5958. doi:10.1016/j.ceramint.2015.01.031

Xu, L., Wang, X., Guan, C., Wu, W., & Zhang, L. (2020). The effect of activators on the mechanical properties and microstructure of alkali-activated nickel slag. *Advances in Civil Engineering*, *2020*, 2020. doi:10.1155/2020/1764108

Yahya, M. Y., Lee Hui, Y., Yassin, A. B. M., Omar, R., Robin, R. O., & Kasim, N. (2019). The Challenges of the Implementation of Construction Robotics Technologies in the Construction. *MATEC Web of Conferences, 266*. doi:10.1051/matecconf/201926605012

Ye, N., Chen, Y., Yang, J., Liang, S., Hu, Y., Hu, J., Zhu, S., Fan, W., & Xiao, B. (2017). Transformations of Na, Al, Si and Fe species in red mud during synthesis of one-part geopolymers. *Cement and Concrete Research*, *101*, 123–130. doi:10.1016/j.cemconres.2017.08.027

Yu, I. T., Li, Y., Wong, T. W., Tam, W., Chan, A. T., Lee, J. H., Leung, D. Y. C., & Ho, T. (2004). Evidence of airborne transmission of the severe acute respiratory syndrome virus. *The New England Journal of Medicine*, *350*(17), 1731–1739. doi:10.1056/NEJMoa032867 PMID:15102999

Zaobao, L. (2021, May 3). *Using technology to allow seniors to age-in-place*. Retrieved March 12, 2022, from https://scis.smu.edu.sg/news/2021/may/03/using-technology-allow-seniors-age-place

Zhang, H. Y., Kodur, V., Qi, S. L., & Wu, B. (2015). Characterizing the bond strength of geopolymers at ambient and elevated temperatures. *Cement and Concrete Composites*, *58*, 40–49. doi:10.1016/j.cemconcomp.2015.01.006

Zhang, H.-Y., Liu, J.-C., & Wu, B. (2021). Mechanical properties and reaction mechanism of one-part geopolymer mortars. *Construction & Building Materials*, *273*, 121973. doi:10.1016/j.conbuildmat.2020.121973

Zhang, J., He, Y., Wang, Y., Mao, J., & Cui, X. (2014). Synthesis of a self-supporting faujasite zeolite membrane using geopolymer gel for separation of alcohol/water mixture. *Materials Letters*, *116*, 167–170. doi:10.1016/j.matlet.2013.11.008

Zhang, P., Gao, Z., Wang, J., Guo, J., Hu, S., & Ling, Y. (2020). Properties of fresh and hardened fly ash/slag based geopolymer concrete: A review. *Journal of Cleaner Production*, *270*, 122389. doi:10.1016/j.jclepro.2020.122389

Zhang, S., Gong, K., & Lu, J. (2004). Novel modification method for inorganic geopolymer by using water soluble organic polymers. *Materials Letters*, *58*(7), 1292–1296. doi:10.1016/j.matlet.2003.07.051

Zhang, Y., Liu, H., & Chen, B. (2013). Comprehensive evaluation of the structural characteristics of an urban metabolic system: Model development and a case study of Beijing. *Ecological Modelling*, *252*, 106–113. doi:10.1016/j.ecolmodel.2012.08.017

Zhang, Z., Provis, J. L., Reid, A., & Wang, H. (2014). Geopolymer foam concrete: An emerging material for sustainable construction. *Construction & Building Materials*, *56*, 113–127. doi:10.1016/j.conbuildmat.2014.01.081

Zhang, Z., Provis, J. L., Reid, A., & Wang, H. (2015). Mechanical, thermal insulation, thermal resistance and acoustic absorption properties of geopolymer foam concrete. *Cement and Concrete Composites*, *62*, 97–105. doi:10.1016/j.cemconcomp.2015.03.013

Zhang, Z., & Sundaresan, S. (2010). Knowledge markets in firms: Knowledge sharing with trust and signalling. *Knowledge Management Research and Practice*, *2010*(8), 322–339. doi:10.1057/kmrp.2010.22

Ziada, M., Erdem, S., Tammam, Y., Kara, S., & Lezcano, R. A. (2021). The Effect of Basalt Fiber on Mechanical, Microstructural, and High-Temperature Properties of Fly Ash-Based and Basalt Powder Waste-Filled Sustainable Geopolymer Mortar. In Sustainability (Vol. 13, Issue 22). doi:10.3390u132212610

Ziada, M., Tammam, Y., Erdem, S., & Lezcano, R. A. (2022). Investigation of the Mechanical, Microstructure and 3D Fractal Analysis of Nanocalcite-Modified Environmentally Friendly and Sustainable Cementitious Composites. In Buildings (Vol. 12, Issue 1). doi:10.3390/buildings12010036

About the Contributors

Roberto Alonso González-Lezcano is a PhD in Industrial Engineering, Postgraduate Diploma in Plant Engineering, Postgraduate Diploma in University Didactics, Master in Mathematical Engineering Extraordinary Doctorate Award. Assistant Professor of the Department of Architecture and Design, in the area of Facilities at the Higher Polytechnic School of the CEU San Pablo University. He has been teaching at Building Installations, Electromechanics and Materials, Physics, Mathematics, Mechanical Engineering and Mechanics of Continuous Media and Theory of Structures. Accredited Professor by the ACAP in the figures of Professor Hired Doctor, Professor Doctor of Private University and Tenured Professor. A six-year research period recognized by the CNEAI. He has taught in undergraduate and master's degrees at the Carlos III University of Madrid, Camilo José Cela University, European University of Madrid, Antonio de Nebrija University, Technical University of Madrid and Alfonso X El Sabio University. He has published several books and articles concerning the areas of Energy, Installations and Continuous Media Mechanics.

* * *

Aarón A. Abad Alva is an Architect, Urban Planner and Researcher, with a postgraduate degree in Urban Planning with recognition for academic excellence, Master in Urban Territorial Planning - Urban Studies, Diploma in Residence, Supervision and Safety in Public Works. He has worked in multiple fields such as: Architecture, Urban Planning, Public Management, Construction, Real Estate and Research. He has worked in several public administrations in different municipalities in Peru, as well as assistant and supervisor in public works. He participated as a peer reviewer for Asian Journal of Biology - India. He worked as editor of the journal Territorios en Formación in Polytechnic University of Madrid. He also worked in real estate in Madrid. He has conducted several researches in the field of urbanism and has published several articles and co-authored a book, related to Green Infrastructure and Sustainability in Cities. Orcid: https://orcid.org/0000-0002-4855-4059.

Vinod Anand Bijlani is an AI and IoT expert. He has contributed to the design and implementation of smart city projects in India, Singapore, and the US. He also has over two decades of experience in creating solutions and is a distinguished inventor with 25 patents in AI and ML technologies. Vinod envisions a radically safer, healthier, and more sustainable planet with the democratization of AI, Industry 5.0, and 5G technologies.

Rolando-Arturo Cubillos-González is a full professor lecturer for the following courses at the RIBA certified Universidad Católicade Colombia: BIM, Principles of Innovation and Technology, Construction Design. Urban Design, Sustainable Design, Resilient design. 2011- 2016: Director of the research group Sustainability, Environment and Technology at the Universidad Católica de Colombia. Five major research projects have been developed by this research group under his direction. 1993-2006: Key roles in more than twenty design, construction and urbanization projects for some of Colombia's leading housing, institutional and industrial companies.

Fernando del Ama is an Assistant Professor of Architecture at Keene State College. Fernando de Ama possesses a Ph.D. in Architectural Acoustics from the School of Architecture of Madrid, Spain. He studied spatial perception of an architectural project and empirical parameters to qualify a Concert Hall. He has a seventeen year teaching experience related to construction technology, building physics, acoustics, energy management in buildings, building construction and architectural design. His research interests are related to energy simulation and monitoring systems in net zero energy buildings; new materials for regenerative buildings; Active Noise Control and acoustical components and geometric considerations for spatial perception of an architectural project.

Juan Echeverría has a degree in Architecture (U.P. Madrid), M.S. in Architecture and Building Design (Columbia U.), PhD in Architecture (U. Navarra), G.C. in Fire Protection Engineering (WPI). Architect in practice since 1988, has developed, in parallel, an academic career at the University San Pablo-CEU and the University of Navarra. Has worked for several academic institutions and professional organizations. His work is focused on Building Regulations, Building Services and Fire Safety. He is currently secretary of the Guipuzoa branch of the Basque and Navarre College of Architects and vice president of the SFPE Spanish chapter.

Savaş Erdem, after receiving his PhD from the Department of Civil Engineering at the University of Nottingham, joined the Department of Civil Engineering at the University of Istanbul in 2012. He progressed rapidly through the academic ranks, in 2016 becoming Associate Professor of Civil Engineering, one of the youngest associate professors at that time. His doctoral research has created a new understanding of the micro- and nanomechanical damage and fracture behaviour of sustainable concrete under short-duration extreme dynamic loads. The results of his study have been published in the most prestigious journals in civil engineering.

María Fernández-Vigil is a MS Architect from the University of Navarra since 2016, when she was awarded with the Training University Lecturers Scholarship from the Spanish Ministry of Education to carry out her PhD in the doctoral program "Environmental and Technological Design in Architecture". Along these years, she published several papers in indexed journals, she was a lecturer in several international conferences and I was a visiting researcher at the NFPA. In December 2020 she defended her PhD "Fire Safety in dwellings occupied by elderly people in Spain: Characterization of the problem and proposal of solutions", obtaining the maximum qualification. Currently, she combines teaching at the School of Architecture of the University of Navarra, with the practice of the profession in the field of self-protection and building services.

Juan Antonio Hernandez Ramos is an associate professor of Aerospace Engineering at the Polytechnic University of Madrid, Spain. He possesses a PhD. in Numeric Simulation from the School of Aeronautics and Space Engineering, UPM. He has been the director of the CES CEU Architecture (UPM) and co-founder of the Technology Based Company, InTelliGlass S.L. His research experience is related to the development of high-energy performance of water-flow glazed façades for buildings and numerical simulations and high order numerical schemes for thermo-fluid-dynamic problems. As a result of his research, he has published papers in high impact factor journals, along with patents.

Belen Moreno is a PhD in Architecture specialist in nearly zero energy building (nZEB). She is a lecturer of the Master of Technological Facades and Sustainable Envelopes at the Technical School of Architecture of Madrid. Over the last ten years she has been working as a Senior Researcher at the School of Aeronautics and Space Engineering in the Department of Applied Mathematics at the UPM. Founder of her own architecture office, her professional career has been enriched in the field of research and development of new technologies focused on industrialized construction and energy efficiency in building at National and International level.

Estela Karem Samamé Zegarra is an architect with a Master's degree in Environmental Managment and Building design from Navarra University. She has been teaching the course of Environmental physical conditioning of buildings to undergraduate students from diferent universities from Peru since 2013. She has recently published two chapters of book related to water efficiency in buildings and an article related to the participatory design for sustainability. In 2020 Now, she is a teacher at Santiago Antunez de Mayolo University (UNASAM) in Huaraz city. Orcid:https://orcid.org/0000-0003-2317-5076.

Yosra Tammam is a writer and researcher. She completed her master's degree from Istanbul Technical University and her PhD in civil engineering from Istanbul University-Cerrahpasa. She has publications in various fields in international journals.

José G. Vargas-Hernández, M.B.A. Ph.D. Member of the National System of Researchers of Mexico and a research professor at Posgraduate and Research Department, Tecnológico Mario Molina, Unidad Zapopan. Professor Vargas-Hernández has a Ph. D. in Public Administration and a Ph.D. in Organizational Economics. He has undertaken studies in Organisational Behaviour and has a Master of Business Administration, published four books and more than 200 papers in international journals and reviews (some translated to English, French, German, Portuguese, Farsi, Chinese, etc.) and more than 300 essays in national journals and reviews. He has obtained several international Awards and recognition.

Allison Zannoni is a student at Keene State College in the class of 2022. By graduations, she will have received a BS in Architecture as well as a BA in Spanish. She plans on working in an architectural firm after graduation as she works towards becoming a certified architect.

Mahmoud Ziada is a writer and researcher born in 1991. He completed his bachelor's and master's degrees from Ondokuz Mayıs University. Also, he is about to finish his PhD in civil engineering at Istanbul University-Cerrahpasa. He has publications in various fields in international journals.

Index

M

N

O

P

R

S

T

U

W

www.ingramcontent.com/pod-product-compliance
Lightning Source LLC
LaVergne TN
LVHW062154300926
849161LV00010B/2663